Mexican Mathematicians Abroad: Recent Contributions

CONTEMPORARY MATHEMATICS

657

APORTACIONES MATEMÁTICAS
SOCIEDAD MATEMÁTICA MEXICANA

Mexican Mathematicians Abroad: Recent Contributions

First Workshop
Matemáticos Mexicanos Jóvenes en el Mundo
August 22–24, 2012
Centro de Investigación en Matemáticas, A.C.,
Guanajuato, Mexico

Noé Bárcenas
Fernando Galaz-García
Mónica Moreno Rocha
Editors

American Mathematical Society
Providence, Rhode Island

2010 *Mathematics Subject Classification.* Primary 13D45, 13F60, 14C40, 18-01, 20C32, 37E45, 46G20, 46H05, 46L54.

Library of Congress Cataloging-in-Publication Data

Mexican mathematicians abroad : recent contributions : first workshop, Matemáticos Mexicanos Jóvenes en el Mundo, August 22–24, 2012, Centro de Investigación en Matemáticas, A.C., Guanajuato, Mexico / Noé Bárcenas, Fernando Galaz-García, Mónica Moreno Rocha, editors.

pages cm. – (Contemporary mathematics : volume 657) (Aportaciones matemáticas / Sociedad Matemática Mexicana)

Includes bibliographical references and index.

ISBN 978-1-4704-2192-2 (alk. paper)

1. Mathematics–Mexico–Congresses. 2. Mathematicians–Mexico–Congresses. 3. Women mathematicians–Mexico–Congresses. I. Bárcenas, Noé, 1984– editor. II. Galaz-García, Fernando, 1979– editor. III. Moreno Rocha, Mónica, 1970– editor.

QA27.M49.M49 2016

510–dc23 2015036424

DOI: `http://dx.doi.org/10.1090/conm/657`

∞ The paper used in this book is acid-free and falls within the guidelines
established to ensure permanence and durability.
Visit the AMS home page at `http://www.ams.org/`

10 9 8 7 6 5 4 3 2 1 21 20 19 18 17 16

Contents

Prologue

As we well know, ever since the modern study of mathematics began in Mexico a little over seventy years ago, many Mexican mathematicians have studied abroad at many different universities in many different countries. Interestingly, however, to date the vast majority of us have returned to Mexico, and are occupying a variety of positions, mostly academic. Like the Mexican mathematicians currently scattered around the world, when we embarked on the adventure and enterprise of carrying out our graduate studies abroad, we wanted to stay connected with the Mexican mathematical community and with our fellow students abroad.

As a background to the "Mexican Mathematicians in the World Meeting" in Guanajuato in 2012, and to these proceedings, I have vague memories of Socorro Soberón, José Seade, Elías Micha and me in Oxford in early 1979 talking about how interesting it would be, both academically and socially, to organize a meeting of Mexican mathematicians studying in Europe at the time, in which we would talk about our doctoral work and, of course, share our experiences. With the enthusiasm of youth, we decided to organize the meeting for the coming summer and hold it at University of Cambridge. I still remember the astonished face of the famous Professor J. W. S. Cassels, Sadleirian Professor of Pure Mathematics at Cambridge, at the time head of the Department of Pure Mathematics and Mathematical Statistics, when I proposed our idea and asked for the use of their facilities for the meeting. He generously agreed, and a similar process was repeated when I requested reservations for some rooms in Pembroke College, Cambridge, to accommodate visiting participants. The meeting, which was called the "First Congress of Mexican Mathematicians Abroad" was held on June 26 to 29, 1979. There are a number of differences between the original conference and the 2012 meeting that are no doubt a reflection of the change for the better that Mexican mathematics has undergone. At the original Cambridge meeting, we were few in number, the most advanced of us were in our third year of graduate studies, and communication among us was limited to a few short letters and a very small number of telephone calls between those who were at Oxford (Soberón, Seade and Micha) and me, at Cambridge. The congress had only twelve presentations, on topics in analysis, geometry, and topology. Besides those already mentioned, the participants were Amira Acosta (Leipzig), Marcelo Aguilar (Warwick), Ricardo Berlanga (Cambridge), Berta Gamboa (Paris VI), Carlos Gómez Mont (Harvard), José Alfredo Jiménez (Durham), Guillermo Pastor (Warwick) and Guillermo Sienra (Southampton). I also remember that in addition to the scheduled talks, there was a presentation by Spanish mathematician Lucía Contreras, then a student of José María Montesinos, who was visiting Cambridge at the time. I am able to list these facts thanks to José Seade who took the time to collect some of the proceedings, of which there are still a few copies extant today, although I must confess that when I look at them, sometimes I can hardly believe that it actually happened. I should also note that in 1980 the Mexican students in Warwick, who now also included Ricardo Berlanga, held a second such conference, but I have little documented information about it. A notable point about our first congress is that at least ten of the twelve presenters did return to Mexico.

Another piece of the prior history leading up to the 2012 "Mexican Mathematicians in the World Meeting" took place earlier the same year at the 2012 Joint Meeting of the SMM–RSME at Malaga, Spain in January. Fernando Galaz García approached Luis Montejano, who was president-elect of the SMM at the time, about the plans that he, Noé Bárcenas and Mónica Moreno were making for the meeting that gave rise to these proceedings. Upon learning about these plans, Montejano offered them all the support that the SMM could give. After the Málaga conference, during a visit to the Alhambra, Montejano told me about the enormous momentum he observed in the new generations of Mexican mathematicians. More than thirty years on, I noted, much and little have changed at the same time. Since the first meeting, the Mexican mathematical community has grown in number and in diversity of areas, and there are representatives of the young generations in many more universities and countries. Moreover, our research is rather more internationally visible. In my opinion, the new generations are also more aware that their talent and training makes them competitive in the profession anywhere in the world. I also say little, because of the enormous challenge we have in our country to palpably affect their development, progress in this direction has been slight, and continues to be something which we all owe to our country.

I feel honored that the editorial committee of Aportaciones has asked me to write a brief prologue to these conference proceedings. The objectives proposed for the meeting have much to tell us; the conference aims to bring Mexican mathematicians abroad together with mathematicians in Mexico, to disseminate the work of young Mexican mathematicians abroad, to enable graduate students abroad to share and exchange experiences, to analyze opportunities for returning to Mexico, and to build and strengthen global networks. Another notable feature is that these proceedings include not only reports of work in analysis, geometry and topology as in the old days but also topics in algebra, applied mathematics and probability. I am confident that the current generation of young Mexican mathematicians, particularly those who are now outside the country, will lead Mexican mathematics to new frontiers. It is commendable that in addition to the hard work they do to advance in their profession, they make the effort to stay in touch with each other and above all not to lose their ties and commitment to our nation. These proceedings present just a sample of what they can achieve, and my desire for them is that their enthusiasm and optimism may long endure.

José Carlos Gómez Larrañaga
Centro de Investigación en Matemáticas, A. C. and
Centro de Investigación y Docencia Económicas, A.C.

Preface

The present volume contains a sample of research work by young Mexican mathematicians currently working abroad, on the occasion of the workshop "Matemáticos Mexicanos Jóvenes en el Mundo", which took place at CIMAT, in Guanajuato, Mexico, on August 22–24, 2012. The workshop had three main goals: fostering the collaboration and exchange of ideas among young Mexican mathematicians working abroad and their peers in Mexico; exchanging experiences concerning studying and working overseas; and designing strategies to strengthen links between mathematical research institutions in Mexico and Mexican mathematicians around the world.

The history of professional mathematics in Mexico is relatively recent and spans less than 75 years. The Instituto de Matemáticas of the Universidad Nacional Autónoma de México (UNAM) was created in 1942 in Mexico City and became the first center fully devoted to mathematical research in the country. The following year saw the foundation of the Sociedad Matemática Mexicana (SMM) and the first issue of its research journal: the *Boletín de la Sociedad Matemática Mexicana*. Nowadays, there are at least six mathematics research centers (with several units around the country) in addition to at least 19 schools and departments of mathematics at public and private universities, according to a list maintained by the SMM.

The development of mathematics as an active research discipline in Mexico has gone through several phases. Among these, one can identify a foundational period centered at the Instituto de Matemáticas at UNAM in the 1940's and 1950's which extended and deepened mathematical activity originated in the Facultad de Ciencias at UNAM. This period experienced an important influence from Solomon Lefschetz and culminated with the creation of the Centro de Investigación y Estudios Avanzados (CINVESTAV) and the Escuela Superior de Física y Matemáticas at the Instituto Politécnico Nacional (IPN) in 1961.

A subsequent period of expansion was marked by the creation of the Centro de Investigaciones en Matemáticas Aplicadas, Sistemas y Servicios (CIMASS) at UNAM, in 1970, and the Universidad Autónoma Metropolitana, in 1974. CIMASS eventually evolved into the actual Instituto de Investigaciones en Matemáticas Aplicadas y en Sistemas (IIMAS), in 1976. In 1980 the Centro de Investigación en Matemáticas (CIMAT) was founded and became one of the first public research centers outside Mexico City. In the last twenty years, the need for decentralization and the growth in some of the already established institutions have led to the creation of associated representations and units (e.g. Cuernavaca, Juriquilla and Oaxaca, for IMUNAM, and Aguascalientes, Mérida, Monterrey and Zacatecas, for CIMAT) and autonomous research centers (e.g. Centro de Ciencias Matemáticas at UNAM, in Morelia) all around Mexico. Parallel to these events, mathematical research has also found a place in public and private universities, such as the Instituto Tecnológico Autónomo de México, Universidad Autónoma de Yucatán, Universidad de Guanajuato, Universidad Juárez Autónoma de Tabasco, Universidad Michoacana de San Nicolás de Hidalgo, Universidad de Sonora and Universidad Veracruzana, to name a few.

The emergence of competitive undergraduate and graduate programs in mathematics around the country has been one of the consequences of the aforementioned development. These programs have been producing a steady number of mathematicians that complete their professional education with doctoral studies or postdoctoral experience abroad, integrating themselves into academic life in their host countries and extending the Mexican mathematical community internationally. The first known effort towards connecting Mexican mathematicians in foreign countries was embodied in the meeting "Primer Congreso de Matemáticos Mexicanos en el Extranjero", held in Cambridge, Great Britain, on June 26–29, 1979. This conference had twelve speakers and brought together Mexican mathematicians living in four different countries and working in analysis, geometry and topology.

The constant growth in the number of Mexican mathematicians abroad in the last decades made a new effort at assembling our colleagues out of the country long overdue, prompting the organization of the workshop "Matemáticos Mexicanos Jóvenes en el Mundo". The meeting gathered 56 participants from 35 universities distributed in 14 countries and it managed to attract graduate students coming from Argentina, Chile, Colombia, and Costa Rica. There was a Q&A session for graduate students interested in studying abroad, along with several informal discussion sessions. Our 17 speakers were young Mexican mathematicians (in the early postdoctoral stage or at the end of their Ph.D. studies) working at institutions in the United States, Canada, Europe, and South America. The talks were grouped into several areas: analysis and algebra, probability and applied mathematics, and geometry and topology, with each group roughly corresponding to a day of activities.

We thank CIMAT, the Sociedad Matemática Mexicana, the Centro de Ciencias Matemáticas at UNAM and the Departamento de Matemáticas of the Universidad de Guanajuato for their support in organizing this meeting. We also thank the speakers and the participants for their enthusiastic response, which helped to bring the workshop to a successful conclusion. Finally, we thank the authors, the referees, and the editorial committees and staff of Aportaciones Matemáticas and Contemporary Mathematics for their support in the preparation of these proceedings.

Noé Bárcenas
Fernando Galaz-García
Mónica Moreno Rocha

Collection of Socorro Soberón

Participants at the First Congress of Mexican Mathematicians Abroad, University of Cambridge, June 26–29, 1979.

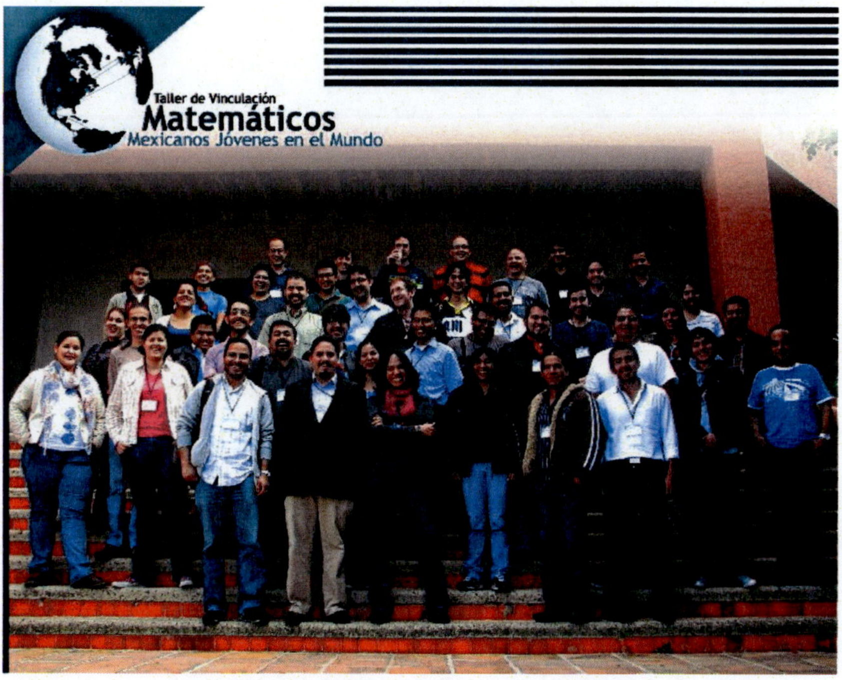

Photo by CIMAT A.C.

Participants at the First Workshop Matemáticos Mexicanos Jóvenes en el Mundo, CIMAT, August 22–24, 2012.

List of Participants

María José Aceves Sepúlveda
CUCEI, Universidad de Guadalajara, Mexico.

Sergio Ángel Almada
University of North Carolina-Chapel Hill, USA.

Octavio Arizmendi
Universität des Saarlandes, Germany.

Samuel Uriel Armendariz Hernández
Universidad Autónoma de Nuevo León, Mexico.

Noé Bárcenas Torres
Hausdorff Center for Mathematics, Universität Bonn, Germany.

Ada Borjas López
Universidad Autónoma Metropolitana, Mexico.

Victor Isidoro Bravo Reyna
Universidad Autónoma de México, Mexico.

Victoria Cantoral Farfán
Université Pierre et Marie Curie Paris VI, France.

Juan Carlos Castro Contreras
CINVESTAV, Instituto Politécnico Nacional, Mexico

Jorge Alberto Coleote Domínguez
Benemérita Universidad Autónoma de Puebla, Mexico.

Román Contreras Morales
Universidad Autónoma de México, Mexico.

José Ezequiel Contreras Hernández
Benemérita Universidad Autónoma de Puebla, Mexico.

Pablo Dávalos
Universidade de São Paulo, Brazil.

Edgar Duéñez
Harvard University, USA and K.U. Leuven, Belgium.

Fernando Galaz García
WWU Münster, Germany.

José Luis García Arias
Universidad Juárez Autónoma de Tabasco, Mexico.

Nayeli González
Centro de Investigación en Matemáticas, A.C., Mexico.

David Guerrero Sánchez
Universidad de Murcia, Spain.

Luis Ángel Gutiérrez Méndez
Benemérita Universidad Autónoma de Puebla, Mexico.

Jesús Hernández Hernández
University of Galway, Ireland and Universidad Veracruzana, Mexico.

Iván Hernández Orzuna
Benemérita Universidad Autónoma de Puebla, Mexico.

Anabel Hernández Ramírez
Benemérita Universidad Autónoma de Puebla, Mexico.

Gabriel Kantú Montiel
Centro de Investigación en Matemáticas, A.C., Mexico.

Gonzalo Jiménez
Pontificia Universidad Católica de Chile, Chile.

Daniel Labardini
Universität Bonn, Germany.

Simón Lunagomez
Harvard University, USA.

José Malagón
Ottawa University, Canada.

Cristabel Martínez
Centro de Investigación en Matemáticas, A.C., Mexico.

Mariana Mireles González
Universidad Autónoma de Mexico, Mexico.

Mónica Moreno Rocha
Centro de Investigación en Matemáticas, A.C., Mexico.

Isidro Munive
Purdue University, USA.

Luis Núñez Betancourt
University of Michigan-Ann Arbor, USA.

Jesús Ángel Núñez Zimbrón
Universidad Autónoma de Mexico, Mexico.

Sofía Ortega Castillo
Texas A&M University, USA.

Daniela Osorio Mendoza
Universidad Iberoamericana, Mexico.

Brenda Aide Peña Cantú
Universidad Autónoma de Nuevo León, Mexico.

Raquel Perales
SUNY-Stony Brook, USA.

Reyna María Pérez Tiscareño
University of Tartu, Estonia.

Emilene Carmelita Pliego Pliego
Benemérita Universidad Autónoma de Puebla, Mexico.

Carlos Ramos
Max Planck Institut für Mathematik, Germany.

Jesús Adrián Ríos Osornio
Universidad de Guanajuato, Mexico.

Rodolfo Ríos Zertuche
Princeton University, USA.

Eli Roblero Méndez
Centro de Investigación en Matemáticas, A.C., Mexico.

Javier Sáenz
Centro de Investigación en Matemáticas, A.C., Mexico.

Emilio Salcedo Martínez
Centro de Investigación en Matemáticas, A.C., Mexico.

Jaime Santos Rodríguez
Instituto de Matemáticas, Universidad Autónoma de México, Mexico.

Carlos Segovia
Universidad de los Andes, Colombia.

Salvador Sierra Murillo
Freie Universität Berlin, Germany.

Pedro Solórzano
University of California-Riverside, USA.

Iván Martín Suárez Barraza
CINVESTAV, Instituto Politécnico Nacional Zacatenco, Mexico.

Porfirio Toledo Hernández
Universidad Veracruzana, Mexico.

Abraham Toriz Cruz
Universidad Veracruzana, Mexico.

Roberto Antonio Ulloa Esquivel
Universidad de Costa Rica, Costa Rica.

Gabriel Uribe
Universidad de Antioquia, Colombia.

Vanesa Noelia Vansteenkieste
Universidad Nacional de Rosario, Argentina.

Oscar Zamora Luna
Universidad de Costa Rica, Costa Rica.

Contemporary Mathematics
Volume 657, 2016
http://dx.doi.org/10.1090/conm/657/13087

Locally pseudoconvex inductive limit of locally pseudoconvex Q-algebras

Mati Abel and Reyna María Pérez-Tiscareño

ABSTRACT. A $\mathcal{LF}pg$-algebra ($\mathcal{LF}p$-algebra) E was defined in Abel and Pérez-Tiscareño (2013) as a locally pseudoconvex inductive limit (respectively, a locally pseudoconvex inductive limit of an increasing sequence) of locally pseudoconvex F-algebras, which satisfies certain conditions. The case when every locally pseudoconvex F-algebra in the inductive limit E is a Q-algebra is considered in the present paper. Conditions for E to be also a Q-algebra are found and examples of such inductive limits of topological algebras are given.

1. Introduction

1. Let E be a unital topological algebra over $\mathbb{K}$ (the field of real numbers $\mathbb{R}$ or complex numbers $\mathbb{C}$) with separately continuous multiplication (in short, a topological algebra). If the underlying linear topological space of E is locally pseudoconvex (see [**8**], [**9**], [**13**], [**14**], [**16**] or [**17**]), then E is called a *locally pseudoconvex algebra*. In this case E has a base $\mathcal{U} = \{U_\lambda : \lambda \in \Lambda\}$ of neighborhoods of zero consisting of balanced ($\mu U_\lambda \subset U_\lambda$, when $|\mu| \leqslant 1$) and pseudoconvex sets ($U_\lambda + U_\lambda \subset \mu U_\lambda$ for a $\mu \geqslant 2$). This base defines a set of numbers $\{k_\lambda : \lambda \in \Lambda\}$ in $(0, 1]$ such that

$$U_\lambda + U_\lambda \subset 2^{\frac{1}{k_\lambda}} U_\lambda$$

and

$$\Gamma_{k_\lambda}(U_\lambda) \subset 2^{\frac{1}{k_\lambda}} U_\lambda$$

for each $\lambda \in \Lambda$, where

$$\Gamma_k(U) =$$

$$= \Big\{ \sum_{\nu=1}^{n} \mu_\nu u_\nu : n \in \mathbb{N}, u_1, \cdots, u_n \in U \text{ and } \mu_1, \cdots, \mu_n \in \mathbb{K} \text{ with } \sum_{\nu=1}^{n} |\mu_\nu|^k \leqslant 1 \Big\}$$

is the *absolutely k-convex hull* of a subset $U \subset E$ for $k \in (0, 1]$. Herewith, a subset $U \subset E$ is *absolutely k-convex*, if $U = \Gamma_k(U)$, and *absolutely pseudoconvex*, if

2010 *Mathematics Subject Classification*. Primary 46H05; Secondary 46H20.

Key words and phrases. Topological algebras, locally pseudoconvex algebras, F-algebra, Q-algebras, locally pseudoconvex inductive limits of locally pseudoconvex algebras, $\mathcal{LF}p$-algebras and $\mathcal{LF}pg$-algebras.

Research is in part supported by the Estonian Targeted Financing Project SF0180039s08 and by the European Union through the European Social Fund (MOBILITAS grant No. MJD247).

$U = \Gamma_k(U)$ for some $k \in (0,1]$. When $k_\lambda = k$ for each $\lambda \in \Lambda$ (that is, k does not depend on U_λ) or

$$\inf\{k_\lambda : \lambda \in \Lambda\} = k > 0,$$

then E is a *locally k-convex algebra.*

The topology of a locally pseudoconvex algebra E can be defined by a family $\mathcal{P} = \{p_\lambda : \lambda \in \Lambda\}$ of k_λ-homogeneous seminorms (that is, $p_\lambda(\mu a) = |\mu|^{k_\lambda} p_\lambda(a)$ for $\lambda \in \Lambda$ and $a \in E$), where $k_\lambda \in (0,1]$ for each $\lambda \in \Lambda$, and

$$p_\lambda(a) = \inf\{|\mu|^{k_\lambda} : a \in \mu\Gamma_{k_\lambda}(U_\lambda)\}$$

for each $a \in A$ and $\lambda \in \Lambda$ (see, for example, [**17**, pp. 3–6] or [**8**, pp. 189 and 195]).

2. We recall first what means an inductive limit of algebras. For it, let I be a (non-empty) directed set with the order "$\prec$". So, for any $\alpha, \beta \in I$ there is a $\gamma \in I$ such that $\alpha \prec \gamma$ and $\beta \prec \gamma$. Let $(E_\alpha)_{\alpha \in I}$ be a family of algebras and for every $\alpha, \beta \in I$ with $\alpha \prec \beta$ let

$$\{f_{\beta\alpha} : E_\alpha \to E_\beta\}$$

be a family of homomorphisms such that the following properties have been satisfied:

1) $f_{\alpha\alpha} = id_{E_\alpha}$ for every $\alpha \in I$.

2) $f_{\gamma\alpha} = f_{\gamma\beta} \circ f_{\beta\alpha}$ for any $\alpha, \beta, \gamma \in I$ such that $\alpha \prec \beta \prec \gamma$.

In case, when every E_α is a unital algebra with unit element e_α, then it is assumed that $f_{\beta\alpha}(e_\alpha) = e_\beta$ whenever $\alpha \prec \beta$. The family of algebras $(E_\alpha)_{\alpha \in I}$ with maps $f_{\beta\alpha}$, defined above, is an *inductive* (or *directed*) *system of algebras* E_α and it is denoted by $(E_\alpha, f_{\beta\alpha})$.

Let $E_0 = \sum_\alpha E_\alpha$ (a disjoint union). Elements $x, y \in E_0$ (then $x \in E_\alpha$ and $y \in E_\beta$ for some $\alpha, \beta \in I$) are called *equivalent* (in short $x \sim y$), if there exists $\gamma \in I$ such that $\alpha \prec \gamma$, $\beta \prec \gamma$ and

$$f_{\gamma\alpha}(x) = f_{\gamma\beta}(y).$$

The quotient set $(E_0/\!\sim)$ is the *inductive* (or *direct*) *limit* of the inductive system $(E_\alpha, f_{\beta\alpha})$. We shall denote this, as usual, by $\varinjlim(E_\alpha, f_{\beta\alpha})$ or simply by $\varinjlim E_\alpha$.

For every $\alpha \in I$, let $i_\alpha : E_\alpha \to E_0$ be the inclusion and $\pi : E_0 \to E_0/\!\sim$ the quotient map. Then,

$$f_\alpha = \pi \circ i_\alpha : E_\alpha \to E = \varinjlim E_\alpha \quad \text{for every} \quad \alpha \in I$$

is a homomorphism (later on, it will be called a *canonical map* from E_α to E). When the inductive limit $\varinjlim E_\alpha$ has the unit element e, then it is assumed that $f_\alpha(e_\alpha) = e$ for each $\alpha \in I$.

It is known that

$$E = \bigcup_{\alpha \in I} f_\alpha(E_\alpha) \tag{1}$$

(see [**15**, p. 110]). Moreover, $f_\beta \circ f_{\beta\alpha} = f_\alpha$ whenever $\alpha \prec \beta$ and $f_\alpha(E_\alpha) \subseteq f_\beta(E_\beta)$ for any $\alpha, \beta \in I$ with $\alpha \prec \beta$. Therefore, the algebraic operations can be given on $\varinjlim E_\alpha$ (see [**15**, p. 111]).

If inductive limits of topological algebras (E_α, τ_α) are considered, it is assumed that the homomorphisms $f_{\beta\alpha} : E_\alpha \to E_\beta$ ($\alpha, \beta \in I$ with $\alpha \prec \beta$) are continuous and $E = \varinjlim E_\alpha$ is endowed with the final topology or the *inductive limit topology* $\tau_{\varinjlim E_\alpha}$ defined by the canonical maps f_α, i.e.

$$\tau_{\varinjlim E_\alpha} = \{U \subset E : f_\alpha^{-1}(U) \in \tau_\alpha \text{ for every } \alpha \in I\}.$$

Then all canonical maps $f_\alpha : E_\alpha \to E$ $(\alpha \in I)$ are continuous homomorphisms (see [**15**, p. 113]).

We shall consider now inductive limits of locally pseudoconvex algebras E_α. Since the inductive limit topology on E is not necessarily locally pseudoconvex, the final locally pseudoconvex topology τ is defined on E by giving the base of neighborhoods at $x \in E$ on the form

$$\mathcal{L}_x = \{x + U : U \text{ is absolutely pseudoconvex in } E \text{ and } f_\alpha^{-1}(U) \in \mathcal{N}_{\tau_\alpha}\}$$

where $\mathcal{N}_{\tau_\alpha}$ denotes the set of all neighborhoods of zero in E_α. Hence, (E, τ) is a locally pseudoconvex algebra, because $\mathcal{L}_\theta$ (here θ denotes the zero element in E) is a base of absolutely pseudoconvex neighborhoods of zero in E. Herewith, the topology τ is the finest locally pseudoconvex topology on E such that f_α is continuous for every $\alpha \in I$. Later on we shall say in this case that τ is the *locally pseudoconvex inductive limit topology* on the inductive limit of locally pseudoconvex algebras.

In case, when $E = \varinjlim E_\alpha$, where every E_α is a subset of E,

$$E = \bigcup_{\alpha \in I} E_\alpha$$

and for every $\alpha, \beta \in I$ there exists $\gamma \in I$ such that $E_\alpha \subseteq E_\gamma$ and $E_\beta \subseteq E_\gamma$, then the notation $\underrightarrow{\bigcup} E_\alpha$ instead of $\varinjlim E_\alpha$ is used.

Next we recall several classes of topological algebras introduced in [**4**]. A topological algebra (E, τ) is a *generalized $\mathcal{LF}$-algebra* (in short, *$\mathcal{LF}g$-algebra*), if E is an inductive limit of F-algebras[1] (E_α, τ_α) such that

$$E = \underrightarrow{\bigcup} E_\alpha \tag{2}$$

(as sets), the topology τ on E coincides with the inductive limit topology, defined by canonical maps, and the topology, induced on every E_α by τ, coincides with the original topology τ_α on E_α. Moreover, (E, τ) is an $\mathcal{LF}$-algebra, if E is an inductive limit of an increasing sequence of F-algebras (E_n, τ_n) such that holds

$$E = \underrightarrow{\bigcup} E_n \tag{3}$$

(as sets), the topology τ coincides with the inductive limit topology on E, defined by canonical maps, and for each $n \in \mathbb{N}$ the topology induced on E_n by E_{n+1} and by the topology τ are identical to the original topology τ_n on E_n. In particular case, when every E_α in (2) (E_n in (3)) is a locally pseudoconvex F-algebra and τ on E coincides with the locally pseudoconvex inductive limit topology, defined by canonical maps, then $\mathcal{LF}pg$-algebra ($\mathcal{LF}p$-algebra) instead of $\mathcal{LF}g$-algebra (respectively $\mathcal{LF}$-algebra) is used. Moreover, when all E_α (respectively, E_n) and E are locally k-convex algebras for some $k \in (0, 1]$, then the term *k-$\mathcal{LF}g$-algebra* (respectively, *k-$\mathcal{LF}$-algebra*) are used.

---

[1] An F-algebra is a metrizable and complete algebra.

A topological algebra (E, τ) is a *Q-algebra* if the set Qinv$E$ of quasi-invertible elements[2] (in case when E is a unital algebra, then the set Inv$E$ of invertible elements) of E is open in the topology τ.

We shall say that an $\mathcal{LF}pg$-algebra ($\mathcal{LF}p$-algebra) (E, τ) is a *Q-$\mathcal{LF}pg$-algebra* (respectively, *Q-$\mathcal{LF}p$-algebra*) if (E, τ) and all algebras (E_α, τ_α) in the inductive limit are Q-algebras. Similar topological algebras for locally m-pseudoconvex case[3] have been considered in [**10**] and for locally m-convex case in [**5**] and [**11**].

Examples of Q-$\mathcal{LF}pg$-algebras and Q-$\mathcal{LF}p$-algebras are given in the present paper. It is shown that (E, τ) is a Q-algebra if it is an inductive limit of Q-algebras and τ is the inductive limit topology on E. Conditions, in order a locally pseudoconvex algebra (E, τ) (which is an inductive limit of locally pseudoconvex Q-algebras and τ coincides with the locally pseudoconvex inductive limit topology on E, defined by canonical maps) is also a Q-algebra are found.

Moreover, conditions in order that $\mathcal{LF}pg$-algebra ($\mathcal{LF}p$-algebra) is a Q-$\mathcal{LF}pg$-algebra (respectively, a Q-$\mathcal{LF}p$-algebra) are given.

2. Q-$\mathcal{LF}pg$-algebra and Q-$\mathcal{LF}g$-algebras

First we give some examples of Q-$\mathcal{LF}pg$-algebras and Q-$\mathcal{LF}g$-algebras.

2.1. Examples of Q-$\mathcal{LF}pg$-algebras and Q-$\mathcal{LF}g$-algebras. Let (A, τ) be a unital commutative locally pseudoconvex F-algebra with continuous inversion which is also a Q-algebra, where τ is defined by k_n-homogeneous seminorms p_n with $k_n \in (0, 1]$ for each $n \in \mathbb{N}$, and let $\mathcal{K}(\mathbb{R}, A)$ be the algebra of continuous functions from $\mathbb{R}$ to A with compact support, in which the algebraic operations are defined by

$$(f+g)(x) = f(x) + g(x), (\lambda f)(x) = \lambda(f(x)) \text{ and } (fg)(x) = f(x)g(x)$$

for each $f, g \in \mathcal{K}(\mathbb{R}, A)$, $\lambda \in \mathbb{K}$ and $x \in X$. Moreover, for any integer $i \geq 1$ let $\mathcal{K}(\mathbb{R}, A; [-i, i])$ be the subalgebras of $\mathcal{K}(\mathbb{R}, A)$, which support of every function is contained in $[-i, i]$. The topology τ_i on $\mathcal{K}(\mathbb{R}, A; [-i, i])$ for each i is defined by k_n-homogeneous seminorms $q_{i,n}$, where

$$q_{i,n}(f) = \sup_{x \in [-i,i]} p_n(f(x))$$

for each $n \in \mathbb{N}$ and $f \in \mathcal{K}(\mathbb{R}, A; [-i, i])$. Then $(\mathcal{K}(\mathbb{R}, A; [-i, i]), \tau_i)$ is a metrizable locally pseudoconvex algebra for every i. To show that $(\mathcal{K}(\mathbb{R}, A; [-i, i]), \tau_i)$ is complete, let $i \in \mathbb{N}$ be fixed and (f_k) be a Cauchy sequence in $(\mathcal{K}(\mathbb{R}, A; [-i, i]), \tau_i)$. Then for every $\varepsilon > 0$ and $n \in \mathbb{N}$ there exists $N > 0$ such that

$$q_{i,n}(f_k - f_m) = \sup_{x \in [-i,i]} p_n(f_k(x) - f_m(x)) < \varepsilon$$

whenever $m > k > N$. Hence, $p_n(f_k(x) - f_m(x)) < \varepsilon$ for every $x \in [-i, i]$ (also for every $x \in \mathbb{R}$, because the support of function f_k is contained in $[-i, i]$ for each $k \in \mathbb{N}$) and every $n \in \mathbb{N}$ whenever $m > k > N$. It means that for every $x \in \mathbb{R}$

[2] An element a is *quasi-invertible* in an algebra A, if there exists another element $b \in A$ such that $a + b = ab$.

[3] A topological algebra A is *locally m-pseudoconvex* (*locally m-convex*), if it has a base of neighborhoods of zero which consists of absolutely pseudoconvex (respectively, absolutely convex) and idempotent sets U (that is, $UU \subset U$).

the sequence $(f_k(x))$ is a Cauchy sequence in A. Since A is complete, then $(f_k(x))$ converges for every $x \in \mathbb{R}$ to an element $y_x \in A$. Therefore

$$p_n(f_k(x) - y_x) = \lim_{m\to\infty} p_n(f_k(x) - f_m(x)) \leqslant \varepsilon$$

for every $x \in \mathbb{R}$ and $n \in \mathbb{N}$ whenever $k > N$. Let $f : \mathbb{R} \to A$ be a function such that $f(x) = y_x$ for each $x \in \mathbb{R}$. Hence, (f_k) converges to f in the topology τ_i.

We shall prove now that f is continuous on $\mathbb{R}$. For this, let $x \in \mathbb{R}$ and (x_l) be a sequence in $\mathbb{R}$ such that $\lim\limits_{l\to\infty} x_l = x$. Then $\lim\limits_{l\to\infty} f_k(x_l) = f_k(x)$ for each k because f_k is continuous on $\mathbb{R}$. Hence, for any $\varepsilon > 0$ and $k, n \in \mathbb{N}$ there are $M > 0$ such that

$$p_n(f_k(x_l) - f_k(x)) < \frac{\varepsilon}{3}$$

whenever $l > M$ and $N > 0$ such that

$$p_n(f_k(x) - f(x)) < \frac{\varepsilon}{3}$$

for each $x \in \mathbb{R}$ whenever $k > N$. Let now $k > N$. Then

$$p_n(f(x_l) - f(x)) = p_n(f(x_l) + f_k(x) - f_k(x) + f_k(x_l) - f_k(x_l) - f(x)) \leqslant$$
$$\leqslant p_n(f_k(x_l) - f(x_l)) + p_n(f_k(x) - f(x)) + p_n(f_k(x_l) - f_k(x)) < \frac{\varepsilon}{3} + \frac{\varepsilon}{3} + \frac{\varepsilon}{3} = \varepsilon$$

for each $n \in \mathbb{N}$ whenever $l > M$. This means, that f is continuous on $\mathbb{R}$ and its support is compact (because $f(x) = \theta_A$ for every $x \in \mathbb{R} \setminus [-i, i]$). Consequently, $\mathcal{K}(\mathbb{R}, A; [-i, i])$ is a locally pseudoconvex F-algebra.

Moreover,

$$\mathcal{K}(\mathbb{R}, A; [-i, i]) \subset \mathcal{K}(\mathbb{R}, A; [-(i + 1), i + 1])$$

for each $i \in \mathbb{N}$ and

$$\mathcal{K}(\mathbb{R}, A) = \bigcup_{i\in\mathbb{N}} \mathcal{K}(\mathbb{R}, A; [-i, i]).$$

Therefore $\mathcal{K}(\mathbb{R}, A)$ is a $\mathcal{LF}p$-algebra[4] without the unit element if we endow $\mathcal{K}(\mathbb{R}, A)$ with the locally pseudoconvex inductive limit topology, defined by the canonical maps φ_i from $\mathcal{K}(\mathbb{R}, A; [-i, i])$ into $\mathcal{K}(\mathbb{R}, A)$. Now, from

$$\mathcal{K}(\mathbb{R}, A; [-i, i]) \times A \subset \mathcal{K}(\mathbb{R}, A; [-(i + 1), i + 1]) \times A$$

for each $i \in \mathbb{N}$ and

$$\mathcal{K}(\mathbb{R}, A) \times A = \Big(\bigcup_{i\in\mathbb{N}} \mathcal{K}(\mathbb{R}, A; [-i, i]\Big) \times A = \bigcup_{i\in\mathbb{N}} (\mathcal{K}(\mathbb{R}, A; [-i, i]) \times A)$$

follows that

$$\mathcal{K}(\mathbb{R}, A) \times A = \underset{\longrightarrow}{\bigcup} (\mathcal{K}(\mathbb{R}, A; [-i, i]) \times A).$$

We consider for each i the topology τ_i' on $\mathcal{K}(\mathbb{R}, A; [-i, i]) \times A$ which is defined by k_n-homogeneous seminorms $Q_{i,n}$, where

$$Q_{i,n}(f, a) = q_{i,n}(f) + p_n(a)$$

for each $f \in \mathcal{K}(\mathbb{R}, A; [-i, i])$, $a \in A$ and $n \in \mathbb{N}$.

[4]The topology on $\mathcal{K}(\mathbb{R}, A; [-i, i])$, induced by τ_{i+1} and by τ, coincides with the topology τ_i because $q_{i,n}(f) = q_{(i+1),n}(f)$ for each $f \in \mathcal{K}(\mathbb{R}, A; [-i, i])$ and $n \in \mathbb{N}$ and every neighborhood O of zero in $(\mathcal{K}(\mathbb{R}, A; [-i, i]), \tau_i)$ defines a neighbourhood U of $(\mathcal{K}(\mathbb{R}, A), \tau)$ such that $U \cap \mathcal{K}(\mathbb{R}, A) = O$.

Following, we prove that $(\mathcal{K}(\mathbb{R}, A; [-i, i]) \times A, \tau_i')$ is complete. For this, for every fixed $i \in \mathbb{N}$ let $((f_k, a_k))$ be a Cauchy sequence in $(\mathcal{K}(\mathbb{R}, A; [-i, i]) \times A, \tau_i')$. Then, for each fixed $n \in \mathbb{N}$ and arbitrary $\varepsilon > 0$, there exists $N > 0$ such that

$$Q_{i,n}((f_k, a_k) - (f_l, a_l)) < \varepsilon \quad \text{whenever} \quad k > l > N.$$

Hence,

$$q_{i,n}(f_k - f_l) < \varepsilon \text{ and } p_n(a_k - a_l) < \varepsilon \text{ whenever } k > l > N.$$

So (f_k) and (a_k) are Cauchy sequences in the complete algebras $\mathcal{K}(\mathbb{R}, A; [-i, i])$ and A respectively. Let $f \in \mathcal{K}(\mathbb{R}, A; [-i, i])$ and $a \in A$ be such that $\lim_{k\to\infty} f_k = f$ and $\lim_{k\to\infty} a_k = a$. Then, $((f_k, a_k))$ converges to (f, a). Indeed, for each fixed $n \in \mathbb{N}$ and arbitrary $\varepsilon > 0$ there exists N such that $q_{i,n}(f_k - f) < \frac{\varepsilon}{2}$ and $p_n(a_k - a) < \frac{\varepsilon}{2}$ whenever $k > N$. Thus $Q_{i,n}((f_k, a_k) - (f, a)) < \varepsilon$ for each $n \in \mathbb{N}$ whenever $k > N$. Consequently, $(\mathcal{K}(\mathbb{R}, A; [-i, i]) \times A, \tau_i')$ is complete and hence a locally pseudoconvex F-algebra. Therefore $\mathcal{K}(\mathbb{R}, A) \times A$ in the locally pseudoconvex inductive limit topology τ' is a $\mathcal{LF}p$-algebra[5].

To show that $\mathcal{K}(\mathbb{R}, A) \times A$ is a Q-$\mathcal{LF}p$-algebra, let τ'' be the topology on $\mathcal{K}(\mathbb{R}, A) \times A$ given by k_n-homogeneous seminorms Q_n, where

$$Q_n(\psi, a) = q_n(\psi) + p_n(a) \quad \text{and} \quad q_n(\psi) = \sup_{x\in\mathbb{R}}(p_n(\psi(x)))$$

for every $n \in \mathbb{N}$, $\psi \in \mathcal{K}(\mathbb{R}, A)$ and $a \in A$. Then $(\mathcal{K}(\mathbb{R}, A) \times A, \tau'')$ is a metrizable locally pseudoconvex algebra.

We prove that the inclusion

$$\Psi_i : (\mathcal{K}(\mathbb{R}, A; [-i, i]) \times A, \tau_i') \to (\mathcal{K}(\mathbb{R}, A) \times A, \tau'')$$

is continuous in this topology for each i. For this, let

$$U_{rn} = \{\psi \in \mathcal{K}(\mathbb{R}, A) : q_n(\psi) < r\} \text{ and } W_{rn} = \{a \in A : p_n(a) < r\}$$

for each $r > 0$ and $n \in \mathbb{N}$. Then $V_{rn} = U_{rn} \times W_{rn}$ is a neighborhood of (θ, θ_A) (θ is the zero element of $\mathcal{K}(\mathbb{R}, A)$ and θ_A the zero element of A) in $(\mathcal{K}(\mathbb{R}, A) \times A, \tau'')$ and

$$V_{rn} \cap (\mathcal{K}(\mathbb{R}, A; [-i, i]) \times A) = \{\psi \in \mathcal{K}(\mathbb{R}, A; [-i, i]) : q_{i,n}(\psi) < r\} \times W_{rn}$$

is a neighborhood of (θ, θ_A) in $\mathcal{K}(\mathbb{R}, A; [-i, i]) \times A$ in the topology τ_i' (θ is the zero element also in $\mathcal{K}(\mathbb{R}, A; [-i, i])$). Hence Ψ_i is continuous at (θ, θ_A) for every i. Therefore Ψ_i is continuous. So $\tau'' \prec \tau'$ by the definition of τ'. Thus, it is enough to show that $(\mathcal{K}(\mathbb{R}, A) \times A, \tau'')$ and $(\mathcal{K}(\mathbb{R}, A; [-i, i]) \times A, \tau_i')$ are Q-algebras for every $i \in I$. By assumption, (A, τ) is a Q-algebra. Then there are $n \in \mathbb{N}$ and $\varepsilon > 0$ such that $e_A + W_{\varepsilon n} \subset \operatorname{Inv}(A)$ and

$$V_{\varepsilon n} = U_{\varepsilon n} \times (e_A + W_{\varepsilon n})$$

is a neighborhood of (θ, e_A) in $(\mathcal{K}(\mathbb{R}, A) \times A, \tau'')$, hence in $(\mathcal{K}(\mathbb{R}, A) \times A, \tau')$ also because $\tau'' \prec \tau'$. To see that $V_{\frac{\varepsilon}{2}n} \subset \operatorname{Inv}(\mathcal{K}(\mathbb{R}, A) \times A)$, let $(\psi, a) \in V_{\frac{\varepsilon}{2}n}$. Then

$$p_n(\psi(x) + a - e_A) \leq p_n(\psi(x)) + p_n(a - e_A) < \frac{\varepsilon}{2} + \frac{\varepsilon}{2} = \varepsilon$$

[5]We consider on $\mathcal{K}(\mathbb{R}, A) \times A$ and on $\mathcal{K}(\mathbb{R}, A; [-i, i]) \times A$ the separately continuous multiplication defined by $(f, a)(f', a') = (ff' + fa' + af', aa')$, where $fa'(x) = f(x)a'$ and $af'(x) = a(f'(x))$. In that case (θ, e_A) is the unit element in $\mathcal{K}(\mathbb{R}, A) \times A$ and $\mathcal{K}(\mathbb{R}, A; [-i, i]) \times A$, where θ is the zero element in $\mathcal{K}(\mathbb{R}, A; [-i, i])$ and e_A is the unit element in A.

for every $x \in \mathbb{R}$. Hence $\psi(x) + a \in e_A + W_{\varepsilon n} \subset \operatorname{Inv}(A)$. It is easy to check that the inverse of (ψ, a) is (ϕ_a, a^{-1}), where $\phi_a : \mathbb{R} \to A$ is a function, defined by

$$\phi_a(x) = -(\psi(x) + a)^{-1}\psi(x)a^{-1}$$

for every $x \in \mathbb{R}$ and $a \in e_A + W_{\frac{\varepsilon}{2}n}$. Since the invertion and the multiplications in A by an element from the right and from the left are continuous, then ϕ_a is continuous for every $a \in e_A + W_{\frac{\varepsilon}{2}n}$. Moreover, $\phi_a(x) = \theta_A$ if and only $\psi(x) = \theta_A$. Hence $\phi_a \in \mathcal{K}(\mathbb{R}, A)$. Therefore, $(\psi, a)^{-1}$ exists in $\mathcal{K}(\mathbb{R}, A) \times A$ for any $(\psi, a) \in V_{\frac{\varepsilon}{2}n}$. So we have proved that $(\theta, e_A) \in V_{\frac{\varepsilon}{2}n} \subset \operatorname{Inv}(\mathcal{K}(\mathbb{R}, A) \times A)$. Consequently, $(\mathcal{K}(\mathbb{R}, A) \times A, \tau'')$ is a Q-algebra. Thus, $(\mathcal{K}(\mathbb{R}, A) \times A, \tau')$ is also a Q-algebra.

Moreover,

$$U_{\frac{\varepsilon}{2}n} \cap \mathcal{K}(\mathbb{R}, A; [-i, i]) = \left\{\psi \in \mathcal{K}(\mathbb{R}, A; [-i, i]) : q_{in}(\psi) < \frac{\varepsilon}{2}\right\}$$

is a neighborhood of zero in $\mathcal{K}(\mathbb{R}, A; [-i, i])$ in the topology τ_i for every i. Therefore $V_{\frac{\varepsilon}{2}n} \cap (\mathcal{K}(\mathbb{R}, A; [-i, i]) \times A)$ is a neighborhood of (θ, e_A) in $\mathcal{K}(\mathbb{R}, A; [-i, i]) \times A$. Similarly as above,

$$(\theta, e_A) \in V_{\frac{\varepsilon}{2}n} \cap (\mathcal{K}(\mathbb{R}, A; [-i, i]) \times A) \subset \operatorname{Inv}(\mathcal{K}(\mathbb{R}, A; [-i, i]) \times A).$$

Hence $\mathcal{K}(\mathbb{R}, A; [-i, i])) \times A$ is also a Q-algebra for every $i \in I$. Thus, $(\mathcal{K}(\mathbb{R}, A) \times A, \tau')$ is a Q-$\mathcal{LF}p$-algebra.

To give other examples, we use the following result.

PROPOSITION 2.1. Let A be an unital Q-algebra and B a Q_l-algebra[6] which is an A-module with separate continuous module multiplication. Then $B \times A$ in the product topology is a unital Q_l-algebra. Moreover, if B is a Q-algebra which is a commutative[7] A-module with separate continuous module multiplication, then $B \times A$ in the product topology is an unital Q-algebra.

PROOF. Let first B be a Q_l-algebra, then $\operatorname{Qinv}_l B$ is a neighborhood of zero in B. By the assumption, the map $f_a : B \to B$, defined by $f_a(b) = ab$ for every $a \in A$ and $b \in B$, is continuous. Therefore, there exists a neighborhood O_B of zero in B such that $f_{-a^{-1}}(O_B) \subseteq \operatorname{Qinv}_l B$. Thus, $-a^{-1}b \in \operatorname{Qinv}_l B$ for each $a \in \operatorname{Inv} A$ and $b \in O_B$. Hence, $-a^{-1}b$ has the left quasi-inverse $(-a^{-1}b)_q^{-1}$ in B. Therefore, (b, a) with $a \in \operatorname{Inv} A$ and $b \in O_B$ is left invertible in $B \times A$ and its left inverse is $(-(-a^{-1}b)_q^{-1}a^{-1}, a^{-1})$. So[8] $O_B \times \operatorname{Inv} A \subset \operatorname{Inv}_l(B \times A)$. As $O_B \times \operatorname{Inv} A$ is a neighborhood of (θ_B, e_A) in the product topology on $B \times A$, then $\operatorname{Inv}_l(B \times A)$ is also a neighborhood of (θ_B, e_A) in the product topology on $B \times A$. Consequently, $B \times A$ is a Q_l-algebra.

The proof is similar if B is a Q-algebra and a commutative A-module. $\square$

Using the Proposition 2.1, we give a family of examples of Q-$\mathcal{LF}pg$ algebras.

Let A be a unital locally pseudoconvex F-algebra, which is also a Q-algebra, and B a Q-$\mathcal{LF}pg$-algebra, defined by a family of locally pseudoconvex F-algebras B_α with $\alpha \in I$. Then

$$B = \underset{\longrightarrow}{\bigcup} B_\alpha.$$

[6]A topological algebra is a Q_l-*algebra* if the set of left quasi-invertible elements in A is open.
[7]That is $ab = ba$ for each $a \in A$ and $b \in B$.
[8]Here $\operatorname{Inv}_l(B \times A)$ denotes the set of left invertible elements in $B \times A$.

In addition, let B be a commutative A-bimodule with separate continuous module multiplication. Then

$$B \times A = \underset{\longrightarrow}{\bigcup}(B_\alpha \times A)$$

(where $B_\alpha \times A$ has the product topology τ_α for each $\alpha \in I$) is Q-$\mathcal{LF}pg$ algebra in the locally pseudoconvex inductive limit topology τ_{ind}. Indeed, $(B_\alpha \times A, \tau_\alpha)$ is a locally pseudoconvex F-algebra for every $\alpha \in I$ (see [**13**, p. 59, Prop. 6] and [**16**, p. 6]). Moreover, $B \times A$ in the product topology τ and $B_\alpha \times A$ in the product topology τ_α are unital Q-algebras by Proposition 2.1 and the inclusion

$$\phi_\alpha : B_\alpha \times A \to (B \times A, \tau)$$

is continuous for each $\alpha \in I$ since

$$(U \times V) \cap (B_\alpha \times A) = (U \cap B_\alpha) \times V \in \tau_\alpha$$

for every $U \in \tau_B$ and $V \in \tau_A$ (the topology of B_α coincides with the topology, induced on B_α by B). Hence $\tau \subset \tau_{ind}$ and therefore $B \times A$ is a Q-algebra also in the topology τ_{ind}. Consequently, $B \times A$ is a Q-$\mathcal{LF}pg$-algebra.

2.2. Cases, when the inductive limit of topological algebras is a Q-algebra. 1. First we consider the case when the inductive limit of topological algebras is a Q-algebra in the inductive limit topology.

THEOREM 2.2. *Let (E, τ) be a topological algebra, which is an inductive limit of Q-algebras (E_α, τ_α) with $\alpha \in I$, and τ is the inductive limit topology on E. Then, (E, τ) is Q-algebra.*

PROOF. It is known that (E, τ) is a Q-algebra if and only if Qinv$E$ is a neighborhood of zero in E (see, for example, [**15**, p. 44]). To prove that Qinv$E$ is a neighborhood of zero in (E, τ), let $f_\alpha : E_\alpha \to E$ be the canonical map for $\alpha \in I$. Since every (E_α, τ_α) is a Q-algebra, then for every $\alpha \in I$ there exists an open neighbourhood O_α of zero in E_α such that $O_\alpha \subseteq \mathrm{Qinv}E_\alpha$. So

$$f_\alpha(O_\alpha) \subseteq f_\alpha(\mathrm{Qinv}E_\alpha) \subseteq \mathrm{Qinv}E$$

for every $\alpha \in I$. Hence

$$E_0 = \bigcup_{\alpha \in I} f_\alpha(O_\alpha) \subseteq \mathrm{Qinv}E.$$

Since $O_\alpha \subseteq f_\alpha^{-1}(E_0)$ for every $\alpha \in I$, then $f_\alpha^{-1}(E_0)$ is a neighborhood of zero in E_α for every $\alpha \in I$. Hence E_0 is a neighborhood of zero in E. So Qinv$E$ is also a neighborhood of zero in (E, τ). □

2. Let now A be an algebra over $\mathbb{C}$ and let $a \in A$. The *spectrum* $sp_A(a)$ of a is defined by

$$sp_A(a) = \left\{\lambda \in \mathbb{C} \setminus \{0\} : \frac{a}{\lambda} \notin \mathrm{Qinv}A\right\} \cup \{0\}$$

and the spectral radius $\rho_A(a)$ of a by

$$\rho_A(a) = \sup\{|\lambda| : \lambda \in sp_A(a)\}.$$

For every topological algebra (A, τ) let hom A denote the set of all nontrivial continuous linear multiplicative functionals on A. Then

$$\{\varphi(a) : \varphi \in \ hom A\} \subseteq sp_A(a)$$

for every $a \in A$. When

$$sp_A(a) = \{\varphi(a) : \varphi \in \hom A\}$$

for each $a \in A$, then (A, τ) is called a *topological algebra with functional spectrum* or (A, τ) *has functional spectrum* (see [**2**]). In this case

$$\rho_A(a) = \sup_{\varphi \in homA} \mid \varphi(a) \mid .$$

Therefore, in this case ρ_A is a seminorm on A. It is known (see [**2**, pp. 22–23]) that every commutative unital invertibly complete[9] locally m-pseudoconvex Hausdorff algebra over $\mathbb{C}$ has functional spectrum. Moreover, there are given equivalent conditions for a commutative unital locally m-pseudoconvex algebra to be a topological algebra with functional spectrum.

LEMMA 2.3. *Let (E, τ) be a topological algebra. Then, the spectral radius ρ_E is upper semicontinuous on (E, τ) if and only if (E, τ) is a Q-algebra.*

PROOF. See [**3**, p. 66] and for unital algebras see [**5**, p. 118], [**6**, pp. 159–160], [**7**, p. 13] or [**12**, p. 60]. □

PROPOSITION 2.4. 1. Let E be a topological algebra over $\mathbb{C}$ such that $E = \underrightarrow{\bigcup} E_\alpha$, where E_α are topological algebras over $\mathbb{C}$.

a) If $sp_E(x)$ and $sp_{E_\alpha}(x)$ (at least for one) $\alpha \in I$ are proper subsets of $\mathbb{C}$, then for such $x \in E$ are true[10]

(1) $sp_E(x) = \bigcap_{\alpha \in I_x} sp_{E_\alpha}(x)$;

(2) $\rho_E(x) \leqslant \inf_{\alpha \in I_x} \rho_{E_\alpha}(x)$. Moreover, if $(0, \rho_{E_\alpha}(x)) \subseteq sp_{E_\alpha}(x)$ for every $\alpha \in I_x$, then $\rho_E(x) = \inf_{\alpha \in I_x} \rho_{E_\alpha}(x)$;

b) $\text{Qinv}E = \bigcup_{\alpha \in I} \text{Qinv}E_\alpha$ and $\text{Inv}E = \bigcup_{\alpha \in I} \text{Inv}E_\alpha$, if E is a unital algebra.

2. Let (E, τ) be a locally pseudoconvex algebra over $\mathbb{C}$ such that $E = \underrightarrow{\bigcup} E_\alpha$, where every (E_α, τ_α) is a locally pseudoconvex Q-algebra and τ the locally pseudoconvex inductive limit topology on E. If, in addition, one of the following statements holds:

(3) $\text{Qinv}E_\alpha \in \tau$ for each $\alpha \in I$;

(4) the spectral radius ρ_{E_α} of E_α is a seminorm on E_α for each $\alpha \in I$;

(5) I has the minimal element α_0 and $f_{\beta\alpha_0}$ is an open map for each $\beta \in I$,

then (E, τ) is a Q-algebra.

PROOF. To see that the statement (1) holds, let first $\lambda \neq 0$ be a complex number and

$$\lambda \notin \bigcap_{\alpha \in I_x} sp_{E_\alpha}(x).$$

[9] A topological algebra (A, τ) is *invertibly complete* (see [**2**, p. 15]) if every invertibly convergent Cauchy net $(a_\lambda)_{\lambda \in \Lambda}$ (that is, there exists an element $a \in A$ such that $(aa_\lambda)_{\lambda \in \Lambda}$ and $(a_\lambda a)_{\lambda \in \Lambda}$ converge to e_A) converges in A. It is known (see [**15**, p. 45]) that every complete unital algebra and every unital Q-algebra are invertibly complete.

[10] Here $I_x = \{\alpha \in I : x \in E_\alpha\}$ for each $x \in E$.

Then $\frac{x}{\lambda}$ is quasi-invertible in E_α for some $\alpha \in I_x$. Since $E_\alpha \subset E$, then $\frac{x}{\lambda}$ is quasi-invertible also in E. Hence, $\lambda \notin sp_E(x)$. Consequently,

$$sp_E(x) \subseteq \bigcap_{\alpha \in I_x} sp_{E_\alpha}(x).$$

Now, we suppose that $\lambda \in \mathbb{C}\backslash\{0\}$ is such that $\lambda \notin sp_E(x)$. Then $\frac{x}{\lambda}$ is quasi-invertible in E.

If $\frac{x}{\lambda} \in E_\alpha$ and $y = (\frac{x}{\lambda})_q^{-1} \in E_\beta$, then there exists $\gamma \in I$ such that $E_\alpha \subseteq E_\gamma$ and $E_\beta \subseteq E_\gamma$. Then, $\frac{x}{\lambda}$, $y \in E_\gamma$ and hence $\frac{x}{\lambda}$ is quasi-invertible in E_γ. This means that $\lambda \notin sp_{E_\gamma}(x)$. Moreover, $\gamma \in I_x$ because $\frac{x}{\lambda} \in E_\gamma$. Hence,

$$\lambda \notin \bigcap_{\alpha \in I_x} sp_{E_\alpha}(x).$$

This means that

$$\bigcap_{\alpha \in I_x} sp_{E_\alpha}(x) \subseteq sp_E(x).$$

Then statement (1) holds.

By (1) is true that $sp_E(x) \subseteq sp_{E_\alpha(x)}$ for every $\alpha \in I_x$. So $\rho_E(x) \leqslant \rho_{E_\alpha}(x)$ for every $\alpha \in I_x$. Hence $\rho_E(x) \leqslant \inf\limits_{\alpha \in I_x} \rho_{E_\alpha}(x)$. Suppose that $\rho_E(x) < \inf\limits_{\alpha \in I_x} \rho_{E_\alpha}(x)$. Then there exists a λ such that $\rho_E(x) < \lambda < \inf\limits_{\alpha \in I_x} \rho_{E_\alpha}(x)$. Hence, $\lambda \notin sp_E(x)$ and $\lambda \in (0, \rho_{E_\alpha}(x)) \subset sp_{E_\alpha}(x)$ for every $\alpha \in I_x$, which is not possible. Consequently, $\rho_E(x) = \inf\limits_{\alpha \in I_x} \rho_{E_\alpha}(x)$ in this case.

b) Let $x \in \text{Qinv}E$. Then there is an element $y \in E$ such that y is the quasi-inverse of x. Similarly as above we can assume that $x, y \in E_\gamma$ for some $\gamma \in I$. Therefore x is quasi-invertible in E_γ. Thus

$$x \in \text{Qinv}E_\gamma \subset \bigcup_{\alpha \in I} \text{Qinv}E_\alpha.$$

Let now

$$x \in \bigcup_{\alpha \in I} \text{Qinv}E_\alpha.$$

Then there exists an index $\beta \in I$ such that $x \in \text{Qinv}E_\beta$. Therefore, x is quasi-invertible in $E_\beta \subset E$. Hence, $x \in \text{Qinv}E$.

The proof for invertible elements in E is similar.

2. Let (E_α, τ_α) be a Q-algebra for every $\alpha \in I$. Since holds (3), then (E, τ) is a Q-algebra by the statement b).

Let now the spectral radius ρ_{E_α} of E_α be a seminorm on E_α for each $\alpha \in I$. Since E_α is a Q-algebra, then ρ_{E_α} is upper semicontinuous at zero by Lemma 2.3, that is, for each $\varepsilon > 0 = \rho_{E_\alpha}(\theta)$ and $\alpha \in I$ there is a neighborhood U_α of zero in (E_α, τ_α) such that $\varepsilon > \rho_{E_\alpha}(x)$ for every $x \in U_\alpha$. Then the absolutely k-convex hull $V_k = \Gamma_k(U)$ of $U = \bigcup\limits_{\alpha \in I} U_\alpha$ for some $k \in (0, 1]$ is a neighborhood of zero[11] in (E, τ). If $x \in V_k$, then

$$x = \sum_{\nu=1}^{n} \mu_\nu u_\nu$$

[11]Because $U_\beta \subset U \cap E_\beta \subset V_k \cap E_\beta$ for every $\beta \in I$.

with $u_1, \ldots, u_n \in U$, $\mu_1, \ldots, \mu_n \in \mathbb{K}$ and $\sum_{\nu=1}^{n} |\mu_\nu|^k \leqslant 1$. Since I is a directed set, then there is an $\alpha \in I$ such that $u_\nu \in E_\alpha$ for every ν with $1 \leqslant \nu \leqslant n$. Hence $x \in E_\alpha$ and

$$\rho_E(x) \leqslant \rho_{E_\alpha}(x) \leqslant \sum_{\nu=1}^{n} |\mu_\nu| \rho_{E_\alpha}(u_\nu) < \varepsilon \sum_{\nu=1}^{n} |\mu_\nu| \leqslant \varepsilon \sum_{\nu=1}^{n} |\mu_\nu|^k \leqslant \varepsilon$$

for every $x \in V_k$. So, ρ_E is upper semicontinuous at zero in E. Taking this into account, the set $U' = \{x \in E : \rho_E(x) < 1\}$ is an open neighborhood of zero in E. Suppose that $U' \not\subset \mathrm{Qinv}E$. Then there is an element $u \in U'$ such that $u \notin \mathrm{Qinv}E$. Now $1 \in sp_E(u)$, but this is not possible, because $\rho_E(u) < 1$. Thus, $U' \subset \mathrm{Qinv}E$ and E is a Q-algebra.

To prove (5), let O_α be for every $\alpha \in I$ an absolutely k_α-convex neighborhood of zero in E_α such that $O_\alpha \subseteq \mathrm{Qinv}E_\alpha$, since $\mathrm{Qinv}E_\alpha$ is a neighborhood of zero in E_α.

Then $O = \Gamma_{k_{\alpha_0}}(f_{\alpha_0}(O_{\alpha_0}))$ is a neighborhood of zero in the topology τ. Indeed, $f_{\beta\alpha_0}(O_{\alpha_0})$ is a neighborhood of zero in E_β by hypothesis and

$$f_{\beta\alpha_0}(O_{\alpha_0}) \subseteq f_\beta^{-1}[f_\beta \circ f_{\beta\alpha_0}(O_{\alpha_0})] = f_\beta^{-1}[f_{\alpha_0}(O_{\alpha_0})] \subseteq f_\beta^{-1}(O)$$

for every $\beta \in I$.

To show that $O \subset \mathrm{Qinv}E$, let $x \in O$. Then

$$x = \sum_{\nu=1}^{n} \mu_\nu x_\nu$$

with

$$n \in \mathbb{N}, \quad x_1, \ldots, x_n \in f_{\alpha_0}(O_{\alpha_0}), \quad \mu_1, \cdots, \mu_n \in \mathbb{K} \text{ and } \sum_{\nu=1}^{n} |\mu_\nu|^{k_{\alpha_0}} \leqslant 1.$$

Now there exist $o_1, \ldots, o_n \in O_{\alpha_0}$ such that $x_\nu = f_{\alpha_0}(o_\nu)$ for each ν. Hence

$$x = f_{\alpha_0}\Big(\sum_{\nu=1}^{n} \mu_\nu o_\nu\Big) \subseteq f_{\alpha_0}(\Gamma_{k_{\alpha_0}}(O_{\alpha_0})) = f_{\alpha_0}(O_{\alpha_0}) \subset \mathrm{Qinv}E.$$

It means that $O \subseteq \mathrm{Qinv}E$. Hence $\mathrm{Qinv}E$ is a neighborhood of zero in E in the locally pseudoconvex inductive limit topology on E. □

Remark 1. The proof of the case 2 – from (5) follows that (E, τ) is a Q-algebra – does not need the restriction $E = \underset{\longrightarrow}{\bigcup} E_\alpha$ to E_α.

COROLLARY 2.5. Let (E, τ) be a $\mathcal{LF}pg$-algebra defined by locally pseudoconvex F-algebras E_α with $\alpha \in I$ and let ρ_{E_α} be a seminorm on E_α for each $\alpha \in I$. If, in addition, every (E_α, τ_α) is a Q-algebra, then (E, τ) is also a Q-$\mathcal{LF}pg$-algebra.

PROPOSITION 2.6. Let (E, τ) be a $\mathcal{LF}p$-algebra defined by locally k_n-convex F-algebras (E_n, τ_n) and let

$$k = inf\{k_n : n \in \mathbb{N}\} > 0.$$

If, in addition, every (E_n, τ_n) is a Q-algebra, then (E, τ) is a Q-$\mathcal{LF}p$-algebra.

PROOF. Since every (E_n, τ_n) is a Q-algebra, then $\mathrm{Qinv}E_n$ is a neighborhood of zero in E_n for each $n \in \mathbb{N}$. Hence, there is an absolutely k_n-convex neighborhood O_n of zero such that $O_n \subseteq \mathrm{Qinv}E_n$ for each $n \in \mathbb{N}$. We can assume that

$$O_1 \supseteq O_2 \supseteq \cdots \supseteq O_n \cdots,$$

otherwise, instead of neighborhoods O_n, we consider in E_n new neighborhoods O'_n of zero such that $O'_1 = O_1$ and $O'_n = O_n \cap O'_{n-1}$ for $n > 1$. Let

$$V = \bigcup_{n \in \mathbb{N}} O_n.$$

Then V is a neighborhood of zero in (E, τ). To show that V is absolutely k-convex in E, let $x, y \in V$ and $\lambda, \mu \in \mathbb{K}$ be such that $|\lambda|^k + |\mu|^k \leqslant 1$. Then there exists $n_0 \in \mathbb{N}$ such that $x, y \in O_{n_0}$ and $\lambda x + \mu y \in O_{n_0}$ because O_{n_0} is absolutely k_{n_0}-convex and

$$|\lambda|^{k_{n_0}} + |\mu|^{k_{n_0}} \leqslant |\lambda|^k + |\mu|^k \leqslant 1.$$

Then V is an absolutely k-convex neighborhood of zero in E such that

$$V \subseteq \bigcup_{n \in \mathbb{N}} \operatorname{Qinv} E_n = \operatorname{Qinv} E$$

by Proposition 2.4 b). It means that Qinv$E$ is a neighborhood of zero in E in the topology τ. Consequently, (E, τ) is a Q-algebra. □

Remark 2. For a locally convex Fréchet Q-algebra A it is shown in [**6**, p. 166] that $\mathcal{K}(\mathbb{R}, A) \times A$ is a Q-$\mathcal{LF}$-algebra. Proposition 2.1 in [**1**, Proposition 2] has been proved in case when $A = \mathbb{C}$; Theorem 2.2 is in [**15**, p. 123] for complete locally m-convex case and Corollary 2.5 and Proposition 2.6 (with some misprints) in [**6**, p. 163] for locally m-convex case.

References

[1] M. Abel, *Dense subalgebras in noncommutative Jordan topological algebras*, Acta Comment. Univ. Tartu. Math. **1** (1996), 65–70. MR1711652 (2000e:46061)

[2] M. Abel, *Advertive topological algebras*, General topological algebras (Tartu, 1999), Math. Stud. (Tartu), vol. 1, Est. Math. Soc., Tartu, 2001, pp. 14–24. MR1853508 (2002j:46055)

[3] M. Abel, *Characterization of associative Q-algebras*, Vestn. Kyrg. Gos. Nats. Univ., Mat. Nauk 3 (2001), 5, 64–68.

[4] M. Abel and R. M. Pérez-Tiscareño, *Locally pseudoconvex inductive limit of topological algebras*, Mediterr. J. Math. **10** (2013), no. 4, 1949–1963, DOI 10.1007/s00009-013-0294-9. MR3119343

[5] M. Akkar and C. Nacir, *Continuité automatique dans les limites inductives localement convexes de Q-algèbres de Fréchet* (French, with English and French summaries), Ann. Sci. Math. Québec **19** (1995), no. 2, 115–130. MR1365821 (97a:46060)

[6] M. Akkar and C. Nacir, *Unicité de structure d'algèbres limites inductives localement convexes de suites de Q-algèbres de Fréchet* (French, with Italian summary), Boll. Un. Mat. Ital. A (7) **10** (1996), no. 1, 157–168. MR1386253 (97f:46067)

[7] H. Arizmendi and V. Valov, *Some characterizations of Q-algebras*, Comment. Math. (Prace Mat.) **39** (1999), 11–21. MR1739013 (2001c:46094)

[8] V. K. Balachandran, *Topological algebras*, North-Holland Mathematics Studies, vol. 185, North-Holland Publishing Co., Amsterdam, 2000. Reprint of the 1999 original. MR1836752 (2002d:46043)

[9] A. Bayoumi, *Foundations of complex analysis in non locally convex spaces*, North-Holland Mathematics Studies, vol. 193, Elsevier Science B.V., Amsterdam, 2003. Function theory without convexity condition. MR2017739 (2004m:46102)

[10] J. Bensalem and C. Nacir, *Uniqueness of the topology in some classes of semisimple locally m-pseudoconvex algebras and their inductive limits*, General topological algebras (Tartu, 1999), Math. Stud. (Tartu), vol. 1, Est. Math. Soc., Tartu, 2001, pp. 48–58. MR1853833 (2003f:46067)

[11] M. Fragoulopoulou, *Uniqueness of topology for semisimple* LFQ-*algebras*, Proc. Amer. Math. Soc. **117** (1993), no. 4, 963–969, DOI 10.2307/2159523. MR1101981 (93e:46052)

[12] T. G. Honary and M. Najafi Tavani, *Upper semicontinuity of the spectrum function and automatic continuity in topological Q-algebras*, Note Mat. **28** (2008), no. 2, 57–62 (2010). MR2680990 (2012a:46082)
[13] H. Jarchow, *Locally convex spaces*, B. G. Teubner, Stuttgart, 1981. Mathematische Leitfäden. [Mathematical Textbooks]. MR632257 (83h:46008)
[14] G. Köthe, *Topological vector spaces. I*, Translated from the German by D. J. H. Garling. Die Grundlehren der mathematischen Wissenschaften, Band 159, Springer-Verlag New York Inc., New York, 1969. MR0248498 (40 #1750)
[15] A. Mallios, *Topological algebras. Selected topics*, North-Holland Mathematics Studies, vol. 124, North-Holland Publishing Co., Amsterdam, 1986. Notas de Matemática [Mathematical Notes], 109. MR857807 (87m:46099)
[16] P. Turpin, *Sur une classe d'algèbres topologiques* (French), C. R. Acad. Sci. Paris Sér. A-B **263** (1966), A436–A439. MR0206749 (34 #6566)
[17] L. Waelbroeck, *Topological vector spaces and algebras*, Lecture Notes in Mathematics, Vol. 230, Springer-Verlag, Berlin-New York, 1971. MR0467234 (57 #7098)

Institute of Mathematics, University of Tartu, 2 J. Liivi Str., Room 614, 50409 Tartu, Estonia
E-mail address: mati.abel@ut.ee

Institute of Mathematics, University of Tartu, 2 J. Liivi Str., Room 221, 50409 Tartu, Estonia
E-mail address: reyna.maria.perez.tiscareno@ut.ee

A whirlwind tour of the world of $(\infty, 1)$-categories

Omar Antolín Camarena

ABSTRACT. This introduction to higher category theory is intended to a give the reader an intuition for what $(\infty, 1)$-categories are, when they are an appropriate tool, how they fit into the landscape of higher category, how concepts from ordinary category theory generalize to this new setting, and what uses people have put the theory to. It is a rough guide to a vast terrain, focuses on ideas and motivation, omits almost all proofs and technical details, and provides many references.

CONTENTS

2010 *Mathematics Subject Classification.* Primary 18-01.
Key words and phrases. $(\infty, 1)$-categories, higher category theory.

1. Introduction

An $(\infty, 1)$-category is a category-like thing that besides objects and morphisms has 2-morphisms between morphisms, 3-morphisms between 2-morphisms, and so on all the way to ∞; but in which all k-morphisms for $k > 1$ are "invertible", at least up to higher invertible morphisms. This is the sort of invertibility that homotopies have: the composition or concatenation of any homotopy with its reverse is not actually the identity but it is homotopic to it. So we can picture an $(\infty, 1)$-category as a "homotopy theory": a kind of category with objects, morphisms, homotopies between morphisms, higher homotopies between homotopies and so on.

Any context where there is a notion of homotopy, can benefit from the use of $(\infty, 1)$-categories in place of ordinary categories. This includes homotopy theory itself, of course, but also homological algebra and more generally wherever Quillen's version of abstract homotopy theory, the theory of model categories, has been used. Notions of homotopy are perhaps more common than one might expect since the philosophy of model categories shows that simply specifying a class of "weak equivalences" in a category, a collection of morphisms which we wish to treat as if they were isomorphisms, produces a notion of homotopy. The theory of $(\infty, 1)$-categories plays a prominent role in derived algebraic geometry, as can be expected from the very rough description of the subject as being what is obtained by replacing the notion of commutative rings in algebraic geometry by, say, commutative differential graded algebras but only caring about them up to quasi-isomorphism.

There are now several different formalizations or models of the notion of $(\infty, 1)$-category, detailed comparison results relating the different definitions and for one particular model of $(\infty, 1)$-category, quasi-categories, a detailed account of how ordinary category theory generalizes to the $(\infty, 1)$ context [**Joy08b**], [**Lur09b**], [**Lur12b**]. (Many definitions and *statements* of results from ordinary category theory generalize straightforwardly to $(\infty, 1)$-categories, often simply by replacing bijections of Hom-sets with weak homotopy equivalences of mapping spaces, but with current technology the traditional *proofs* do not generalize, and instead often require delicate model specific arguments: most of this work has been done using the model of quasi-categories.)

Surveying the applications of (ordinary) category theory is an impossible task: categories, and categorical constructions such as products and adjoint functors appear in very many fields of mathematics. Such a survey would turn into a survey of much of mathematics. Writing an overview of the applications of $(\infty, 1)$-categories could potentially be similarly doomed. This paper attempts it anyway only because $(\infty, 1)$-categories are still relatively new and have not fully caught on yet, making it possible to list a reasonable portion of the current literature. Even so, this is just a small entry point into the world of $(\infty, 1)$-categories and the broader context of higher category theory.

The ideal reader of this survey is someone who has heard about $(\infty, 1)$-categories (perhaps under the name ∞-categories), is interested in reading some work that uses them (such as the derived algebraic geometry literature), or is simply curious about them but wishes to have a better idea of what they are and how they are used before committing to read a rigorous treatment such as [**Joy08b**] or [**Lur09b**]. We will not assume any prior knowledge of $(\infty, 1)$-categories, or even more than a cursory knowledge of 2-categories, but we will assume the reader is comfortable with notions of ordinary category theory such as limits, colimits, adjoint functors (but it's fine if the reader can't give a precise statement of Freyd's Adjoint Functor Theorem, for example). We also assume the reader is acquainted with simplicial sets; if that's not the case we recommend reading [**Fri12**] as a gentle introduction that gives the basic definitions and properties and focuses on conveying geometrical intuition. Anyone attempting to use the theory of $(\infty, 1)$-categories will need to know much more about simplicial sets, and [**Cur71**] is a good place to continue learning about them.

We will begin by briefly exploring the landscape of higher category theory to give a context for $(\infty, 1)$-categories and describe some basic guiding principles and requirements for the theory. Then we'll go on a quick tour of all the different models available for $(\infty, 1)$-categories and discuss the problem of comparing different definitions; an exciting recent development is Barwick and Schommer-Pries's axiomatic characterization of higher categories [**BSP12**]. The next section deals with practical aspects of working with $(\infty, 1)$-categories and describes how concepts from ordinary category theory such as isomorphisms, limits and colimits, adjunctions, monads, monoidal categories and triangulated categories generalize to the $(\infty, 1)$ setting. The final section consists of (very!) brief descriptions of some of the work that applies the theory of $(\infty, 1)$-categories.

2. The idea of higher category theory

The first hint of higher category theory comes from the category Cat of categories. It not only has objects, which are categories, and morphisms between them, functors, but there are also natural transformations between functors. Indeed, Cat is the basic example of a (strict[1]) 2-category, just as Set is the basic example of a category. Of course, once we've imagined having another layer of things we'll call 2-morphisms connecting the morphisms (in the way natural transformations connect functors), there is no reason to stop at 2.

This gives us our first blurry picture of higher categories: an n-category will have a collection of objects, and collections of k-morphisms for $1 \leq k \leq n$ with specified identity morphisms and composition operations for morphisms satisfying appropriate associativity and unit axioms; an ∞-infinity category will be a similar structure having k-morphisms for all $k \geq 1$.

REMARK 2.1. We are being very vague and purposefully so: there is a large design space to explore. There are many possible forms for composition laws and many ways of making the axioms precise, and there are even many choices for

[1] We'll explain what this means and have much more to say about it in Section 2.1.

the "shape" of morphisms, that is, choices for what data specifies the analogue of domain-and-codomain of a k-morphism. We won't have much to say about different shapes for morphisms, so that discussion is postponed to section 2.3.

Another way to visualize this idea is also already present in the 2-category Cat: given two categories $\mathcal{C}$ and $\mathcal{D}$, Cat doesn't just have a *set* of morphisms from $\mathcal{C}$ to $\mathcal{D}$, it has a whole category $\mathrm{Fun}(\mathcal{C}, \mathcal{D})$ whose objects are functors $\mathcal{C} \to \mathcal{D}$ and whose morphisms are natural transformations. Note the funny re-indexing that takes place:

(1) functors $\mathcal{C} \to \mathcal{D}$ are 1-morphisms in Cat but are 0-morphisms (objects) in $\mathrm{Fun}(\mathcal{C}, \mathcal{D})$,
(2) natural transformations are 2-morphisms in Cat but are 1-morphisms in $\mathrm{Fun}(\mathcal{C}, \mathcal{D})$.

This gives us an alternative inductive way to think of higher categories: an n-category is like a 1-category but instead of having a Hom-*sets* between any pair of objects, its Hom-things are $(n-1)$-categories. Readers familiar with *enriched*[2] category theory will recognize that this is similar to defining an n-category as a category enriched over $(n-1)$-categories. That actually defines what is known as a *strict* n-category and we will argue in section 2.1 that this notion does not capture the interesting examples that one would want in higher category, so we really want some kind of category "weakly enriched" over $(n-1)$-categories. But before we discuss that, notice that even brushing aside the issue of strictness, this perspective does not by itself define ∞-categories, as the inductive definition becomes circular in case $n = n - 1 = \infty$. However, if we restrict our attention to higher categories in which above a certain level the morphisms behave like homotopies, we can use the inductive perspective again.

Let's say an (n, k)-category is an n-category in which all j-morphisms for $j \geq k+1$ are invertible in the sense homotopies are: *not* that every j-morphism $\alpha : x \to y$ has an inverse β for which $\beta \circ \alpha$ and $\alpha \circ \beta$ are exactly *equal* to the identity j-morphisms id_x and id_y, but only that there is a β for which those composites have invertible[3] $(j+1)$-morphisms connecting them to id_x and id_y. Of course, if $j + 1 > n$, we *do* require that $\beta \circ \alpha = \mathrm{id}_x$ and $\alpha \circ \beta = \mathrm{id}_y$. This convention for $j+1 > n$ is consistent with viewing any n-category as an $(n+1)$-category where all the $(n+1)$-morphisms are identities. Finally, we can similarly talk about (∞, k)-categories (where reaching the top degree for morphisms is not an issue), and the bulk of this survey will focus on the $(\infty, 1)$ case.

REMARK 2.2. A useful metaphor has us think of an invertible morphism between two objects as a proof that they are the "same". Just as with proofs of theorems in mathematics, sometimes one can argue that two proofs are "really the same proof"; such an argument corresponds to an invertible 2-morphism between two 1-morphisms. Then we can think of proofs establishing that two ways of showing that two proofs that two objects are the same are the same are the same, and so on[4]. In other words: an $(\infty, 0)$-category, usually called an *∞-groupoid*, is what a set is forced to become if we are never satisfied to just note that two things can be proven to be the same, but instead we write down the proof and contemplate the

[2]The definition of enriched category is recalled in section 2.1.

[3]Invertible in this same sense, so this definition is recursive.

[4]Limited only by the number of times we are willing to say "are the same" in a row.

possibility that different looking proofs can be proven to be the same. This is what people mean when they say higher category theory *systematically replaces equality by isomorphism.*

For these (n, k)-categories and (∞, n)-categories, the inductive perspective says that an (n, k)-category has Hom-things which are $(n-1, k-1)$-categories (which does not buy us anything new), but also that an (∞, k)-category has $(\infty, k-1)$-categories as Hom-things. To start picturing (∞, n)-categories, we need to know how to visualize $(\infty, 0)$-categories, which is the next topic on our agenda.

2.1. The homotopy hypothesis and the problem with strictness. The 2-category of categories is *strict*, meaning that the composition of both its 1-morphisms and 2 morphisms is associative and has units (the identity 1- and 2-morphisms), as opposed to being just something like "associative up to homotopy". One says that in Cat composition is *strictly* associative. As mentioned above, it is easy to define n-categories with strictly associative and unital compositions inductively using the notion of enriched category. Recall that given a monoidal category $\mathcal{V}$ with tensor product given by a functor $\otimes : \mathcal{V} \times \mathcal{V} \to \mathcal{V}$, a category $\mathcal{C}$ enriched over $\mathcal{V}$ (sometimes called a $\mathcal{V}$-category) consists of

- a collection of objects,
- Hom-objects $\mathcal{C}(X, Y) \in \mathcal{V}$, for every pair of objects of $\mathcal{C}$,
- composition morphisms $\mathcal{C}(Y, Z) \otimes \mathcal{C}(X, Y) \to \mathcal{C}(X, Z)$ of $\mathcal{V}$, for every triple of objects of $\mathcal{C}$,
- identities given as morphisms $I \to \mathcal{C}(X, X)$ in $\mathcal{V}$, for every object of $\mathcal{C}$ (where I is the tensor unit in $\mathcal{V}$),

and this data is required to satisfy obvious unit and associativity axioms (whose precise statement requires using the unit and associativity constraints of $\mathcal{V}$). When $\mathcal{V}$ is a category with finite products, we can take the tensor product to be the categorical product (and I to be the terminal object); when equipped with this tensor product, $\mathcal{V}$ is said to be a *Cartesian* monoidal category. There is also a notion of $\mathcal{V}$-enriched functor between two categories $\mathcal{C}$ and $\mathcal{D}$ enriched over $\mathcal{V}$: a function associating to every object $X \in \mathcal{C}$ an object $FX \in \mathcal{D}$, plus a collection of morphisms of $\mathcal{V}$, $\mathcal{C}(X, Y) \to \mathcal{D}(FX, FY)$ compatible with identities and composition.

We can now give the inductive definition of strict n-categories:

DEFINITION 2.3. A *strict n-category* is a category enriched over the Cartesian monoidal category StrCat_{n-1}. The category StrCat_n whose objects are all strict n-categories and whose morphisms are StrCat_{n-1}-enriched functors is easily seen to have finite products, making the recursion well defined. The base case can be taken to be StrCat_1, the (1-)category of categories and functors or even $\mathrm{StrCat}_0 = \mathrm{Set}$

The only higher category we've mentioned so far is Cat, and it *is* a strict 2-category, but that's more or less it for naturally occurring examples of strict 2-categories, in the sense that almost all natural examples are based on functions and composition of functions.

EXAMPLE 2.4. A monoid M can be regarded as a category that has a single object x for which $\mathrm{Hom}(x, x) = M$ with composition given by the monoid multiplication. In a similar way one cat try to turn a monoidal category $\mathcal{V}$ into a 2-category with one object x for which $\mathrm{Hom}(x, x) = \mathcal{V}$ with composition given by the tensor product in $\mathcal{V}$. This does not produce a strict 2-category unless the tensor product is strictly associative and unital. The point of this example is that most naturally occurring monoidal categories are *not* strict. For example, the tensor product of vector spaces is only defined up to canonical isomorphism, and while $(U \otimes V) \otimes W \cong U \otimes (V \otimes W)$, it is exceedingly unlikely that any actual choice of specific vector spaces for all tensor products would render both sides exactly *equal*. Similar remarks apply to products for Cartesian monoidal categories.

REMARK 2.5. There is a standard notion of non-strict 2-category: the notion of *bicategory* due to Bénabou [**Bén67**] (or see [**Lac10**]), that has a definition very similar to the usual definition of monoidal category and which reduces to it in the case of a bicategory with a single object.

While we have given what we feel are natural examples of 2-categories that fail to be strict, maybe they do not make a conclusive case for the need to weaken the associativity and unitality axioms: MacLane's coherence theorem for monoidal categories shows that any monoidal category is (monoidally) equivalent to one where the tensor product is strictly associative. And more generally any 2-category is equivalent to a strict one[5] (see [**Lei98**] for an expository account). But once we get to 3-categories, the situation is different: there are examples that cannot be made strict. We'll give an explicit example in section 2.2, namely, the fundamental 3-groupoid of S^2; but first we will discuss fundamental higher groupoids and their role in higher category theory.

Higher groupoids are special cases of higher categories, namely an n-groupoid, in the terminology explained above, is an $(n, 0)$-category and an ∞-groupoid is a $(\infty, 0)$-category. Before we explain what higher fundamental groupoids should be, recall that the fundamental groupoid packages the fundamental groups of a space X at all base points into a single category $\pi_{\leq 1}X$ whose objects are the points of X and whose morphisms $x \to y$ are endpoint-preserving homotopy classes of paths from x to y. Composition is given by concatenation of paths (which is not strictly associative and unital before we quotient by homotopy[6]). For a space X that has some non-zero higher homotopy groups, $\pi_{\leq 1}X$ clearly does not contain all the homotopical information of X, but for 1-*types* it does.

DEFINITION 2.6. A space[7] X is called an n-type if $\pi_k(X, x) = 0$ for all $k > n$ and all $x \in X$.

The homotopy theory of 1-types is completely captured by groupoids:

(1) The fundamental groupoid functor induces an equivalence between (a) the homotopy category of 1-types, where the morphisms are homotopy classes

[5]On the other hand maybe we do have a conclusive case for considering more general notions than strict *functors*: not every functor of bicategories between strict 2-categories is equivalent to a strict 2-functor! See [**Lac07**, Lemma 2] for an example. Bénabou has expressed the view that the point of bicategories is not that they are non-strict themselves, but that they are the natural home for non-strict functors.

[6]We could use Moore paths, which are maps $[0, \ell] \to X$ for some $\ell \geq 0$ called the length of the path. When concatenating Moore paths, the lengths add. This operation *is* strictly associative and unital, but (1) the category of Moore paths is not a groupoid, since the reversal of a path only is an inverse up to homotopy, and (2) there is no analogue of Moore paths for the fundamental n-groupoid when $n > 1$.

[7]For technical reasons, space here shall mean "space with the homotopy type of a CW-complex", otherwise some of the statements need homotopy equivalences replaced by weak homotopy equivalence.

of continuous functions between 1-types[8], and (b) the homotopy category of groupoids, whose morphisms are equivalence classes of functors between groupoids, two functors being equivalent if there is a natural isomorphism between them.

(2) The inverse of the equivalence described above can be given by a *classifying space* functor B that generalizes the well-known construction for groups and is defined before passing to homotopy categories, i.e., is a functor from the category of groupoids to the category of 1-types. Any groupoid G is equivalent to $\pi_{\leq 1} BG$, and any 1-type X is homotopy equivalent to $B\pi_{\leq 1} X$.

(3) Given two 1-types X and Y (or more generally, an arbitrary space X and a 1-type Y), the space of maps $\mathrm{Map}(X, Y)$ is a 1-type and its fundamental groupoid is the category of functors $\mathrm{Fun}(\pi_{\leq 1} X, \pi_{\leq 1} Y)$ (which is automatically a groupoid too).

This means that homotopy theoretic questions about 1-types can be translated to questions about groupoids which thus provide complete algebraic models for 1-types. This is the simplest case of perhaps the main guiding principle in the search for adequate definitions in higher category theory: the *homotopy hypothesis* proposed by Grothendieck in [**Gro83**]. As is common now, we interpret (and phrase!) it as stating desired properties of a theory of higher categories.

The homotopy hypothesis: Any topological space should have a fundamental *n-groupoid* for each n (including $n = \infty$). These should furnish all examples of n-groupoids in the sense that every n-groupoid should be equivalent to the fundamental n-groupoid of some space. Furthermore, the theory of n-groupoids should be the "same" as the homotopy theory of n-types (where if $n = \infty$, "the homotopy theory of n-types" is just "homotopy theory").

Notice that this only puts requirements on (n, k)-categories for $k = 0$, so it certainly does not tell the whole story of higher category theory, but it is enough to rule out basing the theory on strict n-categories as we'll see in the next section. This means that we must search for definitions of higher categories that are *non-strict* or *weak*, in the same sense that monoidal categories are weak: instead of associativity meaning that given three k-morphisms f, g and h, the composites $(f \circ g) \circ h$ and $f \circ (g \circ h)$ are *equal*, we should only require them to be linked by an invertible $(k+1)$-morphism $(f \circ g) \circ h \to f \circ (g \circ h)$ that could be called an *associator*. The reader familiar with the definition of monoidal category will know that these associators should satisfy a condition of their own. Given four k morphisms f, g, h and k, we can relate the composites $((fg)h)k$ and $f(g(hk))$ in two different ways

[8] This definition of the category is correct because we took 1-types to have the homotopy type of a CW-complex; we could instead consider the category obtained from 1-types by inverting weak homotopy equivalences.

(we've dropped the $\circ$ for brevity):

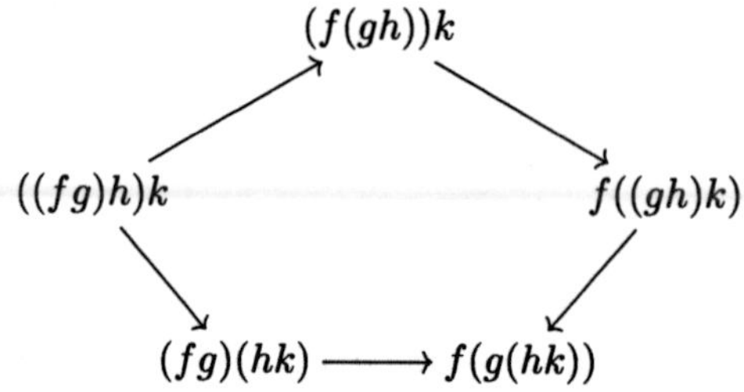

For the case of monoidal categories (where f, g, h and k are objects, $\circ = \otimes$, and the associator is a 1-morphism) we've reached the top level already and we require this diagram to commute; but in a higher category we can instead requires this to commute only up to an invertible $(k+2)$-morphism we could call a *pentagonator*. This pentagonator must satisfy its own condition, but only up to a higher morphisms and so on. This kind of data —the associators, pentagonators, etc.— are what is meant to exist when saying an operation is associative up to *coherent homotopy*.

Clearly, drawing these diagrams gets complicated very quickly and indeed, definitions of n-categories along these lines have only been written down for n up to 4 —for a definition of tricategories see [**GPS95**] or [**Gur06**], for tetracategories see [**Tri06**] or [**Hof11**]. Instead people find ways of implicitly providing all these higher homotopies in a clever roundabout way. We'll see some examples in the section on models of $(\infty, 1)$-categories.

2.2. The 3-type of S^2. We will show that the fundamental 3-groupoid of S^2 is not equivalent to a strict 3-groupoid, or, in other words, that there is no strict 3-groupoid that models the 3-type of S^2, which is commonly denoted P_3S^2 in the theory of Postnikov towers.[9] What we mean by "models" is that we assume the existence of classifying space functors (with certain properties we'll spell out later that *are* satisfied for the "standard realization functors", see the discussion after Theorem 2.4.2 of [**Sim12**]) that produce an n-type BG for a strict n-groupoid, and we say G models a space X if BG is homotopy equivalent to X. The argument shows, more generally, that if X is a simply connected n-type modeled by a strict n-groupoid G, X is in fact an infinite loop space and even a product of Eilenberg-MacLane spaces.

Let's investigate when we can *deloop* a given strict n-groupoid G, i.e., when G can be realized as $\mathrm{Hom}_H(x,x)$ for some strict $(n+1)$-groupoid H with a single object x. That's easy enough: if there exists such an H, G inherits from it a composition $\mu : G \times G \to G$ which makes it into a monoid object in StrCat_n, and clearly for each such monoid structure we can form a delooping H. If we want to deloop more than once, we need a monoid structure on H. And here something remarkable happens: a monoid structure $\nu : H \times H \to H$, in particular restricts to a new monoid structure $\nu_G : G \times G \to G$ on $G = \mathrm{Hom}_H(x,x)$, and, since ν is a StrCat_n-enriched functor, this ν_G must be compatible with composition in H, that is, with μ. The end result is that G has *two* monoid structures one of which is homomorphism for the other.

[9]Recall that the 3-type of S^2 can be obtained, say, by building S^2 as a CW-complex and then inductively attaching cells of larger and larger dimension to kill all homotopy groups π_i for $i \geq 4$.

The classical Eckmann-Hilton argument[10] implies that $\nu_G = \mu$ and that they are commutative. Also, conversely, if μ is a commutative monoid structure for G, H does in fact become a commutative monoid under $\nu(x) = x$, $\nu_G = \mu$.

This means that in the world of strict n-groupoids, being able to deloop twice already implies that you can deloop arbitrarily many times! Using this it is easy to see why you can't find a strict 3-groupoid that models P_3S^2. If there were such a 3-groupoid G, without loss of generality we could assume G had a single object x and a single 1-morphism id_x: otherwise just take the sub-strict-3-groupoid consisting of x, id_x and the groupoid $\mathrm{Hom}_G(\mathrm{id}_x, \mathrm{id}_x)$. But then G is the second delooping of $\mathrm{Hom}_G(\mathrm{id}_x, \mathrm{id}_x)$, which shows that G in turn can be delooped arbitrarily many times. If we had classifying spaces for groupoids that were compatible with looping (by which we mean we had an n-type BG for each strict n-groupoid G such that if G has a single object x, ΩBG is weakly homotopy equivalent to $B(\mathrm{Hom}_G(x, x))$), it would follow that P_3S^2 is an infinite loop space, which it is not. In fact, if classifying spaces preserved products (i.e., $B(G \times H) \cong BG \times BH$), we'd have that P_3S^2 would be a topological abelian monoid and thus homotopy equivalent to a product of Eilenberg-MacLane spaces. It would then have to be $K(\mathbb{Z}, 2) \times K(\mathbb{Z}, 3)$, but it is not, since, for example, the Whitehead product $\pi_2S^2 \times \pi_2S^2 \to \pi_3S^2$ is non-zero.

REMARK 2.7. Vanishing of the Whitehead product $\pi_2 \times \pi_2 \to \pi_3$ does not guarantee that a 3-type can be modeled by a strict 3-groupoid. Consider the space $X = P_3QS^2 = P_3 \operatorname{colim} \Omega^n\Sigma^nS^2$. That Whitehead product for X agrees with the corresponding product for QS^2 but since QS^2 is an infinite loop space its Whitehead products of all degrees vanish. We have that $\pi_2X = \mathbb{Z}$ and $\pi_3X = \mathbb{Z}/2$, the first stable homotopy group of spheres, but one can see X is not homotopy equivalent to $K(\mathbb{Z}, 2) \times K(\mathbb{Z}/2, 3)$ by looking at the operation $\pi_2W \to \pi_3W$ given by composing (maps representing homotopy classes) with the generator of $\pi_3(S^2)$: this operation is non-zero for $W = X$, but is zero for the product of Eilenberg-MacLane spaces. By the argument above, X is not modeled by a strict 3-groupoid.

FURTHER READING. Carlos Simpson [**Sim12**, Section 2.7] proved that there is no classifying space functor for strict 3-groupoids such that BG is homotopy equivalent to P_3S^2 under weaker assumptions than we sketched above: he does not assume that classifying spaces are compatible with looping, in fact, he does not require there to be a family of classifying space functors for strict n-groupoids for all n at all; just a single functor for $n = 3$ satisfying the minimal requirements that BG be a 3-type and that the homotopy groups of BG are functorially isomorphic to algebraically defined ones for G. The simpler argument we sketched (under the stronger assumption of compatibility with looping) can be found in [**Sim12**, Section 2.6]. Clemens Berger proved a stronger result characterizing all connected 3-types (not necessarily simply connected) that can be modeled by strict 3-groupoids [**Ber99**, Corollary 3.4].

---

[10] This says that if a set M has two different monoid structures given by products $\cdot$ and $*$, and we have $(a \cdot b) * (c \cdot d) = (a * c) \cdot (b * d)$ —which says that $* : (M, \cdot) \times (M, \cdot) \to (M, \cdot)$ is a monoid homomorphism— then $\cdot = *$ and M is commutative. Here we are actually using an extension to strict n-categories instead of sets, which is essentially obtained by applying the classical statement to each degree of morphism separately.

2.3. Shapes for cells. There are many possibilities for the shapes of the morphisms in an n-category, of which we'll give a brief representative list. Here we will call the morphisms *cells*, since the word "morphism" is a little awkward when a k-morphism does not simply go from one $(k-1)$-morphism to another.

In the case of the 2-category of categories, the 2-morphisms, which are natural transformations, go between two 1-morphisms (functors) that are parallel, i.e., that share their domain and share their codomain. This pattern can be generalized for higher morphisms and is called *globular*, because drawings of such morphisms looks like topological balls, or more precisely like one of the usual CW-complex structures on disks: the one in which the boundary of the disk is divided into hemispheres meeting along a sphere, which is also divided into hemispheres and so on. Here's a picture of a globular 2-cell:

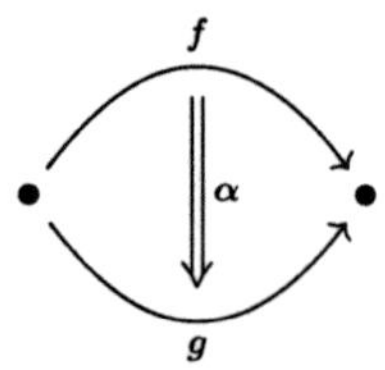

For another example of a shape for morphisms think of homotopies, homotopies between homotopies, and so on. As we mentioned in the introduction, this is one of the examples we are trying to capture. Such higher homotopies are maps $X \times [0,1]^n \to Y$, and so naturally have a *cubical* shape. A cubical 2-morphism looks like a square, and its analogue of domain-and-codomain, the boundary of the square, has four objects and four 1-morphisms:

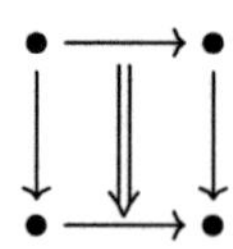

When we get to discussing models for $(\infty, 1)$-categories, and specifically the model of quasi-categories (which are simplicial sets satisfying some condition), we will encounter another shape for morphisms: *simplicial*. A 2-morphism is shaped like a triangle:

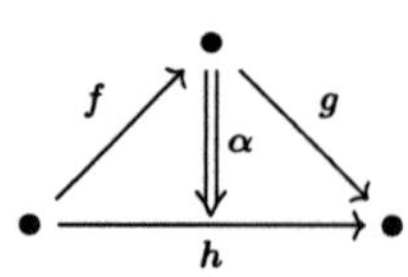

We can interpret α as a homotopy between $g \circ f$ and h (or alternatively we can interpret composition as multivalued, in which case, h is some composite of g and f, and α is a witness to that fact). Similarly we can think of a higher dimensional simplex as being a coherent collection of homotopies between composites of a string of 1-morphisms. See section 3.2.

There are more elaborate cell shapes as well, such as *opetopes*, introduced by John Baez and James Dolan [**BD98**] (see also [**Bae97**] which besides describing opetopes and the proposed definition of higher category based on them, is a nice introduction to n-categories generally). These can be interpreted as being a homotopy between the result of evaluating a *pasting diagram* and a specified target morphism. This is analogous to the above interpretation of simplices, but allowing for more general pasting diagrams than those given by strings of 1-morphisms.

FURTHER READING. To look at pictures of the zoo of higher categories, we recommend the illustrated guidebook by Eugenia Cheng and Aaron Lauda [**CL04**]. (In particular, the above description of opetopes is meaningless without pictures, which can be seen there.) For a concise list of many of the available definitions for n-category and ∞-category see [**Lei02**]. See also the book [**Lei04**], particularly Chapter 10.

Those sources concentrate on definitions attempting to capture n-categories without any requirement of invertibility of morphisms. Thanks to the homotopy hypothesis and to the availability of topological spaces, simplicial sets and homotopy theory, it has turned out somewhat easier in practice to work with notions of (∞, n)-categories (which of course include n-categories as a special case). As John Baez said about climbing up the categorical ladder from 1-groupoids to ∞-groupoids [**Bae05**][11]:

> [...] the n-category theorists meet up with the topologists — and find that the topologists have already done everything there is to do with ∞-groupoids... but usually by thinking of them as spaces, rather than ∞-groupoids!
>
> It's sort of like climbing a mountain, surmounting steep cliffs with the help of ropes and other equipment, and then finding a Holiday Inn on top and realizing there was a 4-lane highway going up the other side.

For the homotopical perspective and a focus on (∞, n)-categories see [**Sim12**]. The rest of this survey will mostly focus on $(\infty, 1)$-categories.

2.4. What does (higher) category theory do for us? The reader might be asking now how exactly higher category theory is useful in mathematics. Here is one possible answer, a purely subjective and personal answer, and should be disregarded if the reader does not find it convincing. It is now widely recognized that category theory is a highly versatile and profitable organizing language for mathematics. Many fields of mathematics have objects of interest and distinguished maps between them that form categories, many comparison procedures between different kinds of objects can be represented as functors and, perhaps, most importantly, basic notions from category theory such as products, coproducts (or more general limits and colimits) and adjoint functors turn out to be well-known important constructions in the specific categories studied in many fields. While it is not reasonable to expect that category theory will swoop in and solve problems from other fields of mathematics, phrasing things categorically does help spot analogies between different fields and to pinpoint where the hard work needs to happen: often arguments are a mix of "formal" parts, which depend very little on the detailed structure of the objects being studied, and "specific" parts which involve understanding their distinguishing properties; categorical language makes short work of many formal arguments, thus highlighting the remainder, the "essential mathematical content" of an argument. Higher category theory promises to extend the scope of such formal methods to encompass situations where we wish to consider objects up to a

[11] John Baez's web column *This Week's Finds* is a highly recommended source for intuition about higher categories.

weaker notion of equivalence than isomorphism; for example, we almost always wish to consider categories up to equivalence, in homological algebra we consider chain complexes up to quasi-isomorphism, and in homotopy theory we consider space up to homotopy equivalence or weak homotopy equivalence.

3. Models of $(\infty, 1)$-categories

This sections gives a list of the main models of $(\infty, 1)$-categories and attempts to motivate each definition. We spend more time discussing quasi-categories than the other models, and in later sections we'll mostly just use quasi-categories whenever we need particular models. The reader will notice an abundance of simplicial sets appearing in the definitions, and is warned again that some basic knowledge of them will be required.

Ideally we would describe for each model, say,

- the definition of $(\infty, 1)$-category,
- the corresponding notion of functor and natural transformation or, even better, given two $(\infty, 1)$-categories $\mathcal{C}$ and D, the $(\infty, 1)$-category $\mathrm{Fun}(\mathcal{C}, \mathcal{D})$ whose objects are functors and whose 1-morphisms are natural transformations,
- how to retrieve the Hom-∞-groupoid, or *mapping space* $\mathrm{Map}_{\mathcal{C}}(X, Y)$ between two objects of a given $(\infty, 1)$-category,
- the homotopy category $\mathrm{ho}\,\mathcal{C}$ of a given $(\infty, 1)$-category, which is the ordinary category with the same objects as $\mathcal{C}$ and whose morphisms correspond to homotopy[12] classes of morphisms in $\mathcal{C}$.

Sadly, for reasons of space we will not do all of those for each model, but we hope to mention enough of these to give an idea of how the story goes.

One excellent feature of the $(\infty, 1)$ portion of higher category theory is that the problem of relating different definitions has a satisfactory answer which will be described in the following section.

FURTHER READING. For a more detailed introduction to the different models and the comparison problem, we recommend [**Ber10**], [**JT06**] or [**Por04**].

3.1. Topological or simplicial categories. As we mentioned above, we can think of an $(\infty, 1)$-category as a category weakly enriched in ∞-groupoids, and to satisfy the homotopy hypothesis we could "cheat" and define ∞-groupoids as topological spaces or simplicial sets (whose homotopy theory is well-known to be equivalent to that of topological spaces). It turns out that one can always "strictify" the enrichment in ∞-groupoids, meaning that we can model $(\infty, 1)$-categories using:

DEFINITION 3.1. A *topological category* is a category enriched over the category of topological spaces[13]. A *simplicial category* is similarly a category enriched over the category of simplicial sets.

[12]Recall that since in an $(\infty, 1)$-category 2-morphisms and higher are invertible, we tend to think of them as homotopies

[13]Instead of the category of all topological spaces it is better to use a so-called "convenient category of spaces" [**Ste67**], such as compactly generated weakly Hausdorff spaces (see, for instance, [**Str**]). This is to make the comparison with other models smoother and is a technical point the reader can safely ignore. One should be aware though, that when using a different category of spaces, the product —which appears in the domain of the composition map $\mathcal{C}(Y, Z) \times \mathcal{C}(X, Y) \to \mathcal{C}(X, Z)$— does not carry the usual product topology.

These models of $(\infty, 1)$-categories are perhaps the easiest to visualize and are a great psychological aid but are inconvenient to work with in practice because, among other problems, enriched functors do not furnish all homotopy classes of functors between the $(\infty, 1)$-categories being modeled, unless the domain and codomain satisfy appropriate conditions[14]. And even when those conditions are met, the category of enriched functors might not correctly model the $(\infty, 1)$-functor category, essentially because enriched functors correspond to functors that preserve composition strictly while $(\infty, 1)$-functors are allowed to preserve it only up to coherent homotopy.

Finally, notice that although we "cheated" by putting the homotopy hypothesis into the definition, there is a sense in which we don't trivially get it back out! We obtained a definition of $(\infty, 1)$-category through enrichment from a definition of ∞-groupoid, but having done so we now have a second definition of ∞-groupoid: an $(\infty, 1)$-category in which all 1-morphisms are invertible (up to higher morphisms, as always). In terms of the homotopy category $\mathrm{ho}\,\mathcal{C}$, this definition of ∞-groupoid that accompanies any notion of $(\infty, 1)$-category is simply: an $(\infty, 1)$-category $\mathcal{C}$ for which $\mathrm{ho}\,\mathcal{C}$ is a groupoid.

For topological or simplicial categories it is easy to construct $\mathrm{ho}\,\mathcal{C}$: take the set of morphisms between X and Y in $\mathrm{ho}\,\mathcal{C}$ to be $\pi_0(\mathcal{C}(X, Y))$; since π_0 preserves products, the composition law in $\mathcal{C}$ induces a composition for $\mathrm{ho}\,\mathcal{C}$. Now, given a topological[15] category $\mathcal{C}$ for which $\mathrm{ho}\,\mathcal{C}$ is a groupoid, what is the space C that this ∞-groupoid is supposed to correspond to? Think first of the case when $\mathcal{C}$ has a single object X. Then $M := \mathcal{C}(X, X)$ is a topological monoid and $\mathrm{ho}\,\mathcal{C}$ being a groupoid just says that $\pi_0(M)$ is a group under the operation induced by the multiplication in M. The topological category $\mathcal{C}$ is a delooping[16] of M, so we should have $\Omega C \cong M$, and there is such a space: the classifying space $C = BM$ of M; when $\pi_0(M)$ is a group, the unit map $M \to \Omega BM$ is a weak homotopy equivalence. For general groupoids $\mathrm{ho}\,\mathcal{C}$, the space C corresponding to $\mathcal{C}$ will be a disjoint union of classifying spaces of the monoid of endomorphisms of an object chosen from each component of $\mathrm{ho}\,\mathcal{C}$.

3.2. Quasi-categories. There are two classes of examples we certainly wish to have in any theory of $(\infty, 1)$-categories: (a) ordinary categories (just add identity morphisms in all higher degrees), and (b) ∞-groupoids, which by the homotopy hypothesis we can take to be anything modeling all homotopy types of spaces. After spaces themselves, the best known models for homotopy types are *Kan complexes*, simplicial sets X that satisfy the *horn filler* condition: that every map $\Lambda^n_k \to X$ extends to a map $\Delta^n \to X$. (Recall that Λ^n_k is obtained from the boundary $\partial\Delta^n$ of Δ^n by removing the k-th face.) Also, every category $\mathcal{C}$ has a nerve which is a simplicial set whose n-simplices are indexed by strings of n composable morphisms of $\mathcal{C}$; and the nerve functor $N : \mathrm{Cat} \to \mathrm{sSet}$ is fully faithful. So inside the category sSet of simplicial sets we find both ordinary categories and Kan complexes and so we might expect to find a good definition of an $(\infty, 1)$-category as a special kind of simplicial set. The following easy characterization of those simplicial sets which arise as nerves of categories shows what to do:

[14]Namely, that the domain be cofibrant and the codomain be fibrant in the model structures discussed in section 4.

[15]The simplicial case is analogous.

[16]See section 2.2.

PROPOSITION 3.2. *A simplicial set X is isomorphic to the nerve of some category if and only if every map $\Lambda^n_k \to X$ with $0 < k < n$ extends* uniquely *to a map $\Delta^n \to X$.*

The least common generalization of the condition above and the definition of Kan complex is:

DEFINITION 3.3. A *quasi-category* is a simplicial set in which all *inner horns* can be filled, that is, in which every map $\Lambda^n_k \to X$ with $0 < k < n$ extends to a map $\Delta^n \to X$.

Probably the greatest advantage of quasi-categories over other models for $(\infty, 1)$-categories is how straightforward it is to deal with functors. A functor $\mathcal{C} \to \mathcal{D}$ between two quasi-categories is simply a map of simplicial sets: the structure of the quasi-categories makes any such maps behave like a functor. (This is related to the nerve functor being fully faithful.) Moreover, there is a simple way to obtain the $(\infty, 1)$-category of functors between two quasi-categories: it is just the simplicial mapping space[17] $\mathcal{D}^{\mathcal{C}}$, which is automatically a quasi-category whenever $\mathcal{C}$ and $\mathcal{D}$ are. In fact, more generally, given a quasi-category $\mathcal{C}$, and an arbitrary simplicial set X, $\mathcal{C}^X$ is a quasi-category which we think of as the category of X-shaped diagrams in $\mathcal{C}$.

The definition of quasi-category is very clean, but it may seem mysterious that it does not mention anything like composition of morphisms. Quasi-categories have something like a "multivalued" composition operation. Consider two morphisms $f : X \to Y$ and $g : Y \to Z$ in a quasi-category $\mathcal{C}$ —this really means that X, Y and Z, are vertices or 0-simplices in the simplicial set $\mathcal{C}$ and that f and g are 1-simplices with the specified endpoints. The data (X, f, Y, g, Z) determines a map $\Lambda^2_1 \to \mathcal{C}$, that we display by drawing Λ^2_1 and labeling the simplices by their images in $\mathcal{C}$. A filler for this horn is a 2-simplex α whose third edge h gives a possible composite of g and f. The 2-simplex itself can be considered to be some sort of certificate that h is a composite of g and f. There may be more than one composite h, and for a given h there may be more than one certificate.

This might seem like chaos, but homotopically composition is well-defined in a sense we'll now make precise. The space of composable pairs of 1-simplices in $\mathcal{C}$ is given by the simplicial mapping space $\mathcal{C}^{\Lambda^2_1}$ and the space of "certified compositions" is similarly $\mathcal{C}^{\Delta^2}$. The set of vertices of $\mathcal{C}^{\Lambda^2_1}$ is precisely the set of pairs of composable 1-simplices, and the higher dimensional simplices capture homotopies between diagrams of composable pairs, and homotopies between those, and so on. Similar remarks apply to $\mathcal{C}^{\Delta^n}$.

PROPOSITION 3.4 (Joyal). *For a quasi-category $\mathcal{C}$, the map $\mathcal{C}^{\Delta^2} \to \mathcal{C}^{\Lambda^2_1}$ induced by composition with the inclusion $\Lambda^2_1 \hookrightarrow \Delta^2$ is a trivial Kan fibration, which implies in particular that its fibers are contractible Kan complexes.*

---

[17]This is the internal hom in sSet, its n-simplices are simplicial maps $\mathcal{C} \times \Delta^n \to \mathcal{D}$.

REMARK 3.5. Joyal proved the converse as well: if $\mathcal{C}$ is a simplicial set such that $\mathcal{C}^{\Delta^2} \to \mathcal{C}^{\Lambda^2_1}$ is a trivial Kan fibration, then $\mathcal{C}$ is a quasi-category.

We can think of the map in the proposition roughly as, given a "certified composition", forgetting both the certificate and the composite being certified. That the fibers of this map are contractible says that to a homotopy theorist composition is uniquely defined after all.

This result can be extended to strings of n composable 1-simplices, namely, for a quasi-category $\mathcal{C}$ the canonical map $\mathcal{C}^{\Delta^n} \to \mathcal{C}^{P_n}$ is a trivial Kan fibration. Here, $P_n = \Delta^1 \vee_{\Delta^0} \Delta^1 \vee_{\Delta^0} \cdots \vee_{\Delta^0} \Delta^1$ is the simplicial path of length n obtained by gluing n different 1-simplices end to end. The map $\mathcal{C}^{\Delta^n} \to \mathcal{C}^{P_n}$ is induced by the inclusion of P_n into Δ^n as the *spine of the simplex*, that is, the subsimplicial set generated by the 1-simplices $\{0, 1\}, \{1, 2\}, \ldots, \{n-1, n\}$ in Δ^n. The case $n = 3$ can be interpreted as specifying a precise sense in which composition is associative.

We can generalize even further to say that when defining a functor from the free (ordinary) category on a directed graph X into a quasi-category $\mathcal{C}$, we can choose a diagram of 0-simplices and 1-simplices in $\mathcal{C}$ of shape X arbitrarily: there will always be an extension to a functor, and moreover, the space of all such extension is contractible. Formally, we have:

PROPOSITION 3.6. *Let X be a reflexive*[18] *directed graph which we will think of as a simplicial*[19] *set which has no non-degenerate k-simplices for $k \geq 2$. For any quasi-category $\mathcal{C}$, the canonical map $\mathcal{C}^{NFX} \to \mathcal{C}^X$ is a trivial Kan fibration, where NFX is the nerve of the free category on X.*

For $X = P_n$, the free category on X is the category which objects $0, 1, \ldots, n$ and a unique morphism from i to j when $i \leq j$; its nerve is the n-simplex Δ^n, so we recover the previous statement. As an example of this proposition, take X to be a single loop[20]: an X-shaped diagram in a category is an object together with an endomorphism. The free category on X is just the monoid of natural numbers under addition regarded as a one object category, say, $FX = B\mathbb{N}$. In the world of ordinary categories, once you've chosen an object and an endomorphism f of it, you've uniquely specified a functor out of $B\mathbb{N}$: the functor sends k to $f^k = f \circ f \circ \cdots \circ f$. For quasi-categories, there is no canonical choice of f^k, you must make a choice for each k and then, to specify a functor out of $B\mathbb{N}$ you need to further choose homotopies and higher homotopies showing you made compatible choices of iterates of f. The proposition says then that all these choices (of iterates and homotopies between their composites) can be made and that they are homotopically unique in the sense that the spaces of such choices is contractible.

We haven't yet described how to get at mapping spaces in quasi-categories. One intuitive approach is to use the arrow $(\infty, 1)$-category of $\mathcal{C}$, which is simply the simplicial mapping space $\mathcal{C}^{\Delta^1}$. This has a projection π to $\mathcal{C} \times \mathcal{C} = \mathcal{C}^{\Delta^0 \sqcup \Delta^0}$ which sends each 1-simplex of $\mathcal{C}$ to its source and target. Then, given two objects X and Y in $\mathcal{C}$, we can think of them as being picked out by maps $\Delta^0 \to \mathcal{C}$ and form the

[18] Reflexive means the graph has a distinguished loop at each vertex; these will play the role of the identities in the free category on the graph.

[19] When thought of a simplicial set, it is understood that the degenerate 1-simplices are the distinguished loops in the graph.

[20] Well, a single non-distinguished loop, in addition to the distinguished one.

pullback:

$$\begin{array}{ccc} \mathrm{Map}_{\mathcal{C}}(X,Y) & \longrightarrow & \mathcal{C}^{\Delta^1} \\ \downarrow & & \downarrow \pi \\ \Delta^0 & \xrightarrow[(X,Y)]{} & \mathcal{C}^{\Delta^0 \sqcup \Delta^0} \end{array}$$

This does work, that is, it produces a simplicial set $\mathrm{Map}_{\mathcal{C}}(X,Y)$ with the correct homotopy type, but there are many other descriptions of the mapping spaces that are all homotopy equivalent but not isomorphic as simplicial sets. One such alternative description of the mapping spaces is given by Cordier's *homotopy coherent nerve* [**Cor82**], used in [**Lur09b**] to compare quasi-categories with simplicial categories. Cordier's construction not only provides models for the mapping spaces but is also a procedure for strictifying composition in a quasi-category: that is, constructing a simplicial category (where composition *is* required to be single-valued and strictly associative) that represents the same $(\infty,1)$-category as a given quasi-category. Dan Dugger and David Spivak in [**DS11b**] explain a really nice way to visualize the mapping spaces appearing in the homotopy coherent nerve through "necklaces" of simplices strung together; they also wrote a second paper giving a detailed comparison of the known constructions for mapping spaces in quasi-categories [**DS11a**].

3.3. Segal categories and complete Segal spaces. Segal categories are a different formalization of the idea discussed above for quasi-categories of a multivalued composition that is uniquely defined homotopically. Just as quasi-categories can be motivated by Proposition 3.2, Segal categories can be motivated by the following equally easy result:

PROPOSITION 3.7. *A simplicial set X is isomorphic to the nerve of a category if and only if for each n, the canonical map $X_n \to X_1 \times_{X_0} X_1 \times_{X_0} \cdots \times_{X_0} X_1$ is a bijection.*

This canonical map is the map $X^{\Delta^n} \to X^{P_n}$ we've already met in section 3.2: it sends an n-simplex to its *spine*, the string of 1-simplices connecting vertices 0 and 1, 1 and 2, ..., $n-1$ and n. It is tempting to try to make compositions only defined up to homotopy simply by requiring these canonical maps to be homotopy equivalences instead of bijections, but, of course, that requires working with spaces rather than sets.

DEFINITION 3.8. A Segal category is a simplicial space (or more precisely a simplicial simplicial-set), that is, a functor $\mathbf{\Delta}^{\mathrm{op}} \to \mathrm{sSet}$ such that

(1) the space of 0-simplices X_0 is discrete, and
(2) for each n, the canonical map $X_n \to X_1 \times_{X_0} X_1 \times_{X_0} \cdots \times_{X_0} X_1$ is a weak homotopy equivalence.

Complete Segal spaces, also called Rezk categories, were defined by Charles Rezk in [**Rez01**]; his purpose was explicitly to find a nice model for the "homotopy theory of homotopy theories", i.e., the $(\infty,1)$-category of $(\infty,1)$-categories[21]. Their definition is a little complicated and we'll only describe it informally, but they do have some advantages one of which was worked out by Clark Barwick in his PhD

[21] $(\infty,1)$-categories naturally form an $(\infty,2)$-category, but we can discard non-invertible natural transformations to get an $(\infty,1)$-category.

thesis [**Bar05**]: the construction of complete Segal spaces starting from simplicial sets as a model for ∞-groupoids, can be iterated to provide a model for (∞, n)-categories. These are called *n-fold complete Segal spaces*, see [**BSP12**] or [**Lur09c**] for a definition, if Barwick's thesis proves too hard to get a hold of.

A *Segal space* like a Segal category, is also a simplicial space, but we do not require that the space of objects X_0 be discrete. In that case, the second condition must be modified to use homotopy pullbacks[22] so that it reads: for each n, the canonical map $X_n \to X_1 \times^h_{X_0} X_1 \times^h_{X_0} \cdots \times^h_{X_0} X_1$ is a weak homotopy equivalence. The *completeness* condition has to do with the fact that having a non-discrete space of objects means we have two different notions of equivalence of objects: one is having an invertible morphism between them in the $(\infty, 1)$-category modeled by X, the other is being in the same connected component of X_0. Even better, there are two canonical ∞-groupoids of objects: one is the *core* of the $(\infty, 1)$-category modeled by X, this is the subcategory obtained by throwing away all non-invertible 1-morphisms (all higher morphisms are already invertible); the other is the ∞-groupoid represented by X_0. The core of X can be described as a simplicial set directly in terms of the simplicial space X; the completeness condition then says that it and X_0 are homotopy equivalent.

3.4. Relative categories. Relative categories are based on the intuition that higher category theory is meant for situations where we want to treat objects up to a notion of equivalence that is weaker than isomorphism in the category they live in. The reader should have in mind the examples of equivalence of categories, homotopy equivalence of spaces, quasi-isomorphism of chain complexes, etc. The definition of a relative category couldn't be simpler:

DEFINITION 3.9. A *relative category* is a pair $(\mathcal{C}, \mathcal{W})$ of an ordinary category $\mathcal{C}$ and a subcategory $\mathcal{W}$ of $\mathcal{C}$ required only to contain all the objects of $\mathcal{C}$. Morphisms in $\mathcal{W}$ are called *weak equivalences*.

Implicit in the claim that these somehow provide a model for $(\infty, 1)$-categories is the claim that out of just a collection of weak equivalences we get some sort of notion of homotopy between morphisms, to play the role of 2-morphisms in the $(\infty, 1)$-category represented by a given relative category. To give the first idea of how this happens, let's describe the homotopy category of the $(\infty, 1)$-category modeled by $(\mathcal{C}, \mathcal{W})$: it is $\mathcal{C}[\mathcal{W}^{-1}]$, the *localization* of $\mathcal{C}$ obtained by formally adding inverses for all morphisms in $\mathcal{W}$. Let's see in the example $\mathcal{C} = \text{Top}$, $\mathcal{W} = \{\text{homotopy equivalences}\}$ that homotopic maps become equal[23] as morphisms in $\mathcal{C}[\mathcal{W}^{-1}]$. First, notice that the projection $p : X \times [0, 1] \to X$ is a homotopy equivalence and thus becomes an isomorphism in $\mathcal{C}[\mathcal{W}^{-1}]$. This means that the two maps $i_0, i_1 : X \to X \times [0, 1]$ given by $i_0(x) = (x, 0)$ and $i_1(x) = (x, 1)$ become equal in the localization because $p \circ i_0 = p \circ i_1$. Finally, two maps are homotopic when they can be written in the form $f \circ i_0$ and $f \circ i_1$ for a single map f.

But of course, a satisfactory answer to the question of how higher morphisms appear in the $(\infty, 1)$-category represented by $(\mathcal{C}, \mathcal{W})$ would construct the mapping space between two objects of $\mathcal{C}$, and this is precisely what an enhancement of

[22]The reader unfamiliar with homotopy limits can find a quick introduction in section 5.2.

[23]One's first instinct —at least, if one hasn't localized rings which are not integral domains— might be that *adding* inverses to some morphisms shouldn't force other morphisms to become equal.

localization called *simplicial localization* does. We refer the reader to the classic papers of William Dwyer and Dan Kan **[DK80a]**, **[DK80b]** for details on how simplicial localizations may be constructed. The most intuitive construction is probably the hammock localization, which we'll explain by contrasting it with the (non-simplicial) localization $\mathcal{C}[\mathcal{W}^{-1}]$.

Morphisms in $\mathcal{C}[\mathcal{W}^{-1}]$ can be represented by *zig-zags*:

$$X_0 \xleftarrow{\sim} X_1 \to X_2 \xleftarrow{\sim} \cdots \to X_n$$

where arrows can go either way, but if they point to the left they are required to be in $\mathcal{W}$ (this is typically indicated by placing a $\sim$ on the arrow). [24] To form the mapping space in the hammock localization we add homotopies between zig-zags: the mapping space is constructed as a simplicial sets whose vertices are zig-zags and whose 1-simplices are weak equivalences of zig-zags, by which we mean diagrams of the form:

$$\begin{array}{ccccccccccc} & & X_1 & \xleftarrow{\sim} & X_2 & \longrightarrow & X_3 & \xleftarrow{\sim} & \cdots \longrightarrow & X_{n-1} & & \\ & \nearrow & \downarrow{\scriptstyle\sim} & & \downarrow{\scriptstyle\sim} & & \downarrow{\scriptstyle\sim} & & & \downarrow{\scriptstyle\sim} & \searrow & \\ X_0 & & & & & & & & & & & X_n \\ & \searrow & & & & & & & & & \nearrow & \\ & & Y_1 & \xleftarrow{\sim} & Y_2 & \longrightarrow & Y_3 & \xleftarrow{\sim} & \cdots \longrightarrow & Y_{n-1} & & \end{array}$$

in which all the left pointing morphisms and all the vertical ones are required to be in $\mathcal{W}$. Higher dimensional simplices are similar but have more rows (and look even more like hammocks than 1-simplices do).[25]

FURTHER READING. The papers by Dwyer and Kan on simplicial localization already indicate that relative categories, bare-bones though they may be, can be used to model $(\infty, 1)$-categories. The book **[DHKS04]** develops homotopy theory for relative categories (there called "homotopical categories" and required to satisfy a mild further axiom). More recently, Clark Barwick and Dan Kan, in a series of papers **[BK12c]**, **[BK12a]**, **[BK12b]**, compare relative categories (which they named) to other models of $(\infty, 1)$-categories and define a generalization of them that provides a model for (∞, n)-categories.

3.5. A_∞-categories. A (non-symmetric) *operad* is a collection of spaces $\mathcal{O}(n)$ together with composition maps

$$\mathcal{O}(n) \times \mathcal{O}(k_1) \times \mathcal{O}(k_2) \times \cdots \times \mathcal{O}(k_n) \to \mathcal{O}(k_1 + k_2 + \cdots + k_n),$$

which are required to satisfy associativity conditions[26] that are easy to guess if one thinks of the process of substituting n functions, one of k_1 variables, one of k_2 variables, etc., into a function of n variables to obtain an overall function of $k_1 + k_2 + \cdots + k_n$ variables. As this suggests, the elements of $\mathcal{O}(n)$ are called n-ary operations. They can be used to parametrize all the homotopies required for a composition that is associative up to coherent homotopy:

DEFINITION 3.10. An A_∞-operad is one such that all $\mathcal{O}(n)$ are contractible. Given any such operad, an A_∞-category $\mathcal{C}$ consists of

[24]Really, the morphisms in $\mathcal{C}[\mathcal{W}^{-1}]$ are equivalence classes of zig-zags in the smallest equivalence relation preserved by the operations of (1) removing an identity morphism, (2) composing two consecutive morphisms that point the same way, and (3) canceling a pair of the form $\cdot \xrightarrow{w} \cdot \xleftarrow{w} \cdot$ or $\cdot \xleftarrow{w} \cdot \xrightarrow{w} \cdot$.

[25]In this description some details are missing, see **[DK80b]**.

[26]We are also omitting a couple of conditions on $\mathcal{O}(0)$ and $\mathcal{O}(1)$.

(1) a collection of objects,
(2) a space $\mathcal{C}(X, Y)$ for every pair of objects, and
(3) composition maps

$$\mathcal{O}(n) \times \mathcal{C}(X_{n-1}, X_n) \times \mathcal{C}(X_{n-2}, X_{n-1}) \times \cdots \times \mathcal{C}(X_0, X_1) \to \mathcal{C}(X_0, X_n)$$

for every n and every sequence of objects $X_0, X_1, \ldots, X_n$,

which are required to be compatible with the composition operations of the operad in an obvious sense.[27]

REMARK 3.11. This is not the most common definition of the term "A_∞-category". Usually the term denotes a related concept using chain complexes instead of spaces. See section 3.6.3.

Notice that, just like topological categories, this model provides easy access to homotopy categories: since $\mathcal{O}(n)$ is contractible, applying π_0 to an A_∞-category produces an ordinary category with Hom-sets given by $\pi_0(\mathcal{C}(X, Y))$.

In case the A_∞-operad $\mathcal{O}$ is the operad of Stasheff associahedra, an A_∞-category with a single object is equivalent to the original notion of an A_∞-space introduced by Jim Stasheff in his work on homotopy associative H-spaces [**Sta63**]. The main result of that work can be interpreted as proving the homotopy hypothesis for A_∞-categories with a single object. We'll state a less precise version informally:

PROPOSITION 3.12. *An A_∞-space X is weak homotopy equivalent to a loop space ΩY (in such a way that composition in X corresponds to concatenation of loops) if and only if $\pi_0(X)$ is a group under the operation induced from composition in X.*

This *recognition principle* for loop spaces is part of the original motivation for Peter May's definition of operad[28], which he used to prove a recognition principle in the same spirit for iterated loop spaces $\Omega^n Y$ and infinite loop spaces (which can be thought of as a sequence of spaces $Y_0, Y_1, \ldots$ each of which is equivalent to the loop spaces of the following one). See May's book [**May72**].

FURTHER READING. This model doesn't seem to get used that much in practice. The only example of a paper constructing some $(\infty, 1)$-category as an A_∞-category that the author is aware of is [**AC12**, Proposition 1.4]. Todd Trimble used A_∞-categories as the first step in an inductive definition of (∞, n)-category, see [**Che11**]. A talk given in Morelia by Peter May [**May05**] expressed the hope that a simpler, more general version of the inductive approach would work. The author is unaware of the current status of May's approach, but it seems that Michael Batanin pointed out some difficulty in carrying it out (this is mentioned in Eugenia Cheng's paper just cited).

3.6. Models of subclasses of $(\infty, 1)$-categories. There are also several ways of modeling special classes of $(\infty, 1)$-categories, which, when applicable can be simpler to calculate with. We'll mention model categories and derivators for which it is hard to say exactly which $(\infty, 1)$-categories they can model, but which certainly can only model $(\infty, 1)$-categories that have all small homotopy limits and

[27] In both the description of operad and A_∞-category we've omitted discussing identities. The reader can easily supply the missing details.

[28] Which, we repeat, besides the data mentioned above includes actions of Σ_n on $\mathcal{O}(n)$ and requires the composition maps to be equivariant in an easily guessed sense.

colimits, and linear models, which model $(\infty, 1)$-categories that are enriched over an $(\infty, 1)$-categories of chain complexes.

3.6.1. *Model categories.* Quillen's model categories are the most successful setting for abstract homotopy theory. A *model category* $\mathcal{C}$ is an ordinary category that has all small limits and colimits and is equipped with three collections of morphisms called *cofibrations*, *fibrations* and *weak equivalences* which are required to satisfy axioms that abstract properties that hold of the classes of maps of topological spaces that they are named after. We won't give a precise definition, but refer the reader to standard references such as the introduction [**DS95**], the books [**Hov99**] and [**Hir03**], the fast paced [**Lur09b**, Appendix A.2] or Quillen's original [**Qui67**] (but note that what we are calling model categories where called *closed* model categories there). The $(\infty, 1)$-category modeled by a model category $\mathcal{C}$ is just the one modeled by the relative category $(\mathcal{C}, \text{weak equivalences})$. We mention them separately from relative categories because the extra structure makes them easier to deal with than a random relative category, so they form an eminently practical way to construct particular $(\infty, 1)$-categories. Even on the level of homotopy categories, the fibrations, cofibrations and the axioms make the localization better behaved. For example, in the homotopy category of a model category we do not need to consider zig-zags of arbitrary length, it is enough to look at zig-zags of the form $\cdot \xleftarrow{\sim} \cdot \to \cdot \xleftarrow{\sim}$.

REMARK 3.13. Small homotopy limits and colimits always exist in a model category and thus they can only model $(\infty, 1)$-categories that are complete and cocomplete. It is not known to the author whether or not all such $(\infty, 1)$-categories arise from model categories. There is however a result of Carlos Simpson's under further smallness assumptions, namely he showed that *combinatorial* model categories provide models precisely for the class of *locally presentable* $(\infty, 1)$-categories. See [**Sim99**] (but beware that what are called cofibrantly generated model categories there are what we are calling combinatorial model categories), or [**Lur09b**, Section 5.5.1]. Roughly speaking, a locally presentable category is one that is cocomplete and generated under colimits by a small subcategory of objects which are small or compact in some sense. A combinatorial model category is required to be locally presentable (and to have a model structure which is cofibrantly generated, which is also a condition with the flavor of the whole being determined by a small portion). For information about Jeff Smith's notion of combinatorial model category see [**Bek00**], which is the first published account.

Model categories have been hugely successful in providing workable notions of homotopy theory in many topological and algebraic contexts. A wealth of model structures have been constructed and all provide examples of $(\infty, 1)$-categories that people care about. When performing further constructions based on these $(\infty, 1)$-categories, such as taking categories of diagrams in one of them, functors between two of them or homotopy limits or colimits of them it can be very hard to remain in the world of model categories. In those cases, using model categories to present the inputs to these constructions but carrying them out in the world of $(\infty, 1)$-categories is a very reasonable compromise.

We will meet model categories again in section 4 since the original comparison results between models of $(\infty, 1)$-categories were formulated in that language.

3.6.2. *Derivators.* When working with an $(\infty, 1)$-category $\mathcal{C}$, it might be tempting to do as much as possible in $\operatorname{ho}\mathcal{C}$, since ordinary categories are much simpler and more familiar objects. We can't get very far, though, we run into trouble as soon as we start talking about homotopy limits and colimits[29]. Say we have a small (ordinary) category $\mathcal{I}$ and wish to talk about homotopy limits or colimits of $\mathcal{I}$-shaped diagrams in $\mathcal{C}$. Homotopy limits should be homotopy invariant: if two diagrams $F, G : \mathcal{I} \to \mathcal{C}$ are connected by a natural isomorphism[30], they should have equivalent limits in $\mathcal{C}$. So, taking homotopy limits should induce a functor $\mathrm{ho}(\mathcal{C}^{\mathcal{I}}) \to \mathrm{ho}\,\mathcal{C}$. Now, this homotopy category $\mathrm{ho}(\mathcal{C}^{\mathcal{I}})$ is not something we can construct just from $\mathrm{ho}\,\mathcal{C}$ and $\mathcal{I}$, in particular it is not equivalent to $(\mathrm{ho}\,\mathcal{C})^{\mathcal{I}}$.

EXAMPLE 3.14. Let $\mathcal{I}$ be $\mathbb{Z}/2$ regarded as a category with a single object and let $\mathcal{C}$ be the $(\infty, 1)$-category of ∞-groupoids (or spaces). An $\mathcal{I}$-shaped diagram in $\mathcal{C}$ is just a space with an action of $\mathbb{Z}/2$. Consider the diagrams given by the trivial action and the 180° rotation on S^1. Since the 180° rotation is homotopic to the identity on S^1, these two diagrams become equal in $(\mathrm{ho}\,\mathcal{C})^{\mathcal{I}}$, but are not isomorphic in $\mathrm{ho}(\mathcal{C}^{\mathcal{I}})$ since, for example, they have different homotopy colimits: since the rotation action is free, the homotopy colimit in that case is just $S^1/(\mathbb{Z}/2) \cong S^1$; for the trivial action, we get $(E(\mathbb{Z}/2) \times S^1)/(\text{diagonal action}) = B(\mathbb{Z}/2) \times S^1$.

The idea of derivators is to hold on to not just $\mathrm{ho}\,\mathcal{C}$, but $\mathrm{ho}(\mathcal{C}^{\mathcal{I}})$ for every small (ordinary) category $\mathcal{I}$ as well. This at least allows one to hope to be able to discuss homotopy limits and colimits. Given an $(\infty, 1)$-category $\mathcal{C}$, the construction $\mathcal{I} \to \mathrm{ho}(\mathcal{C}^{\mathcal{I}})$ provides a strict 2-functor $(\mathrm{Cat}_{\mathrm{small}})^{\mathrm{op}} \to \mathrm{Cat}$ where Cat is the strict 2-category of all not necessarily small categories and $\mathrm{Cat}_{\mathrm{small}}$ is the sub-2-category of small ones.[31] By definition, derivators are strict 2-functors $(\mathrm{Cat}_{\mathrm{small}})^{\mathrm{op}} \to \mathrm{Cat}$ satisfying further conditions that guarantee that homotopy limits and colimits (and more generally homotopy versions of the left and right Kan extensions) exist and are well-behaved. As in the case of model categories: (1) the definition directly implies derivators can only model $(\infty, 1)$-categories which are complete and cocomplete, (2) the author does not know if all such $(\infty, 1)$-categories can be modeled, and (3) adding presentability on both sides of the equation balances it, see **[Ren06]**. The later **[Ren09]** deals with representing $(\infty, 1)$-categories coming from left proper model categories by derivators.

FURTHER READING. Derivators were defined by Alexander Grothendieck (the term appears first in **[Gro83]**, a few years later Grothendieck wrote **[Gro91]**) and independently by Alex Heller **[Hel88]** (who called them "homotopy theories"). Good introductions can be found in **[Mal01]**, **[Gro13]**, and the review section of **[GPS12]**.

3.6.3. *dg-categories, A_∞-categories.* Now we'll discuss two "linear" models (or more precisely, models based on chain complexes) for special kinds of $(\infty, 1)$-categories that have seen much use in algebra and algebraic geometry. These are the notions of dg-categories, which are analogous to topological or simplicial categories,

---

[29]See section 5.2 for a quick introduction.

[30]A natural transformation whose components are invertible in the sense we always use for $(\infty, 1)$-categories: invertible up to higher invertible morphisms.

[31]If the reader is not versed in the art of worrying about size issues, we advise not to start until after reading this survey. We do however caution that while it might seem like a merely technical point there is substance to it: for example, it is easy to prove that if the collection of morphisms of a category has size λ and the category has all products $\prod_{i<\lambda} X_i$ then it is a preorder.

and A_∞-categories, which are analogous to the identically named A_∞-categories mentioned in the previous section. In both cases the analogy comes from replacing spaces by chain complexes. We will only discuss the case where the analogy is closest, namely, when the chain complexes are *connective*, that is, concentrated in non-negative degrees; but the definitions can easily be adapted to chain complexes with different boundedness conditions.

DEFINITION 3.15. Let R be a commutative ring. A *differential graded category* or *dg-category* over R is a category enriched in the monoidal category of connective chain complexes of R-modules.

The move from spaces to connective chain complexes can be thought of as going from ∞-groupoids to the abelian, fully strict ∞-groupoids. This philosphy is encoded by the Dold-Kan correspondence which establishes an equivalence of categories between simplicial abelian groups and chain complexes of abelian groups concentrated in non-negative degrees. Given a dg-category $\mathcal{C}$, we can form a category enriched in simplicial sets by applying the Dold-Kan correspondence to all of the hom complexes (and forgetting the abelian group structure), this gives the $(\infty, 1)$-category $\mathcal{C}$ as a simplicial category. There is also a direct construction of the quasi-category corresponding to $\mathcal{C}$, called the dg-nerve of $\mathcal{C}$, see [**Lur12b**, Construction 1.3.1.6].

A_∞-categories can be defined using an operad in chain complexes, analogously to the definition in section 3.5, but there is a more explicit description of them in terms of one n-ary composition operation for each n. We will not repeat the definition here, since we won't have anything to say about it, but the interested reader can look it up in the references.

FURTHER READING. Good introductions to dg-categories include [**Kel06**] and [**Toë10**]. The $(\infty, 1)$-category of all dg-categories is described as a model category in Gonçalo Tabuada's PhD thesis [**Tab07**]. That model structure is used in [**Toë07**] to develop Morita theory for dg-categories. A concise introduction to A_∞-categories can be found in [**Kel06**]; a thorough reference is the book [**BLM08**]. Also see Maxim Kontsevich and Yan Soibelman's notes [**KS09**] which deal mostly with A_∞-algebras which are A_∞-categories with a single object (although that definition is anachronistic, of course). The most conspicuous example of an A_∞-category is the Fukaya category of a symplectic manifold that plays a starring role in homological mirror symmetry. It was first mentioned in [**Fuk93**]. For comprehensive treatments see the books [**FOOO09**] and [**Sei08**], the later of which has a good introduction to A_∞-categories in Chapter 1.

4. The comparison problem

Comparing different definitions of higher categories is harder than it might seem at first, since even what it means to show that two theories are equivalent is not completely clear. Imagine we wish to compare two theories of n-categories, call them red categories and blue categories. Just like ordinary categories form a 2-category, red n-categories should form an $(n+1)$-category, and we'd like to show that this $(n+1)$-category is equivalent in an appropriate sense to the $(n+1)$-category of blue n-categories. But what color are the $(n+1)$-categories and the equivalence? It is reasonable to expect that red n-categories form a red $(n+1)$-category and similarly that the blue n-categories form a blue $(n+1)$-category. But this means we can't easily compare the $(n+1)$-categories before solving the comparison for red and blue higher categories!

We might be able to assemble all blue n-categories into a red $(n+1)$-category by an ad hoc construction, and show that that red $(n+1)$-category is red-equivalent to the red $(n+1)$-category of red n-categories. In that case we could say the red theory regards the two theories as equivalent. But a priori, if that happens, the blue theory might disagree and not consider the two theories equivalent!

The way the comparison problem was first solved for $(\infty,1)$-categories is as follows:

(1) What gets compared are not the $(\infty,2)$-categories of all $(\infty,1)$-categories of particular kind, but rather $(\infty,1)$-categories of $(\infty,1)$-categories (obtained from the $(\infty,2)$-category by throwing away non-invertible natural transformations).
(2) For each model M the $(\infty,1)$-category $\mathrm{Cat}^M_{(\infty,1)}$ of all M-style $(\infty,1)$-categories was described as a model category. [32]
(3) It was shown that these model categories are all connected by zig-zags of *Quillen equivalences*: these are equivalences of categories that preserve enough aspects of the model structure to ensure that two model categories have the same homotopy theory, i.e., model the same $(\infty,1)$-category.

In the non-standard terminology above, this says that the theory of model categories regards all models as equivalent. It can also be shown that any model regards all models as equivalent. For example, take quasi-categories. One can construct a quasi-category from a model category and show that two model categories connected by a zig-zag of Quillen equivalences produce quasi-categories that are equivalent (according to a particular definition of equivalence of quasi-categories). Then we get a quasi-category $\mathrm{Cat}^M_{(\infty,1)}$ for each of the models, and they are all equivalent.

Portions of the program outlined above showing the equivalence of five of the models we discussed (namely quasi-categories, simplicial categories, Segal categories, complete Segal spaces and relative categories) were carried out by Julie Bergner [**Ber07**], Clark Barwick and Dan Kan [**BK12a**], [**BK12a**], André Joyal and Miles Tierney [**Joy08a**], [**JT06**]. For a beautiful diagram showing the Quillen equivalences at a glance and further references see [**BSP12**, Figure 1]. For a summary of the model structures and Quillen equivalences comparing the first four models (i.e., excluding relative categories) see Julie Bergner's survey [**Ber10**].

4.1. Axiomatization. A recent breakthrough in the theory of (∞,n)-categories is the axiomatization by Clark Barwick and Chris Schommer-Pries [**BSP12**] of the $(\infty,1)$-category $\mathrm{Cat}_{(\infty,n)}$ of (∞,n)-categories. As in the direct comparison results we mentioned for $(\infty,1)$-categories, what gets axiomatized is not an $(\infty,n+1)$-category of (∞,n)-categories but rather the $(\infty,1)$-category one obtains from that $(\infty,n+1)$-category by throwing away higher non-invertible morphisms. Their work was inspired by Bertrand Toën's influential [**Toë05**] that similarly axiomatizes *model categories* of $(\infty,1)$-categories. Toën's axioms are closely related to Giraud's

[32] This does not necessarily mean that a model structure was put precisely on some ordinary category category of all M-style $(\infty,1)$-categories; but rather on a larger category in which the objects which are both fibrant and cofibrant are the M-style $(\infty,1)$-categories. For example, for quasi-categories the model structure is on the category of all simplicial sets: every object is cofibrant and the fibrant ones are precisely the quasi-categories.

axioms for toposes, while Barwick and Schommer-Pries's axioms stray a bit further from them. Very roughly, Barwick and Schommer-Pries axioms are as follows:

(1) There is an embedding in $\mathrm{Cat}_{(\infty,n)}$ of the category of *gaunt* n-categories. A gaunt n-category is a strict n-category all of whose invertible k-morphisms are identities, for all k. The images of these gaunt n-categories are required to generate $\mathrm{Cat}_{(\infty,n)}$ under homotopy colimits.
(2) The embedding provides us in particular with (∞, n)-categories that are "walking k-cells": they consist of a single k-morphism and its required (globular) boundary. Certain gaunt n-categories obtained by gluing a few cells together are required to still be obtained by the same gluing process inside $\mathrm{Cat}_{(\infty,n)}$, that is, the embedding of gaunt n-categories is required to preserve a few colimits of diagrams of cells.
(3) $\mathrm{Cat}_{(\infty,n)}$ is required to have internal Homs, and more generally, slices $\mathrm{Cat}_{(\infty,n)}/C_k$ are required to have internal Homs for all cells C_k.
(4) The $(\infty, 1)$-category $\mathrm{Cat}_{(\infty,n)}$ is *universal* with the respect to the above properties, in the sense that any other $(\infty, 1)$-category $\mathcal{D}$ with an embedding of gaunt n-categories satisfying the first three axioms is obtained from $\mathrm{Cat}_{(\infty,n)}$ via a localization functor $\mathrm{Cat}_{(\infty,n)} \to \mathcal{D}$ that commutes with the embeddings.

Even more roughly: the first axiom gives us the ability to present (∞, n)-categories by means of generators and relations, and the second guarantees that at least for presentations of gaunt n-categories we get the expected answer. There are many imprecisions in our description of the axioms and we refer the reader to [**BSP12**] for a correct statement.

REMARK 4.1. One might think that discarding those morphisms loses too much information, but in a sense it doesn't: the $(\infty, 1)$-category $\mathrm{Cat}_{(\infty,n)}$ characterized by the axioms is Cartesian closed, so for any two (∞, n)-categories $\mathcal{C}$ and $\mathcal{D}$ there is an (∞, n)-category $\mathrm{Fun}(\mathcal{C}, \mathcal{D})$ that contains all the non-invertible natural transformations and higher morphisms that are not directly observable in the mapping space $\mathrm{Map}_{\mathrm{Cat}_{(\infty,n)}}(\mathcal{C}, \mathcal{D})$.

Toën also proved in [**Toë05**] that the only automorphism of the theory of $(\infty, 1)$-categories is given by taking opposite categories. It is unique in the strong sense that not only is every automorphism equivalent to that one, but automorphisms themselves have no automorphisms, or more precisely, there is a naturally defined ∞-groupoid of automorphisms of $\mathrm{Cat}_{(\infty,1)}$ and it is homotopy equivalent to the discrete group $\mathbb{Z}/2$. Barwick and Schommer-Pries prove the analogue of this result for $\mathrm{Cat}_{(\infty,n)}$, showing that its space of automorphisms is the discrete group $(\mathbb{Z}/2)^n$ corresponding to the possibility of deciding for each degree k separately whether or not to flip the direction of the k-morphisms.

5. Basic $(\infty, 1)$-category theory

In this section we review how the most basic concepts of category theory generalize to $(\infty, 1)$-categories. The philosophy is that $(\infty, 1)$-categories are much more like ordinary 1-categories than fully weak (∞, ∞)-categories where no morphisms are required to be invertible[33]. A little more precisely, the intuition that $(\infty,1)$-categories have spaces of morphisms and that these spaces only matter up to (weak) homotopy equivalence usually leads to useful definitions and correct statements. We will also frequently point out what happens in the case of quasi-categories, which due both intrinsically to some features they possess and externally to the availability of [**Joy08b**], [**Lur09b**] and [**Lur12b**] are probably the most "practical" model to work with.

FURTHER READING. Besides those systematic treatises by Joyal and Lurie already mentioned, we recommend [**Gro10**], an excellent and well motivated summary of large chunks of [**Lur09b**], [**Lur07a**], [**Lur07b**] and [**Lur07c**] (the last three of which were reworked into the first few chapters of [**Lur12b**]). The availability of [**Gro10**] is why we feel justified in giving very little detail in this section, just giving the flavor of the topic. Also highly recommended is the book [**Rie14**], Part IV of which is about quasi-categories and includes, among other things, (1) a discussion of which aspects of $(\infty,1)$-categories are already captured by the 2-category whose objects are quasi-categories, whose morphisms are functors and whose 2-morphisms are homotopy classes of natural transformations; and (2) plenty of geometrical information about quasi-categories viewed as simplicial sets, such as how to visualize mapping spaces in quasi-categories or the homotopy coherent nerve that produces the equivalence between quasi-categories and simplicial categories.

5.1. Equivalences. There are two things typically called *equivalences* in $(\infty,1)$-category theory: one generalizes isomorphisms in an ordinary category, and the other generalizes equivalences between categories. These are related in the following way: we can ignore non-invertible natural transformations to get $\mathrm{Cat}_{(\infty,1)}$, the $(\infty,1)$-category of $(\infty,1)$-categories, functors and invertible natural transformations. A functor is then an equivalence of $(\infty,1)$-categories if and only if it is an equivalence as a morphism in $\mathrm{Cat}_{(\infty,1)}$.

REMARK 5.1. That $(\infty,1)$-isomorphisms go by the name equivalences is probably due to (1) the case of the $(\infty,1)$-category of spaces where they are just (weak) homotopy equivalences, and (2) there being, in the model of topological categories (resp. simplicial categories), a second, stricter notion of isomorphism, coming from enriched category theory: morphisms such that composition with them induces homeomorphisms (resp. isomorphisms of simplicial sets) between mapping spaces.

DEFINITION 5.2. A morphism $f : X \to Y$ in an $(\infty,1)$-category is an *equivalence* if its image in $\mathrm{ho}\,\mathcal{C}$ is an isomorphism, or, equivalently, if for every object $Z \in \mathcal{C}$, $f \circ \_ : \mathrm{Map}_{\mathcal{C}}(Z,X) \to \mathrm{Map}_{\mathcal{C}}(Z,Y)$ and $\_ \circ f : \mathrm{Map}_{\mathcal{C}}(Y,Z) \to \mathrm{Map}_{\mathcal{C}}(X,Z)$ are weak homotopy equivalences.

André Joyal proved, using the model of quasi-categories, that when a morphism f is an equivalence, one can coherently choose an inverse g for it, 2-morphisms showing f and g are inverses, 3-morphisms showing invertibility of those 2-morphisms, and so on. More precisely, such a coherent system of choices is given by a functor $F : \mathcal{J} \to \mathcal{C}$ where $\mathcal{J}$ is the ordinary category $0 \xrightarrow{\cong} 1$, which has two objects and a unique isomorphism between them. The precise statement then is as follows:

[33]In fact, even 2-categories have some tricky points that that do not arise when dealing with $(\infty,1)$-categories, such as the need to distinguish between several kinds of limits and colimits called pseudo-limits, lax-limits and colax-limits.

PROPOSITION 5.3 ([**Joy02**, Corollary 1.6]). *A morphism f in an $(\infty,1)$-category $\mathcal{C}$ is an equivalence if and only if there is a functor $F : \mathcal{J} \to \mathcal{C}$ with $F(0 \to 1) = f$, where $\mathcal{J}$ is the* walking isomorphism *defined above.*

This result is in the same spirit as the much easier automatic homotopy coherence result in 3.6.

5.1.1. *Further results for quasi-categories.* The idea behind requiring the horn filler condition only for inner horns, Λ^n_k with $0 < k < n$ in the definition of quasi-category is that, for $n = 2$, filling a Λ^2_0 horn requires inverting the edge between vertices 0 and 1:

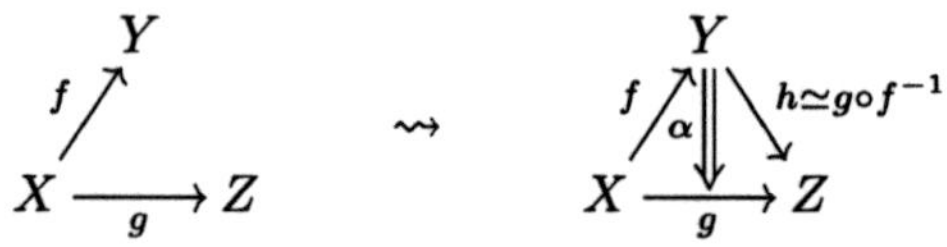

And indeed Joyal proved that this is the only obstacle:

PROPOSITION 5.4 ([**Joy02**, Theorem 1.3]). *In a quasi-category $\mathcal{C}$, a morphism f is an equivalence if and only if there are fillers for every horn $\Lambda^n_0 \to \mathcal{C}$ whose edge joining vertices 0 and 1 is f.*

There is, of course, a dual result about Λ^n_n horns. Putting the two together we get the homotopy hypothesis for quasi-categories:

COROLLARY 5.5. *Every 1-simplex of a quasi-category $\mathcal{C}$ is an equivalence if and only if $\mathcal{C}$ is a Kan complex.*

5.2. Limits and colimits. The notion of limit and colimit in an $(\infty,1)$-category should be thought of as generalizations of *homotopy* limits and colimits, and indeed, reduce to those for the $(\infty,1)$-category of spaces (or more generally, for an $(\infty,1)$-category coming from a model category; there is also a notion of homotopy limit and colimit in a model category). We begin by recalling what homotopy limits and colimits of diagrams of spaces are, by describing a construction of them that provides the correct intuition for $(\infty,1)$-categories. When reading the following description, the reader should think of paths as *invertible* morphisms in an $(\infty,1)$-categories, and homotopies as higher morphisms.

Let's consider a diagram of spaces $F : \mathcal{I} \to \mathrm{Top}$ (where, for now, $\mathcal{I}$ is an ordinary category). Recall that its (ordinary, non-homotopy) limit can be constructed as follows: it consists of the subspace of $\prod_{i\in\mathcal{I}} F(i)$ of points $(x_i)_{i\in\mathcal{I}}$ such that for any $\alpha : i \to j$ in $\mathcal{I}$, we have that $F(\alpha)(x_i) = x_j$. To construct instead the homotopy limit, following the philosophy described in remark 2.2, we replace the strict notion of equality in $F(\alpha)(x_i) = x_j$ by the corresponding homotopical notion: a path from $F(\alpha)(x_i)$ to x_j, this path is witness to the fact that "homotopically speaking, $F(\alpha)(x_i)$ and x_j are the same". For simple diagrams, without composable arrows, this is enough. For example, the homotopy pullback of a diagram $X \xrightarrow{f} Z \xleftarrow{g} Y$ *can* be constructed as the subspace of $X \times Y \times Z \times Z^{[0,1]} \times Z^{[0,1]}$ consisting of 5-tuples $(x, y, z, \gamma, \sigma)$ with $\gamma(0) = f(x), \gamma(1) = z = \sigma(1), \sigma(0) = g(y)$.

For diagrams that have pairs of composable arrows, we need to ensure these paths that act as witnesses compose as well, that is, if we have $i \xrightarrow{\alpha} j \xrightarrow{\beta} k$ somewhere in $\mathcal{I}$, every point of the homotopy limit will have among its coordinates the data of

(1) points $x_i \in F(i), x_j \in F(j), x_k \in F(k)$, and
(2) paths $\gamma_{ij} : F(\alpha)(x_i) \to x_j$, $\gamma_{jk} : F(\beta)(x_j) \to x_k$ and $\gamma_{ik} : F(\beta \circ \alpha)(x_i) \to x_j$.

The paths γ_{ik} and $\gamma_{jk} \cdot (F(\beta) \circ \gamma_{ij})$ both witness that x_i and x_j are in the same path component of $F(k)$, but we shouldn't regard their testimonies as being independent! The points of the homotopy limit should also include a homotopy between these two paths. Clearly, this doesn't stop here: for diagrams with triples of composable arrows we should have homotopies between homotopies and so on. For any given small diagram shape $\mathcal{I}$ it is clear which paths and homotopies are required.

Dually, for homotopy colimits, instead of taking a subspace of a large product where we require the presence of some paths and homotopies, we form a large coproduct of spaces and glue in paths and homotopies that enforce "sameness" of points in the colimits.

REMARK 5.6. There is also a general formula for the required homotopies in homotopy limits due to Pete Bousfield and Dan Kan [**BK72**]: for a functor $F : \mathcal{I} \to \mathrm{Top}$, we have

$$\operatorname{holim} F = \int_{i \in \mathcal{I}} F(i)^{|N(I \downarrow i)|}.$$

Here $|N(I \downarrow i)|$ is the geometric realization of the nerve of the slice category $\mathcal{I} \downarrow i$ whose objects are objects of $\mathcal{I}$ with a map to i, and whose morphisms are commuting triangles. The $\int$ sign denotes a type of limit called an *end*, and can be constructed as a subspace of the product of function spaces $\prod_{i \in \mathcal{I}} F(i)^{|N(\mathcal{I} \downarrow i)|}$ consisting of compatible families of functions $\gamma_i : |N(\mathcal{I} \downarrow i)| \to F(i)$: a morphism $i \xrightarrow{\alpha} j$ induces a functor $\mathcal{I} \downarrow i \to \mathcal{I} \downarrow j$ by composition, and thus gives a map $\alpha_* : |N(\mathcal{I} \downarrow i)| \to |N(\mathcal{I} \downarrow j)|$; the family $\{\gamma_i\}$ is compatible if $\gamma_j \circ \alpha_* = F(\alpha) \circ \gamma_i$. The reader can check that the homotopies mentioned above for composable pairs $i \xrightarrow{\alpha} j \xrightarrow{\beta} k$ are encoded by this formula as maps out of triangles that restrict on the boundary to the paths corresponding to the morphisms α, β and $\beta \circ \alpha$.

All limits and colimits in $(\infty, 1)$-categories can be defined in terms of homotopy limits of spaces. Recall that for a diagram $F : \mathcal{I} \to \mathcal{C}$ in an ordinary category $\mathcal{C}$, the limit and colimit can be defined by requiring the canonical functions of sets

$$\operatorname{Hom}_{\mathcal{C}}(X, \lim_{i \in \mathcal{I}} F(i)) \to \lim_{i \in \mathcal{I}} \operatorname{Hom}_{\mathcal{C}}(X, F(i))$$

and

$$\operatorname{Hom}_{\mathcal{C}}(\operatorname{colim}_{i \in \mathcal{I}} F(i), X) \to \lim_{i \in \mathcal{I}^{\mathrm{op}}} \operatorname{Hom}_{\mathcal{C}}(F(i), X)$$

to be bijections natural in X. The limits occurring in the codomain of these canonical maps are taken in the category of sets.[34]

DEFINITION 5.7. Given a functor $F : \mathcal{I} \to \mathcal{C}$ between two $(\infty, 1)$-categories, its *limit* and *colimit*, if they exist, are determined up to equivalence in $\mathcal{C}$ by requiring that

$$\operatorname{Map}_{\mathcal{C}}(X, \lim_{i \in \mathcal{I}} F(i)) \to \operatorname{holim}_{i \in \mathcal{I}} \operatorname{Map}_{\mathcal{C}}(X, F(i))$$

and

$$\operatorname{Map}_{\mathcal{C}}(\operatorname{colim}_{i \in \mathcal{I}} F(i), X) \to \operatorname{holim}_{i \in \mathcal{I}^{\mathrm{op}}} \operatorname{Map}_{\mathcal{C}}(F(i), X)$$

are weak equivalences of spaces, natural in X.

[34]Note that *co*limits in the category of sets do not appear here.

Note the special case of initial and terminal objects: an object X of an $(\infty,1)$-category $\mathcal{C}$ is initial if $\mathrm{Map}_{\mathcal{C}}(X,Y)$ is contractible for all objects Y, and terminal if $\mathrm{Map}_{\mathcal{C}}(Y,X)$ is contractible for all Y. As expected initial objects are unique when they exist: more precisely, Joyal proved [**Joy02**, Proposition 4.4] that the full subcategory of $\mathcal{C}$ consisting of all initial objects is a contractible ∞-groupoid, that is, any two initial objects are equivalent, any two equivalences between initial objects are homotopic, etc.

Other constructions of limits and colimits in ordinary categories also generalize to the $(\infty,1)$-setting and give the same definition as above. For example, the limit of a functor F can be characterized as a terminal object in the category of cones over F. For concreteness we'll use quasi-categories to describe how this goes for $(\infty,1)$-categories: given a functor $F : \mathcal{I} \to \mathcal{C}$, the quasi-category of cones over F, $\mathrm{Cones}(F)$, is the simplicial set whose n-simplices are given by maps of simplicial sets $\Delta^n \star \mathcal{I} \to \mathcal{C}$ which restrict to F on $\mathcal{I}$. Its vertices are exactly what we'd expect, since $\Delta^0 \star \mathcal{I}$ is $\mathcal{I}$ with a new initial object adjoined. Here $K \star L$ denotes the join of simplicial sets, a geometric operation that can be thought as providing a canonical triangulation of the union of line segments joining all pairs of a point of $|K|$ and a point of $|L|$; each k-simplex of $\mathcal{I}$ contributes an $(n+k+1)$-simplex to $\Delta^n \star \mathcal{I}$.

REMARK 5.8. While it is perfectly fine to define a diagram in a category to be a functor for developing the theory, most people don't actually think of many common diagram shapes, such as diagrams for pullbacks, pushouts, infinite sequences $X_0 \to X_1 \to \cdots$, etc., as being given by a category; instead these shapes are usually thought of as being given by a directed graph. Another advantage of the model of quasi-categories is that there is a very convenient generalization of directed graphs and diagrams shaped like them: simply take arbitrary simplicial sets for shapes and define a K-shaped diagram in $\mathcal{C}$ to be a map of simplicial sets $K \to \mathcal{C}$. Both descriptions given above make sense for these more general types of diagrams. Also, this notion of K-shaped diagram can be used in other model of $(\infty,1)$-categories but is more cumbersome, since essentially one needs to define the free $(\infty,1)$-category on K.

Most classical results about limits and colimits hold for $(\infty,1)$-categories with appropriate definitions, and we hope that the examples shown here give a rough idea of how the definitions are generalized.

5.3. Adjunctions, monads and comonads. As for other concepts, the definition of adjunction in ordinary category theory that uses Hom-sets and bijections generalizes to $(\infty,1)$-categories by using mapping spaces and homotopy equivalences:

DEFINITION 5.9. Given functors $F : \mathcal{C} \to \mathcal{D}$ and $G : \mathcal{D} \to \mathcal{C}$ between $(\infty,1)$-categories, an *adjunction* is specified by a giving a *unit*, a natural transformation $u : \mathrm{id}_{\mathcal{C}} \to g \circ f$ such that the composite map

$$\mathrm{Map}_{\mathcal{D}}(F(C),D) \xrightarrow{G} \mathrm{Map}_{\mathcal{C}}(G(F(C)),G(D)) \xrightarrow{-\circ u_C} \mathrm{Map}_{\mathcal{C}}(C,G(D))$$

is a weak homotopy equivalence.

As in the case of ordinary categories, if F has a right adjoint, the adjoint is uniquely determined up to natural equivalence. The basic continuity properties also hold: left adjoints preserve colimits, and right adjoints preserve limits. For ordinary categories, Freyd's adjoint functor theorem is a partial converse to this result. It says roughly that for $G : \mathcal{D} \to \mathcal{C}$ to have a left adjoint it is sufficient that $\mathcal{D}$ is complete, G preserves limits and satisfies a further condition called the *solution set condition*. The precise form of the solution set condition will not matter for us, only that (1) it is a size condition in the sense that it requires there to exist a small set of morphisms that somehow control all morphisms of a certain form, (2) it is adapted to G: it is not just a condition on the category $\mathcal{C}$.

There is an adjoint functor theorem for $(\infty, 1)$-categories due to Lurie which in some sense is less precise than Freyd's theorem for 1-categories in that its size condition is not adapted to G, but rather is a global condition on $\mathcal{C}$ and $\mathcal{D}$. In practice, this is not a problem: the conditions of Lurie's theorem are usually met when they need to be. What takes the place of the solution set condition is the notion of "*accessibility*"; for $(\infty, 1)$-categories it intuitively means that the $(\infty, 1)$-category is determined by a small subcategory of objects which themselves are small in some sense, for functors between such categories it means it preserves certain colimits which are a generalization of filtered colimits. For precise definitions see [**Lur09b**, Chapter 5] or, for the analogous theory in the case of ordinary categories, [**AR94**]. Finally, an $(\infty, 1)$-category is *presentable* if it has all small colimits and is accessible. Lurie's version of the adjoint functor theorem is as follows:

THEOREM 5.10 ([**Lur09b**, Corollary 5.5.2.9]). *Let $F : \mathcal{C} \to \mathcal{D}$ be a functor between* presentable *$(\infty, 1)$-categories.*

(1) *The functor F has a right adjoint if and only if it preserves small colimits.*
(2) *The functor F has a left adjoint if and only if it is* accessible *and preserves small limits.*

In classical category theory whenever $F : \mathcal{C} \to \mathcal{D}$ and $G : \mathcal{D} \to \mathcal{C}$ are adjoint functors (with F being the left adjoint), the composite $G \circ F$ is a monad and $F \circ G$ is comonad. This is also true in the $(\infty, 1)$-categorical context, but much harder to show since, as the reader expects by now, the concept of monad requires not an associative multiplication but one that is associative up to coherent homotopy. The monad corresponding to an adjunction is constructed in [**Lur12b**, Section 6.2.2]; Lurie uses it to prove an $(\infty, 1)$-analogue of the Barr-Beck theorem characterizing those adjunctions in which $\mathcal{D}$ is equivalent to the category of $G \circ F$-algebras in $\mathcal{C}$ and G is equivalent to the forgetful functor. This result is very useful in the theory of descent in derived algebraic geometry, just as the classical version is useful in algebraic geometry.

Another construction, providing more explicit information, of the monad and comonad corresponding to an adjunction appears in [**RV13**]. There the authors construct "the walking $(\infty, 1)$-adjunction": an $(\infty, 2)$-category $\mathcal{A}$ that has two objects 0 and 1, and two morphisms $f : 0 \to 1$ and $g : 1 \to 0$ which are adjoint to each other and form the free adjunction in the sense that any pair of adjoint functors (F, G) between $(\infty, 1)$-categories arise as the images of f and g under some functor from $\mathcal{A}$ into the $(\infty, 2)$-category of all $(\infty, 1)$-categories. Given such a functor $H : \mathcal{A} \to \mathrm{Cat}_{(\infty,1)}$ (where temporarily, $\mathrm{Cat}_{(\infty,1)}$ is an $(\infty, 2)$-category), the restriction of H to $\mathrm{Hom}_{\mathcal{A}}(0, 0)$ gives the monad $G \circ F$ with its multiplication and all the higher coherence data. This $(\infty, 2)$-category $\mathcal{A}$ is surprisingly just a 2-category, the "walking (ordinary) adjunction" described in [**SS86**].

REMARK 5.11. We mentioned above that if F has a right adjoint G, then G is canonically determined by F. In fact, all of the adjunction data is determined by F in a sense similar to that in proposition 3.4: there is an $(\infty, 1)$-category of adjunction data that has a forgetful functor to the arrow category of $\mathrm{Cat}_{(\infty,1)}$ which just keeps the left adjoint F; this forgetful functor has contractible fibers. Similarly there is a contractible space of adjunction data with a given left adjoint F, right adjoint G and unit u. The results in **[RV13]** describe more generally which pieces of adjunction data determine the rest up to a contractible space of choices.

5.4. Less basic $(\infty, 1)$-category theory. Much more than just the basic notions of category theory have been extended to $(\infty, 1)$-categories. There is a large number of topics we could have included here, and we picked only two that are important in applications: the different sorts of fibrations between $(\infty, 1)$-categories, and stable $(\infty, 1)$-categories. The various fibrations play a larger role in actual use of $(\infty, 1)$-categories, specially when incarnated as quasi-categories, than do the notions in ordinary category theory they generalize. Stable $(\infty, 1)$-categories are widely used in derived algebraic geometry, since they replace triangulated categories. Stable $(\infty, 1)$-categories are also downright pleasant to work with and we have no compunctions about advertising them.

A very important topic we've omitted is the study of monoidal and symmetric monoidal $(\infty, 1)$-categories. As usual in higher category theory they're tricky to define since one must make the tensor product associative up to coherent homotopy (or associative and commutative up to coherent homotopy in the symmetric case). We've decided to omit them since we feel that the coherence issues we'd point out about them would be similar to the ones we've mentioned already in other contexts. Also, there are gentler introductions to them than the one in **[Lur12b]** already available: we recommend Lurie's old version in **[Lur07b]** and **[Lur07c]**, or Groth's nice exposition **[Gro10**, Sections 3 and 4**]**.

Finally, a topic that will surely become increasingly important is the theory of enriched $(\infty, 1)$-categories. Often this is dealt with in a somewhat ad hoc manner when it arises, but recently David Gepner and Rune Haugseng have produced a systematic treatment **[GH14]**. See also Haugseng's follow up paper proving a strictification result for enriched $(\infty, 1)$-categories **[Hau14]**.

5.4.1. *Fibrations and the Grothendieck construction.* The *Grothendieck construction* takes as input a weak 2-functor $F : \mathcal{C}^{\mathrm{op}} \to \mathrm{Cat}$ where $\mathcal{C}$ is an ordinary category (thought of as a 2-category with only identity 2-morphisms), and Cat is the 2-category of categories. Even though Cat is a strict 2-category, it makes sense to consider a weak functor F, that is, one that does not preserve composition on the nose, but rather comes equipped with a natural isomorphism $F(g) \circ F(f) \xrightarrow{\cong} F(f \circ g)$ satisfying a coherence condition for compositions of three morphisms. It produces as output a category $\mathcal{E}$ and a functor $P : \mathcal{E} \to \mathcal{C}$. The functor P produced is always what is called a *Grothendieck fibration*, and the Grothendieck construction provides an equivalence of 2-categories between $\mathrm{Fun}(\mathcal{C}^{\mathrm{op}}, \mathrm{Cat})$ and the 2-category of Grothendieck fibrations over $\mathcal{C}$. We won't give precise definitions here, but we'll illustrate in an example.

EXAMPLE 5.12. Let $\mathcal{C}$ be the category Top of topological spaces and F the functor that assigns to each space X the category of vector bundles on X (the morphisms from a bundle $E_1 \to X$ to a bundle $E_2 \to X$ are maps $E_1 \to E_2$ which form a commuting triangle with the projections to X and which are linear on each fiber). For a continuous function $f : X \to Y$, $F(f)$ is given by pullback of vector bundles. For this functor, the Grothendieck construction produces a category $\mathcal{E}$ whose objects are vector bundles $E \to X$ on arbitrary base spaces X and whose morphisms from a bundle $E_1 \to X_1$ to a bundle $E_2 \to X_2$ are commuting squares

$$\begin{array}{ccc} E_1 & \xrightarrow{e} & E_2 \\ \downarrow & & \downarrow \\ X_1 & \xrightarrow{x} & X_2 \end{array}$$

such that the map e is linear on each fiber. The projection $P : \mathcal{E} \to \mathcal{C}$ simply forgets the bundles and keeps the bases. Now let's spell out what being a Grothendieck fibration means in this example. There are certain morphisms in $\mathcal{E}$ that are distinguished: the *Cartesian* morphisms for which the above square is a pullback. Given any map $x : X_1 \to X_2$ in Top and any vector bundle E_2 on X_2, there is always a Cartesian morphism in $\mathcal{E}$ with codomain $E_2 \to X_2$ that P sends to x: namely, the one in which E_1 is the pullback $x^*(E_2)$. Every morphism e' in $\mathcal{E}$ such that $P(e')$ factors through x can be factored through the morphism $e : x^*(E_2) \to E_2$.

The Grothendieck construction is very versatile. First, it clearly allows one to deal with weak 2-functors to Cat *without* leaving the world of 1-categories. It is used this way in the theory of stacks, for example. There one wants to replace the notion of a sheaf of sets with a 2-categorical version that has values in categories, or more typically, just groupoids. Such a thing can be defined in terms of weak 2-functors to groupoids, but can also be handled as Grothendieck fibration, which is the approach typically taken. See for example Angelo Vistoli's notes [**Vis05**]. (By the way, that also serves as a references for the details on the Grothendieck construction we skipped above.)

But the Grothendieck construction also allows one to calculate limits and colimits of functors to Cat. There is a dual version that given a covariant $F : \mathcal{C} \to \mathrm{Cat}$ produces what is called a Grothendieck opfibration $\mathcal{E} \to \mathcal{C}$, and it's not too hard to show that:

(1) $\lim F$ is given by the category of *coCartesian* sections of $P : \mathcal{E} \to \mathcal{C}$, that is, sections $\sigma : \mathcal{C} \to \mathcal{E}$ such that $P \circ \sigma = \mathrm{id}_{\mathcal{C}}$ and $\sigma(f)$ is a coCartesian morphism in $\mathcal{E}$ for every morphism f of $\mathcal{C}$.
(2) $\mathrm{colim}\, F$ is given by the localization $\mathcal{E}[\mathrm{coCart}^{-1}]$ of $\mathcal{E}$ obtained by inverting all coCartesian morphisms.

Here lim and colim denote what are sometimes called pseudo-limits and pseudo-colimits which are the closest analogues in the world of 2-categories to homotopy limits and colimits. Indeed, the reader who knows how to perform the Grothendieck construction should compare it with the description of homotopy limits and colimits in section 5.2.

All of this is generalized to $(\infty, 1)$-categories in [**Lur09b**, Chapter 2]. (See also Moritz Groth's notes [**Gro10**, Section 3]). Lurie defines *(co)Cartesian fibrations* corresponding to Grothendieck (op)fibrations, and also discusses Joyal's notions of left and right fibrations. As in the classical case, functors $\mathcal{C}^{\mathrm{op}} \to \mathrm{Cat}_{(\infty,1)}$ for an $(\infty, 1)$-category $\mathcal{C}$ are classified by Cartesian fibrations $\mathcal{E} \to \mathcal{C}$; and right fibrations are the subclass of Cartesian fibrations classifying functors that land in the subcategory of $\mathrm{Cat}_{(\infty,1)}$ of ∞-groupoids.

The constructions described above for limits and colimits of categories also generalize to $(\infty,1)$-categories. See [**Lur09b**, Proposition 3.3.3.1] for limits of $(\infty,1)$-categories; Corollary 3.3.3.4 specializes the previous result to left fibrations and thus provides a construction for homotopy limits of spaces. On the colimit side, there are corollaries 3.3.4.3 and 3.3.4.6 which construct colimits in $\mathrm{Cat}_{(\infty,1)}$ and homotopy colimits of spaces respectively.

5.4.2. *Stable $(\infty,1)$-categories.* Stable $(\infty,1)$-categories are a wonderfully practical replacement for the notion of triangulated category that fixes many of the problems in their theory. Many prominent examples of triangulated categories, are given to us almost by definition as homotopy categories of naturally occurring $(\infty,1)$-categories with nice properties that make them homotopical analogues of abelian categories. These $(\infty,1)$-categories are the *stable* ones we will shortly define. For example, the derived category of an abelian category $\mathcal{A}$ is basically defined as the homotopy category of the relative category of chain complexes in $\mathcal{A}$ with weak equivalences given by the quasi-isomorphisms. Let's now give the definition.

DEFINITION 5.13. An $(\infty,1)$-category is *pointed* if it has an object which is both initial and terminal. Such an object is called a *zero object* and denoted 0.

Between any two objects X and Y in a pointed $(\infty,1)$-category there is a unique homotopy class of *zero morphisms*, those that factor through the zero object.

DEFINITION 5.14. A *triangle* in a $(\infty,1)$-category is a commuting square of the form:

$$\begin{array}{ccc} X & \xrightarrow{f} & Y \\ \downarrow & & \downarrow{\scriptstyle g} \\ 0 & \longrightarrow & Z \end{array}$$

That is, a triangle is a homotopy between $g \circ f$ and the zero morphism. If the square is a pushout square, g is called the *cofiber* of f; if the square is a pullback, f is called the *fiber* of g.

Fibers and cofibers of maps play a role similar to kernels and cokernels in abelian categories and to cocones and cones in triangulated categories. In terms of these concepts the definition of stable $(\infty,1)$-category is easy to give:

DEFINITION 5.15. A *stable* $(\infty,1)$-category is a pointed $(\infty,1)$-category where every morphism has a fiber and a cofiber and every triangle is a pushout if and only if it is a pullback.

That's it, that's the whole definition. Of course, the notion of $(\infty,1)$-category is much more complicated than the notion of 1-category, but after making the initial investment in $(\infty,1)$-categories, the definition of stable $(\infty,1)$-category is simpler not only than the definition of triangulated category but even than that of abelian category.

For readers familiar with the definition of triangulated category, we now explain how the homotopy category of any stable $(\infty,1)$-category is canonically triangulated. The triangles are defined to be to diagrams of the form

$$\begin{array}{ccccc} X & \longrightarrow & Y & \longrightarrow & 0 \\ \downarrow & & \downarrow & & \downarrow \\ 0 & \longrightarrow & Z & \longrightarrow & U \end{array}$$

where both squares are pushouts. This makes the outer rectangle a pushout too, and since for spaces homotopy pushouts of that form produce suspensions, we call U the suspension of X and write $U \simeq \Sigma X$. Suspension is the translation functor of the triangulated structure.

The advantage of stable $(\infty, 1)$-categories over triangulated begins to be visible even from here: in a triangulated category the cone of a morphism is determined up to isomorphism but *not* up to a canonical isomorphism, and this is because the universal property we should be asking of the cone is homotopical in nature. In a stable $(\infty, 1)$-category the cofiber is determined, as all colimits are, up to a contractible space of choices which is exactly canonical enough in the $(\infty, 1)$-world to make the constructions functorial.

As further evidence of the simplicity of stable $(\infty, 1)$-categories, notice that proving the octahedral axiom in the homotopy category becomes simply a matter of successively forming pushout squares, and putting down a suspension whenever we see a rectangle with zeros in the bottom left and top right corners:

$$\begin{array}{ccccccccc}
X & \longrightarrow & Y & \longrightarrow & Z & \longrightarrow & 0 & & \\
\downarrow & & \downarrow & & \downarrow & & \downarrow & & \\
0 & \longrightarrow & Y/X & \longrightarrow & Z/X & \longrightarrow & \Sigma X & \longrightarrow & 0 \\
 & & \downarrow & & \downarrow & & \downarrow & & \downarrow \\
 & & 0 & \longrightarrow & Z/Y & \longrightarrow & \Sigma Y & \longrightarrow & \Sigma(Y/X)
\end{array}$$

(Here we've used notation like $X \to Y \to Y/X \to \Sigma X$ for the objects in a distinguished triangle.)

FURTHER READING. We refer the reader to [**Lur12b**, Chapter 1] for the development of the theory, or to Groth's presentation [**Gro10**, Section 5].

6. Some applications

This final section will briefly describe some uses of $(\infty, 1)$-categories outside of higher category theory itself. (The characterization of (∞, n)-categories mentioned in section 4 is an excellent application of $(\infty, 1)$-categories within higher category theory.) Each of these applications is a whole subject in itself and we cannot hope to do any of them justice here, rather, we hope merely to whet the readers' appetite and suggest further reading. Unfortunately many great topics had to be left out, such as algebraic K-theory, where the use of $(\infty, 1)$-categories has finally allowed to produce a universal property of higher K-theory similar to the familiar one satisfied by K^0: approximately, that K-theory is the universal invariant satisfying Waldhausen's additivity theorem. For precise descriptions and proofs of this universal property see [**Bar12**] and [**BGT13**]. Another proof of Waldhausen additivity in the $(\infty, 1)$-setting can be found in [**FL12**] (but without the universal property), and an $(\infty, 1)$-version of Waldhausen's approximation theorem is proved in [**Fio13**].

6.1. Derived Algebraic Geometry. One motivation behind studying stacks in algebraic geometry is a desire to "fix" quotients of schemes by group actions. For example, if an algebraic group G acts freely on a variety X, sheaves on the quotient X/G are the same as G-equivariant sheaves on X. But if the action of G is not free, then the relation between G-equivariant geometry on X and geometry on the quotient X/G is complicated, aside from the difficulty in saying precisely what the quotient means in this case. Passing to stacks "fixes" the quotient, so that sheaves on the quotient stack $[X/G]$ are again G-equivariant sheaves on X.

Just as stacks can be seen as a way to improve certain colimits of schemes so that they always behave as they do in the nice cases, *derived algebraic geometry* can be regarded as fixing a dual problem: correcting certain limits so that they behave more often as they do in good cases. Namely, one motivation for derived algebraic geometry is to "improve" intersections so that they behave more like transverse intersections. Some intersections are already improved by passing from varieties to schemes, for example consider the case of two curves X and Y of degrees m and n in the complex projective plane. If they intersect transversely, according to Bézout's theorem they always have mn points of intersection. When the intersections are not transverse there might be fewer than mn points of intersection, but as long as X and Y have no common irreducible component, there is a natural way to assign multiplicities to the points so that counted with multiplicity there are still mn points of intersection.

We can write the result in the transverse case in terms of cohomology classes as $[X] \cup [Y] = [X \cap Y]$, where $[X], [Y] \in H^2(\mathbb{CP}^2, \mathbb{Z})$ are the fundamental classes of X and Y, and $[X \cap Y] \in H^4(\mathbb{CP}^2, \mathbb{Z}) = \mathbb{Z}$ is the fundamental class of the 0-dimensional variety consisting of the finitely many points of intersection of X and Y. The point of this example is that when the intersection of X and Y is not transverse (but still assuming X and Y do not share a common component), the same formula is true provided we think of $X \cap Y$ as a scheme instead of a variety (and define fundamental classes in an appropriate way). The dimension over $\mathbb{C}$ of the local ring of $X \cap Y$ at the points of intersection gives the relevant multiplicities.

By passing from varieties to schemes we reach a context where computing $[X] \cup [Y]$ does not require full knowledge of the pair (X, Y), just knowledge of the intersection $X \cap Y$, and the intersection is again a scheme, the same kind of object as X and Y. So schemes "fix" the intersection of curves that don't share a component, but they don't fix intersections of curves that do share a component nor do they help intersections in higher dimensions. If X and Y are now subvarieties of $\mathbb{CP}^n$ of complementary dimension that intersect in a finite set of points, the dimension of the local ring of $X \cap Y$ at a point of intersection is no longer the correct multiplicity to make $[X] \cup [Y] = [X \cap Y]$ true; now the multiplicity is given by Serre's formula:

$$(\text{multiplicity at } p \in X \cap Y) = \sum_{i \geq 0} \dim_{\mathbb{C}} \operatorname{Tor}_i^{\mathcal{O}_{\mathbb{CP}^n,p}}(\mathcal{O}_{X,p}, \mathcal{O}_{Y,p}).$$

Notice that this sum *cannot* be computed from just the knowledge of $X \cap Y$ as a scheme, since that only determines the Tor_0 term, but the same formula suggests a fix: we'd be able to compute the correct multiplicities just from $X \cap Y$ if we could arrange for

$$\mathcal{O}_{X \cap Y,p} = \mathcal{O}_{X,p} \overset{\mathbb{L}}{\otimes}_{\mathcal{O}_{\mathbb{CP}^n,p}} \mathcal{O}_{Y,p},$$

where $\otimes^{\mathbb{L}}$ denotes the derived tensor product, a chain complex whose homology gives the relevant Tor groups. We can attempt, then, to define *derived schemes* by replacing commutative rings in the definition of schemes by the chain complex version of commutative algebras: commutative differential graded algebras or cdgas for short, and to compute structure sheaves of intersections using derived tensor products. This suggests jumping into $(\infty, 1)$-waters, since the derived tensor product is only naturally defined up to quasi-isomorphism while sheaves expect to have values determined up to isomorphism.

EXAMPLE 6.1. The higher categorical point of view also suggests itself by considering the other case we mentioned that schemes don't fix: intersecting two curves with a common component. For the simplest possible example let's intersect the y-axis in the affine plane $\mathbb{C}^2$ with itself: the ring of functions we get is $\mathbb{C}[x,y]/(x) \otimes \mathbb{C}[x,y]/(x) = \mathbb{C}[x,y]/(x,x)$. If, as for ordinary rings we had $\mathbb{C}[x,y]/(x,x) = \mathbb{C}[x,y]/(x)$ the intersection would be 1-dimensional and we'd have no hope of a uniform intersection theory. We want instead to have that modding out by x twice is different than doing it once, and makes the quotient ring 0-dimensional again. The philosophy of higher categories suggests that instead of forcing x to be 0, we make it isomorphic to 0; that way if we mod out twice, we can add different isomorphisms each time. Let's do that using cdgas for simplicity. Let's start out with $\mathbb{C}[x,y]$ with x and y in degree zero. To compute $A := \mathbb{C}[x,y]//(x)$ (where we've used a double slash to avoid confusion with the ordinary quotient of rings) we add a generator u in degree one with $du = x$, to make x isomorphic (maybe "homologous" is a better term for isomorphism inside A) to 0. We can readily compute $H_0(A) = \mathbb{C}[x,y]/(x)$ (ordinary quotient), and $H_i(A) = 0$ for $i \neq 0$. Now, we quotient by x again, to get $B := A//(x) = \langle x_0, y_0, u_1, v_1 : du = dv = x \rangle$ (where the subindices on the generators indicate the degree). The non-zero homology is $H_i(B) = \mathbb{C}[x,y]/(x)$ for $i = 0, 1$. This computes the correct Tor terms from Serre's formula: $\mathrm{Tor}_i^{\mathbb{C}[x,y]}(\mathbb{C}[x,y]/(x), \mathbb{C}[x,y]/(x))$ is also given by $\mathbb{C}[x,y]/(x)$ for $i = 0, 1$ and is 0 otherwise. For similar but less trivial examples see the introduction of **[Lur11b]**.

6.1.1. *The cotangent complex.* Another motivation for derived algebraic geometry comes from deformation theory. The cotangent complex $\mathbb{L}_X$ of a scheme X over a field controls its deformation theory through various Ext groups of $\mathbb{L}_X$. Roughly speaking, it can be used to count how many isomorphism classes of first order deformations X has, and to count how many non-isomorphic ways each first order deformation extends to a second order deformation, and so on. But it seems initially to be a purely algebraic creature. Grothendieck asked in 1968 for a geometrical interpretation of the cotangent complex, and derived algebraic geometry provides one. We'll only sketch how the groups $\mathrm{Ext}^i(\mathbb{L}_{X,k}, k)$ are reified through the derived affine schemes.

For $i = 0$, we'll only need ordinary rings, in fact. Let $D_0 = \operatorname{Spec} k[\epsilon]/(\epsilon^2)$ be the walking tangent vector: it is a tiny scheme whose underlying topological space has only one point, but that additionally carries a direction so that morphisms $D_0 \to Y$ correspond to tangent vectors on Y. We have that $\mathrm{Ext}^0(\mathbb{L}_{X,x}, k) \simeq \mathrm{Hom}_*(D_0, (X,x))$ (where Hom_* means base-point preserving morphisms), but to obtain similar representations of the higher Ext, we need to pass to derived rings.

One of the great advantages of admitting nilpotent elements in the rings used for algebraic geometry is that it allows for schemes like D_0. A good intuitive picture to have for derived schemes is that they are to ordinary schemes as those are to reduced schemes, that is, derived schemes can have something like "higher nilpotent elements" in their structure sheaves. We are about to see an example of that now. Let $D_i = \operatorname{Spec} k[\epsilon_i]/(\epsilon_i^2)$ where the ring is now a cdga with generator ϵ_i in degree i (thinking of the chain complex as an abelian higher groupoid, it has a single non-identity i-morphism). As is nicely explained in [**Vez10**], it turns out that $\operatorname{Ext}^i(\mathbb{L}_{X,x}, k) \simeq \pi_0 Map_*(D_i, (X, x))$, where Map_* denotes a mapping space in an $(\infty, 1)$-category of derived pointed schemes.

6.1.2. *Rough sketches of the definitions.* Now we can say approximately what derived schemes are: they are what you get by taking some definition of scheme in terms of commutative rings and performing two replacements: (1) replacing notions from ordinary category theory appearing in the definition with the corresponding definitions in $(\infty, 1)$-category theory, and (2) replacing the category of commutative rings with an $(\infty, 1)$-category of generalized commutative rings (that allow extraction of higher Tor groups from some tensor product operation defined for them).

REMARK 6.2. Since passing to derived schemes and passing to stacks are meant to solve independent problems in the category of schemes, it is entirely possible and even desirable to generalize schemes in both directions at once to obtain the theory of derived stacks. The references we will mention actually treat derived stacks as well. Derived stacks are also higher stacks in the sense that the groupoids appearing in the theory of ordinary stacks are replaced by ∞-groupoids, but it also makes sense to consider *underived* higher stacks as in [**Sim96**].

There are choices for both what definition of schemes to generalize and what sort of rings to use. One can use the definition of schemes as locally ringed spaces locally isomorphic to affine schemes or the point of view of the functor of points. For rings one could use commutative differential graded algebras, simplicial commutative rings or, for application to homotopy theory, E_∞-ring spectra. Also, given any commutative ring R, we can consider R-algebras of each of those three kinds.

Let's talk about the choice of rings first. The three notions we listed are related by functors of $(\infty, 1)$-categories $\mathrm{SCR}_R \to \mathrm{CDGA}_R \to E_\infty$-$R$-Alg. The situation is very simple if R is a $\mathbb{Q}$-algebra: the second functor is an equivalence, the first functor is fully faithful and its image consists of connective cdgas: those whose homology groups are concentrated in non-negative degrees. In general, it gets a little messy: neither functor nor their composite is fully faithful. Comparing free algebras might be illuminating: the free cdga on one generator x of degree 0, say, is just the polynomial ring $\mathbb{Z}[x]$ concentrated in degree 0. On the other hand, the free E_∞-$\mathbb{Z}$-algebra A on x is quite different, since the multiplication on A is not strictly commutative but only commutative up to coherent homotopy. This means for example, that permuting the n factors in the product x^n, doesn't quite fix x^n, rather it produces automorphisms of x^n, and in particular we get a non-trivial homomorphism $\Sigma_n \to \pi_1(A, x^n)$ from the symmetric group on n letters. But in $\mathbb{Z}[x]$ the multiplication *is* strictly commutative and this Σ_n-action is trivial. It seems fair to say that simplicial commutative rings are sufficient for applications of derived algebraic geometry to algebraic geometry itself, while E_∞-ring spectra are mainly for applications to homotopy theory.

Now we'll deal with the choice of definition of scheme. Unlike what happens for the choice of rings, the choice of style of definition does not lead to different notions of derived scheme. Both the locally ringed space point of view and the functor of points view as well as the equivalence between them were described by Jacob Lurie in his PhD thesis or, more recently, in [**Lur11b**]. The latter also contains a very general definition of derived geometric objects that includes derived schemes with any of the above mentioned choices of rings, but also things like derived smooth manifolds studied by David Spivak in his PhD thesis (reworked into [**Spi10**]) and derived analytic spaces (further studied in [**Lur11a**]). The rest of the DAG series of papers, [**Lur11c**], [**Lur11d**], [**Lur11a**], [**Lur11e**], [**Lur11f**], [**Lur11g**], [**Lur11h**], [**Lur12a**] contain a wealth of information about algebraic geometry in the derived setting that we don't have any space to describe.

Bertrand Toën and Gabriele Vezzosi favor the functor of points in their work on homotopical algebraic geometry [**TV05**], [**TV08**]. Their approach is based on an idea of Grothendieck's developed in Hakim's thesis [**Hak72**]: it is possible to construct ordinary stacks, say, by a categorical procedure that receives very little outside input. Namely, one starts with the symmetric monoidal category of abelian groups: commutative monoids for the tensor product there are commutative rings, so we can get our hands on the category of affine schemes. Next, one needs a second piece of information: a topology on the category of affine schemes. Using this one can define purely categorically stacks for the topology. The idea of *relative algebraic geometry* is that one do the same for other symmetric monoidal categories and choices of topology. Toën and Vezzosi develop an $(\infty,1)$-version of this idea starting from a symmetric monoidal $(\infty,1)$-category [**TV01**].[35]

Both approaches require a notion of $(\infty,1)$-version of sheaf and of $(\infty,1)$-topos. A sheaf with values in $\mathcal{C}$ on an $(\infty,1)$-topos $\mathcal{X}$ is just a functor $\mathcal{X}^{\mathrm{op}} \to \mathcal{C}$ that sends colimits in $\mathcal{X}$ to limits in $\mathcal{C}$. Just as in ordinary topos theory, for any topological space X there is a topos $\mathrm{Shv}(X)$ of sheaves on X (in the 1-categorical case these are sheaves of sets, in the $(\infty,1)$-case they are sheaves of ∞-groupoids), which completely determines the space X if it satisfies the mild technical condition of being sober. Sheaves on a space X are just defined to be sheaves on the $(\infty,1)$-topos $\mathrm{Shv}(X)$, but the theory does not require derived schemes or stacks to have underlying topological spaces and is developed for $(\infty,1)$-toposes.

REMARK 6.3. The notion of $(\infty,1)$-topos is due to Charles Rezk who described them as model categories (and called them model toposes) in [**Rez05**]. Lurie developed their theory, using quasi-categories, in his book [**Lur09b**]. Toën and Vezzosi have also written about $(\infty,1)$-toposes using simplicial categories [**TV05**] and Segal categories [**TV02**]. For the reader familiar with the notion of *elementary topos* in ordinary category theory, we should point out that $(\infty,1)$-toposes only generalize the notion of Grothendieck toposes. A Grothendieck topos can be succinctly defined as a localization of a presheaf category, that is as a category $\mathcal{E}$ that admits a functor $F : \mathcal{E} \hookrightarrow \mathrm{Fun}(\mathcal{C}^{\mathrm{op}}, \mathrm{Set})$ for some $\mathcal{C}$ such that F is fully faithful and has left adjoint which preserves finite limits. This definition generalizes to $(\infty,1)$-toposes by replacing Set with ∞-groupoids. Like their 1-categorical counterparts, $(\infty,1)$-toposes can be characterized by analogues of Giraud's axioms. There is also an $(\infty,1)$-version of the notion of site and sheaf on a site, and $(\infty,1)$-categories of sheaves on sites provide examples of $(\infty,1)$-toposes; but unlike the case for ordinary toposes, these examples are not all the $(\infty,1)$-toposes.

FURTHER READING. The idea of derived algebraic geometry can be traced back at least to the 1980s when the *hidden smoothness principle* came out of the work of Alexander Beilinson, Vladimir Drinfeld, Pierre Deligne and Maxim Kontsevich.

[35] They had their work cut out for them, since at the time the best available model for symmetric monoidal $(\infty,1)$-categories were monoidal model categories, which are very rigid: they have an actual monoidal structure, that is, one associative up to coherent *isomorphism* in the ordinary category on which the model structure is defined.

This principle expressed the hope that many moduli spaces in algebraic geometry that are singular were really just truncations of smooth derived moduli spaces. An early example of the use of derived moduli spaces, before all the machinery was in place, can be found in [**Kon95**]. It was also Kontsevich who introduced dg-schemes and these were used by Ionut Ciocan-Fontanine and Mikhail Kapranov to construct derived versions of Quot schemes [**CFK01**] and Hilbert schemes [**CFK02**]. For more recent work on derived moduli spaces see [**TV04**].

For a through overview of higher and derived stacks see [**Toë09**]. More information about the deformation theory aspects of derived algebraic geometry, can be found in the already mentioned [**Lur11e**], but we also recommend Lurie's earlier ICM address [**Lur10**] which gives an expository account. To see derived algebraic geometry in action, see, for example:

(1) David Ben-Zvi and David Nadler's work on derived loop spaces in algebraic geometry, (for example, [**BZN12**]), or
(2) their paper [**BZFN10**], joint with John Francis, about the $(\infty, 1)$-category $QC(X)$ of quasi-coherent sheaves on a derived stack X, where they show that given derived stacks X and Y over k satisfying an appropriate finiteness condition, any k-linear colimit preserving functor $F : QC(X) \to QC(Y)$ is given by a quasi-coherent sheaf $K \in QC(X \times_k Y)$ by means of an *integral transform* $F(S) := (\pi_Y)_*(K \otimes \pi_X^* S)$.

6.1.3. *Topological modular forms.* It would take us too far afield to describe topological modular forms in any kind of detail, but for readers interested in homotopy theory we want to at least mention this application of derived algebraic geometry.

There is a beautiful relation due to Quillen [**Qui69**] between generalized cohomology theories which have an analogue of the theory of Chern classes, properly called complex-oriented cohomology theories, and formal groups laws. A formal group law is[36] a power series $F(x, y) = x + y +$ (higher order terms), with coefficents in some commutative ring A, that satisfies $F(x, F(y, z)) = F(F(x, y), z)$. Given any Lie group or algebraic group, the power series expansion of the multiplication at the origin gives a formal group law[37], but not every such law arises from this construction. There is a notion of isomorphism of formal group laws given by changes of coordinates $x' = g(x)$, $y' = g(y)$ where g is a power series. If A is a field of characteristic 0, all one-dimensional formal group laws are isomorphic to $F(x, y) = x + y$, and there is a classification due independently to Lazard [**Laz55**] and Dieudonné [**Die55**] of one-dimensional formal groups laws over algebraically closed fields of characteristic p.

A complex-oriented cohmology theory has an associated one-dimensional commutative group law F, that gives the formula for the first Chern class of a tensor product of line bundles in terms of the Chern classes of the individual line bundles, $c_1(L_1 \otimes L_2) = F(c_1(L_1), c_1(L_2))$. For ordinary cohomology, the formal group law is the additive one, $F(x, y) = x + y$. For complex K-theory, it is the multiplicative formal group law, $F(x, y) = x + y + xy = (1 + x)(1 + y) - 1$. Both of these come from expanding the product of a one-dimensional commutative algebraic group,

[36] Actually the definition we give is for *one-dimensional* formal groups laws. There are also n-dimensional formal group laws that are defined in the same way but with x, y, z denoting n-tuples of variables and $F(x, y)$ denoting an n-tuple of power series.

[37] Of the same dimension as the group.

the additive group $\mathbb{G}_a$, and the multiplicative group $\mathbb{G}_m$, respectively. Over an algebraically closed field there is only one other kind of connected commutative one-dimensional algebraic group: elliptic curves, whose associated formal group laws correspond to *elliptic cohomology* theories. We can think of assembling all of the elliptic cohomology theories into something like a sheaf of cohomology theories on the moduli space of elliptic curves. If we had such a thing we could imagine that taking global sections of it would produces a new cohomology theory that samples from all elliptic cohomology theories at once. That cohomology theory is (almost[38]) what tmf, the cohomology theory of topological modular forms, is supposed to be.

There are serious technical difficulties in constructing it, since the notion of "sheaf of cohomology theories" isn't well-behaved enough to use for this purpose. Instead one can attempt to represent the cohomology theories by spectra, for which the $(\infty, 1)$-notion of sheaf we mentioned earlier works well. Lifting this sheaf of cohomology theories to a sheaf of spectra turns out to be made easier, paradoxically, by making the problem harder and lifting instead to a sheaf of E_∞-ring spectra, making tmf the global sections of an object in derived algebraic geometry. This lifting was first achieved (and shown to be essentially unique) by Mike Hopkins, Haynes Miller and Paul Goerss using an obstruction theory of E_∞-ring spectra. A second construction due to Jacob Lurie uses more of the theory of derived algebraic geometry: he defines a moduli problem for (derived) elliptic curves that can only be stated in the derived setting and then uses his general representability criterion to show that the moduli problem is represented by a derived Deligne-Mumford stack, whose global sections then give tmf.

FURTHER READING. For a great introduction to the topic see Paul Goerss's Bourbaki Seminar talk [**Goe10**] and the references therein. (That paper is also a good introduction to derived schemes.) Also recommended is Lurie's [**Lur09a**] where he outlines the second construction mentioned above.

6.2. The cobordism hypothesis. Michael Atiyah [**Ati88**] proposed a mathematical definition of *topological quantum field theory* (henceforth TQFT) inspired by Graeme Segal's axioms [**Seg88**][39] for *conformal field theory*. The definition uses the category $\mathrm{Bord}_{\langle n-1,n\rangle}$ whose objects are closed oriented $(n-1)$-manifolds and whose morphisms $M \to N$ are diffeomorphism classes of *bordisms*[40] from M to N, that is, n-manifolds W whose boundary is identified with $\overline{M} \sqcup N$. Here $\overline{M}$ denotes M with the opposite orientation and the diffeomorphisms we consider are those fixing the boundary. This category has symmetric monoidal structure given by taking disjoint unions (both at the level of $(n-1)$-manifolds and n-dimensional bordisms). The brunt of Atiyah's axioms is that a TQFT is a symmetric monoidal functor $F : \mathrm{Bord}_{\langle n-1,n\rangle} \to \mathrm{Vect}$, where Vect carries the monoidal structure given by the tensor product of vector spaces.

This definition implies that one can calculate the value of F on some bordism W by chopping it up into a composition of simpler bordisms and composing the images under F of those pieces. For example, when $n = 2$, we could chop any surface into a

[38]For one thing, we actually need a sheaf on the Deligne-Mumford compactification of the moduli space.

[39]The reader might prefer the more recently published [**Seg04**] which is an expanded version of an unpublished paper that circulated widely for many years.

[40]Perhaps confusingly, there is no difference between bordisms and cobordisms. We are following someone's terminology in each case...

combination of pairs of pants, cylinders and spherical caps. This suggests that such a functor is determined by just a handful of its values. This approach becomes more and more complicated in higher dimensions because the pieces which are required to build up all bordisms increase in number and complexity. One might hope that a simpler theory could result from allowing more general ways to cut up bordisms: Atiyah's definition allows cutting along codimension 1 submanifolds, but if we allowed, say, arbitrary codimensional cuts, we could triangulate every manifold. The cobordism hypothesis doesn't make it quite that simple to cut up manifolds, but we will allow pieces of all dimensions.

John Baez and James Dolan's *cobordism hypothesis* [**BD95**] is about such extended TQFT's and does indeed say that they are determined by very few of their values, namely, that they are determined by where they send single points! This hypothesis has now been proved in the $n = 2$ case by Jacob Lurie and Mike Hopkins and in general by Lurie. A detailed and very readable sketch of the proof is available as [**Lur09c**]. It might be a little unfair to say that the cobordism hypothesis fulfills the promise of a simpler theory mentioned in the previous paragraph, since there were complicated foundational issues involved in both the precise statement in Lurie's paper and in its proof. We'll say a little more about these issues after stating a version of the cobordism hypothesis.

To allow chopping up manifolds into pieces of arbitrary dimension, we replace the domain category of a TQFT by a higher category $\mathrm{Bord}_n^{\mathrm{fr}}$ whose k-morphisms are n-framed[41] k-manifolds with corners, and whose $(k+1)$-morphisms are n-framed bordisms between such. (The framing is used to remove some ambiguity coming from the diffeomorphism group of $\mathbb{R}^m$, it is important to understand and to know what happens without it, but we leave the explanations to the references.) What we've said so far is a little sloppy. Baez and Dolan's original formulation involved an n-category, so the description of the morphisms we gave is only valid for $k < n$; for $k = n$, we must again take diffeomorphism classes of bordisms.

A $\mathcal{C}$-valued framed n-dimensional extended TQFT then is a symmetric monoidal functor from $\mathrm{Bord}_n^{\mathrm{fr}} \to \mathcal{C}$; where $\mathcal{C}$ is an arbitrary symmetric monoidal n-category. We generalize to arbitrary symmetric monoidal n-categories partly because there is no canonical substitute for Vect in this case; categorifying Vect to make an n-category is an interesting question of it own. Fortunately, we don't need any particular symmetric monoidal n-category for the statement:

Baez and Dolan's cobordism hypothesis: A framed n-dimensional extended TQFT with values in $\mathcal{C}$ is completely determined by its value on the point. Moreover, the value on the point is always a *fully dualizable* object of $\mathcal{C}$ and there is a bijection between isomorphism classes of such TQFTs and isomorphism classes of fully dualizable objects in $\mathcal{C}$.

We will now explain what is meant by fully dualizable objects. First, notice that the usual definition of adjoint functors can be phrased in terms of 1-morphisms and 2-morphisms in Cat and thus makes sense in any 2-category, and can even be used for k- and $(k+1)$-morphisms in a higher category to define adjunctions of k-morphisms. Call a k-morphism dualizable if it has both a left and a right adjoint, and call it n-times dualizable if it is dualizable and the $(k+1)$-morphisms that

[41] An n-framing on a k-manifold M is a trivialization of its stabilized tangent bundle, $TM \oplus \underline{\mathbb{R}}^{n-k}$.

show it has adjoints are themselves $(n-1)$-times dualizable[42]. Finally, a monoidal n-category $\mathcal{C}$ can be thought of as an $(n+1)$-category with a single object $B\mathcal{C}$, so we can say that an object of $\mathcal{C}$ is m-times dualizable if as a 1-morphism in $B\mathcal{C}$, it is m-times dualizable. Then a fully dualizable object of a symmetric monoidal n-category $\mathcal{C}$ is just an $(n-1)$-times dualizable object. Notice that this really does depend on n: a symmetric monoidal n-category can also be regarded as a symmetric monoidal $(n+1)$-category (with only identity $(n+1)$-morphisms), but being $(n-1)$-times dualizable is different from being n-times dualizable.

This n-category version of the cobordism hypothesis was proven for $n = 2$ by Chris Schommer-Pries in his PhD thesis [**SP09**], but for higher n a lack of a solid, practical theory of n-categories impeded progress. Lurie's key insight was to prove instead a more general version using the easier theory of (∞, n)-categories and then deduce the original formulation by truncation. So Lurie replaced the n-category described above by one in which

(1) n-morphisms are bordisms, not diffeomorphisms classes of such,
(2) $(n+1)$-morphisms are diffeomorphisms,
(3) $(n+2)$-morphisms are isotopies of diffeomorphisms,
(4) $(n+3)$-morphisms are isotopies of isotopies, and so on.

FURTHER READING. For a more detailed introduction to the cobordism hypothesis and its applications, Dan Freed's survey [**Fre13**] is highly recommended, as is, of course, Lurie's [**Lur09c**] which is not just a detailed outline of the proof but also includes a lot of motivation (and an introduction to higher category theory!), and many other interesting versions of the homotopy hypothesis, and applications. The reader might also be interested in Julie Bergner's survey [**Ber11**] which focuses on models for (∞, n)-categories but does state Lurie's version of the cobordism hypothesis in some detail, describing the construction of $\mathrm{Bord}_n^{\mathrm{fr}}$ as an n-fold complete Segal space.

References

[AC12] Gregory Arone and Michael Ching, *A classification of Taylor towers of functors of spaces and spectra*, Adv. Math. **272** (2015), 471–552, DOI 10.1016/j.aim.2014.12.007. MR3303239

[AR94] Jiří Adámek and Jiří Rosický, *Locally presentable and accessible categories*, London Mathematical Society Lecture Note Series, vol. 189, Cambridge University Press, Cambridge, 1994. MR1294136 (95j:18001)

[Ati88] Michael Atiyah, *Topological quantum field theories*, Inst. Hautes Études Sci. Publ. Math. **68** (1988), 175–186 (1989). MR1001453 (90e:57059)

[Bae97] John C. Baez, *An introduction to n-categories*, Category theory and computer science (Santa Margherita Ligure, 1997), Lecture Notes in Comput. Sci., vol. 1290, Springer, Berlin, 1997, pp. 1–33, DOI 10.1007/BFb0026978. MR1640335 (99h:18008)

[Bae05] John C. Baez, *This Week's Finds in Mathematical Physics (Week 223)*, 2005, `http://math.ucr.edu/home/baez/week223.html`.

[Bar05] Clark Barwick, *(∞, n)-Cat as a closed model category*, Ph.D. thesis, University of Pennsylvania, 2005.

[Bar12] Clark Barwick, *On the algebraic k-theory of higher categories. the universal property of Waldhausen K-theory*, submitted, preprint available as `arXiv:1204.3607 [math.KT]`, 2012.

[BD95] John C. Baez and James Dolan, *Higher-dimensional algebra and topological quantum field theory*, J. Math. Phys. **36** (1995), no. 11, 6073–6105, DOI 10.1063/1.531236. MR1355899 (97f:18003)

[42]Where, of course, "1-times dualizable" just means "dualizable".

[BD98] John C. Baez and James Dolan, *Higher-dimensional algebra. III. n-categories and the algebra of opetopes*, Adv. Math. **135** (1998), no. 2, 145–206, DOI 10.1006/aima.1997.1695. MR1620826 (99j:18005)

[Bek00] Tibor Beke, *Sheafifiable homotopy model categories*, Math. Proc. Cambridge Philos. Soc. **129** (2000), no. 3, 447–475, DOI 10.1017/S0305004100004722. MR1780498 (2001i:18015)

[Bén67] Jean Bénabou, *Introduction to bicategories*, Reports of the Midwest Category Seminar, Springer, Berlin, 1967, pp. 1–77. MR0220789 (36 #3841)

[Ber99] Clemens Berger, *Double loop spaces, braided monoidal categories and algebraic 3-type of space*, Higher homotopy structures in topology and mathematical physics (Poughkeepsie, NY, 1996), Contemp. Math., vol. 227, Amer. Math. Soc., Providence, RI, 1999, pp. 49–66, DOI 10.1090/conm/227/03252. MR1665460 (2000e:55009)

[Ber07] Julia E. Bergner, *Three models for the homotopy theory of homotopy theories*, Topology **46** (2007), no. 4, 397–436, DOI 10.1016/j.top.2007.03.002. MR2321038 (2008e:55024)

[Ber10] Julia E. Bergner, *A survey of $(\infty,1)$-categories*, Towards higher categories, IMA Vol. Math. Appl., vol. 152, Springer, New York, 2010, pp. 69–83, DOI 10.1007/978-1-4419-1524-5_2. MR2664620 (2011e:18001)

[Ber11] Julia E. Bergner, *Models for (∞,n)-categories and the cobordism hypothesis*, Mathematical foundations of quantum field theory and perturbative string theory, Proc. Sympos. Pure Math., vol. 83, Amer. Math. Soc., Providence, RI, 2011, pp. 17–30, DOI 10.1090/pspum/083/2742424. MR2742424 (2012j:55023)

[BGT13] Andrew J. Blumberg, David Gepner, and Gonçalo Tabuada, *A universal characterization of higher algebraic K-theory*, Geom. Topol. **17** (2013), no. 2, 733–838, DOI 10.2140/gt.2013.17.733. MR3070515

[BK72] A. K. Bousfield and D. M. Kan, *Homotopy limits, completions and localizations*, Lecture Notes in Mathematics, Vol. 304, Springer-Verlag, Berlin-New York, 1972. MR0365573 (51 #1825)

[BK12a] C. Barwick and D. M. Kan, *A characterization of simplicial localization functors and a discussion of DK equivalences*, Indag. Math. (N.S.) **23** (2012), no. 1-2, 69–79, DOI 10.1016/j.indag.2011.10.001. MR2877402

[BK12b] C. Barwick and D. M. Kan, *n-relative categories: A model for the homotopy theory of n-fold homotopy theories*, to appear in *Homology, Homotopy, and Applications*, preprint available as `arXiv:1102.0186 [math.AT]`, 2012.

[BK12c] C. Barwick and D. M. Kan, *Relative categories: another model for the homotopy theory of homotopy theories*, Indag. Math. (N.S.) **23** (2012), no. 1-2, 42–68, DOI 10.1016/j.indag.2011.10.002. MR2877401 (2012m:55019)

[BLM08] Yu. Bespalov, V. Lyubashenko, and O. Manzyuk, *Pretriangulated A_∞-categories*, Pratsī Īnstitutu Matematiki Natsīonal′noï Akademīï Nauk Ukraïni. Matematika ta ïï Zastosuvannya [Proceedings of Institute of Mathematics of NAS of Ukraine. Mathematics and its Applications], vol. 76, Natsīonal′na Akademīya Nauk Ukraïni, Īnstitut Matematiki, Kiev, 2008. MR2513383 (2010j:18019)

[BSP12] Clark Barwick and Chris Schommer-Pries, *On the unicity of the homotopy theory of higher categories*, preprint available as `arXiv:1112.0040 [math.AT]`, 2012.

[BZFN10] David Ben-Zvi, John Francis, and David Nadler, *Integral transforms and Drinfeld centers in derived algebraic geometry*, J. Amer. Math. Soc. **23** (2010), no. 4, 909–966, DOI 10.1090/S0894-0347-10-00669-7. MR2669705 (2011j:14023)

[BZN12] David Ben-Zvi and David Nadler, *Loop spaces and connections*, J. Topol. **5** (2012), no. 2, 377–430, DOI 10.1112/jtopol/jts007. MR2928082

[CFK01] Ionuţ Ciocan-Fontanine and Mikhail Kapranov, *Derived Quot schemes* (English, with English and French summaries), Ann. Sci. École Norm. Sup. (4) **34** (2001), no. 3, 403–440, DOI 10.1016/S0012-9593(01)01064-3. MR1839580 (2002k:14003)

[CFK02] Ionuţ Ciocan-Fontanine and Mikhail M. Kapranov, *Derived Hilbert schemes*, J. Amer. Math. Soc. **15** (2002), no. 4, 787–815, DOI 10.1090/S0894-0347-02-00399-5. MR1915819 (2003h:14007)

[Che11] Eugenia Cheng, *Comparing operadic theories of n-category*, Homology Homotopy Appl. **13** (2011), no. 2, 217–248, DOI 10.4310/HHA.2011.v13.n2.a14. MR2854336

[CL04] Eugenia Cheng and Aaron Lauda, *Higher-Dimensional Categories: an illustrated guide book*, draft version available as `http://cheng.staff.shef.ac.uk/guidebook/guidebook-new.pdf`, 2004.

[Cor82] Jean-Marc Cordier, *Sur la notion de diagramme homotopiquement cohérent* (French), Cahiers Topologie Géom. Différentielle **23** (1982), no. 1, 93–112. Third Colloquium on Categories, Part VI (Amiens, 1980). MR648798 (83g:18018)

[Cur71] Edward B. Curtis, *Simplicial homotopy theory*, Advances in Math. **6** (1971), 107–209 (1971). MR0279808 (43 #5529)

[DHKS04] William G. Dwyer, Philip S. Hirschhorn, Daniel M. Kan, and Jeffrey H. Smith, *Homotopy limit functors on model categories and homotopical categories*, Mathematical Surveys and Monographs, vol. 113, American Mathematical Society, Providence, RI, 2004. MR2102294 (2005k:18027)

[Die55] Jean Dieudonné, *Lie groups and Lie hyperalgebras over a field of characteristic $p > 0$. II*, Amer. J. Math. **77** (1955), 218–244. MR0067872 (16,789f)

[DK80a] W. G. Dwyer and D. M. Kan, *Simplicial localizations of categories*, J. Pure Appl. Algebra **17** (1980), no. 3, 267–284, DOI 10.1016/0022-4049(80)90049-3. MR579087 (81h:55018)

[DK80b] W. G. Dwyer and D. M. Kan, *Calculating simplicial localizations*, J. Pure Appl. Algebra **18** (1980), no. 1, 17–35, DOI 10.1016/0022-4049(80)90113-9. MR578563 (81h:55019)

[DS95] W. G. Dwyer and J. Spaliński, *Homotopy theories and model categories*, Handbook of algebraic topology, North-Holland, Amsterdam, 1995, pp. 73–126, DOI 10.1016/B978-044481779-2/50003-1. MR1361887 (96h:55014)

[DS11a] Daniel Dugger and David I. Spivak, *Mapping spaces in quasi-categories*, Algebr. Geom. Topol. **11** (2011), no. 1, 263–325, DOI 10.2140/agt.2011.11.263. MR2764043 (2012e:55026)

[DS11b] Daniel Dugger and David I. Spivak, *Rigidification of quasi-categories*, Algebr. Geom. Topol. **11** (2011), no. 1, 225–261, DOI 10.2140/agt.2011.11.225. MR2764042 (2012f:55025)

[Fio13] Thomas Fiore, *Approximation in K-theory for Waldhausen quasicategories*, preprint available as `arXiv:1303.4029 [math.AT]`, 2013.

[FL12] Thomas Fiore and Wolfgang Lück, *Waldhausen additivity: Classical and quasicategorical*, preprint available as `arXiv:1207.6613 [math.AT]`, 2012.

[FOOO09] Kenji Fukaya, Yong-Geun Oh, Hiroshi Ohta, and Kaoru Ono, *Lagrangian intersection Floer theory: anomaly and obstruction. Part I*, AMS/IP Studies in Advanced Mathematics, vol. 46, American Mathematical Society, Providence, RI; International Press, Somerville, MA, 2009. MR2553465 (2011c:53217)

[Fre13] Daniel S. Freed, *The cobordism hypothesis*, Bull. Amer. Math. Soc. (N.S.) **50** (2013), no. 1, 57–92, DOI 10.1090/S0273-0979-2012-01393-9. MR2994995

[Fri12] Greg Friedman, *Survey article: an elementary illustrated introduction to simplicial sets*, Rocky Mountain J. Math. **42** (2012), no. 2, 353–423, DOI 10.1216/RMJ-2012-42-2-353. MR2915498

[Fuk93] Kenji Fukaya, *Morse homotopy, A^∞-category, and Floer homologies*, Proceedings of GARC Workshop on Geometry and Topology '93 (Seoul, 1993), Lecture Notes Ser., vol. 18, Seoul Nat. Univ., Seoul, 1993, pp. 1–102. MR1270931 (95e:57053)

[GH14] David Gepner and Rune Haugseng, *Enriched ∞-categories via non-symmetric ∞-operads*, preprint available as `arXiv:1312.3178 [math.AT]`, 2014.

[Goe10] Paul G. Goerss, *Topological modular forms [after Hopkins, Miller and Lurie]*, Astérisque **332** (2010), Exp. No. 1005, viii, 221–255. Séminaire Bourbaki. Volume 2008/2009. Exposés 997–1011. MR2648680 (2011m:55003)

[GPS95] R. Gordon, A. J. Power, and Ross Street, *Coherence for tricategories*, Mem. Amer. Math. Soc. **117** (1995), no. 558, vi+81, DOI 10.1090/memo/0558. MR1261589 (96j:18002)

[GPS12] Moritz Groth, Kate Ponto, and Michael Shulman, *The additivity of traces in monoidal derivators*, preprint, available as `arXiv:1212.3277 [math.AT]`, 2012.

[Gro83] Alexander Grothendieck, *Pursuing stacks*, unpublished manuscript, 1983.

[Gro91] Alexander Grothendieck, *Les Dérivateurs*, unpublished manuscript, 1991.

[Gro10] Moritz Groth, *A short course on ∞-categories*, preprint available from author's website as `http://www.math.ru.nl/~mgroth/preprints/groth_scinfinity.pdf`.

[Gro13] Moritz Groth, *Derivators, pointed derivators and stable derivators*, Algebr. Geom. Topol. **13** (2013), no. 1, 313–374, DOI 10.2140/agt.2013.13.313. MR3031644

[Gur06] Michael Nicholas Gurski, *An algebraic theory of tricategories*, ProQuest LLC, Ann Arbor, MI, 2006. Thesis (Ph.D.)–The University of Chicago. MR2717302

[Hak72] Monique Hakim, *Topos annelés et schémas relatifs* (French), Springer-Verlag, Berlin-New York, 1972. Ergebnisse der Mathematik und ihrer Grenzgebiete, Band 64. MR0364245 (51 #500)

[Hau14] Rune Haugseng, *Rectification of enriched ∞-categories*, preprint available as `arXiv:1312.3881 [math.AT]`, 2014.

[Hel88] Alex Heller, *Homotopy theories*, Mem. Amer. Math. Soc. **71** (1988), no. 383, vi+78, DOI 10.1090/memo/0383. MR920963 (89b:55013)

[Hir03] Philip S. Hirschhorn, *Model categories and their localizations*, Mathematical Surveys and Monographs, vol. 99, American Mathematical Society, Providence, RI, 2003. MR1944041 (2003j:18018)

[Hof11] Alexander Hoffnung, *Spans in 2-categories: A monoidal tricategory*, preprint, available as `arXiv:1112.0560 [math.CT]`, 2011.

[Hov99] Mark Hovey, *Model categories*, Mathematical Surveys and Monographs, vol. 63, American Mathematical Society, Providence, RI, 1999. MR1650134 (99h:55031)

[Joy02] A. Joyal, *Quasi-categories and Kan complexes*, J. Pure Appl. Algebra **175** (2002), no. 1-3, 207–222, DOI 10.1016/S0022-4049(02)00135-4. Special volume celebrating the 70th birthday of Professor Max Kelly. MR1935979 (2003h:55026)

[Joy08a] A. Joyal, *Notes on quasi-categories*, in preparation, 2008.

[Joy08b] André Joyal, *The theory of quasi-categories I*, 2008, in progress.

[JT06] André Joyal and Miles Tierney, *Quasi-categories vs Segal spaces*, preprint, available as `arXiv:math/0607820 [math.AT]`, 2006.

[Kel06] Bernhard Keller, *On differential graded categories*, International Congress of Mathematicians. Vol. II, Eur. Math. Soc., Zürich, 2006, pp. 151–190. MR2275593 (2008g:18015)

[Kon95] Maxim Kontsevich, *Enumeration of rational curves via torus actions*, The moduli space of curves (Texel Island, 1994), Progr. Math., vol. 129, Birkhäuser Boston, Boston, MA, 1995, pp. 335–368. MR1363062 (97d:14077)

[KS09] M. Kontsevich and Y. Soibelman, *Notes on A_∞-algebras, A_∞-categories and non-commutative geometry*, Homological mirror symmetry, Lecture Notes in Phys., vol. 757, Springer, Berlin, 2009, pp. 153–219. MR2596638 (2011f:53183)

[Lac07] Stephen Lack, *Bicat is not triequivalent to Gray*, Theory Appl. Categ. **18** (2007), No. 1, 1–3. MR2276246 (2007m:18005)

[Lac10] Stephen Lack, *A 2-categories companion*, Towards higher categories, IMA Vol. Math. Appl., vol. 152, Springer, New York, 2010, pp. 105–191, DOI 10.1007/978-1-4419-1524-5_4. MR2664622 (2011d:18012)

[Laz55] Michel Lazard, *Sur les groupes de Lie formels à un paramètre* (French), Bull. Soc. Math. France **83** (1955), 251–274. MR0073925 (17,508e)

[Lei98] Tom Leinster, *Basic bicategories*, preprint available as `arXiv:math/9810017 [math.CT]`, 1998.

[Lei02] Tom Leinster, *A survey of definitions of n-category*, Theory Appl. Categ. **10** (2002), 1–70 (electronic). MR1883478 (2002i:18008)

[Lei04] Tom Leinster, *Higher operads, higher categories*, London Mathematical Society Lecture Note Series, vol. 298, Cambridge University Press, Cambridge, 2004. MR2094071 (2005h:18030)

[Lur07a] Jacob Lurie, *Derived Algebraic Geometry I: Stable ∞-categories*, preprint available as `arXiv:math/0608228 [math.CT]`, 2007.

[Lur07b] Jacob Lurie, *Derived Algebraic Geometry II: Noncommutative Algebra*, preprint available as `arXiv:math/0702299 [math.CT]`, 2007.

[Lur07c] Jacob Lurie, *Derived Algebraic Geometry III: Commutative Algebra*, preprint available as `arXiv:math/0703204 [math.CT]`, 2007.

[Lur09a] J. Lurie, *A survey of elliptic cohomology*, Algebraic topology, Abel Symp., vol. 4, Springer, Berlin, 2009, pp. 219–277, DOI 10.1007/978-3-642-01200-6_9. MR2597740 (2011f:55009)

[Lur09b] Jacob Lurie, *Higher topos theory*, Annals of Mathematics Studies, vol. 170, Princeton University Press, Princeton, NJ, 2009. MR2522659 (2010j:18001)

[Lur09c] Jacob Lurie, *On the classification of topological field theories*, Current developments in mathematics, 2008, Int. Press, Somerville, MA, 2009, pp. 129–280. MR2555928 (2010k:57064)

[Lur10] Jacob Lurie, *Moduli problems for ring spectra*, Proceedings of the International Congress of Mathematicians. Volume II (New Delhi), Hindustan Book Agency, 2010, pp. 1099–1125. MR2827833 (2012i:55008)

[Lur11a] Jacob Lurie, *Derived Algebraic Geometry IX: Closed Immersions*, preprint available from author's website as `http://math.harvard.edu/~lurie/papers/DAG-IX.pdf`, 2011.

[Lur11b] Jacob Lurie, *Derived Algebraic Geometry V: Structured Spaces*, preprint available from author's website as `http://math.harvard.edu/~lurie/papers/DAG-V.pdf`, 2011.

[Lur11c] Jacob Lurie, *Derived Algebraic Geometry VII: Spectral Schemes*, preprint available from author's website as `http://math.harvard.edu/~lurie/papers/DAG-VII.pdf`, 2011.

[Lur11d] Jacob Lurie, *Derived Algebraic Geometry VIII: Quasi-Coherent Sheaves and Tannaka Duality Theorems*, preprint available from author's website as `http://math.harvard.edu/~lurie/papers/DAG-VIII.pdf`, 2011.

[Lur11e] Jacob Lurie, *Derived Algebraic Geometry X: Formal Moduli Problems*, preprint available from author's website as `http://math.harvard.edu/~lurie/papers/DAG-X.pdf`, 2011.

[Lur11f] Jacob Lurie, *Derived Algebraic Geometry XI: Descent Theorems*, preprint available from author's website as `http://math.harvard.edu/~lurie/papers/DAG-XI.pdf`, 2011.

[Lur11g] Jacob Lurie, *Derived Algebraic Geometry XII: Proper Morphisms, Completions, and the Grothendieck Existence Theorem*, preprint available from author's website as `http://math.harvard.edu/~lurie/papers/DAG-XII.pdf`, 2011.

[Lur11h] Jacob Lurie, *Derived Algebraic Geometry XIII: Rational and p-adic Homotopy Theory*, preprint available from author's website as `http://math.harvard.edu/~lurie/papers/DAG-XIII.pdf`, 2011.

[Lur12a] Jacob Lurie, *Derived Algebraic Geometry XIV: Representability Theorems*, preprint available from author's website as `http://math.harvard.edu/~lurie/papers/DAG-XIV.pdf`, 2012.

[Lur12b] Jacob Lurie, *Higher algebra*, 2012, draft available from author's website as `http://www.math.harvard.edu/~lurie/papers/HigherAlgebra.pdf`.

[Mal01] Georges Maltsiniotis, *Introduction à la théorie des dérivateurs (d'après grothendieck)*, preprint available from author's website, as `http://www.math.jussieu.fr/~maltsin/ps/m.ps`, 2001.

[May72] J. P. May, *The geometry of iterated loop spaces*, Springer-Verlag, Berlin-New York, 1972. Lectures Notes in Mathematics, Vol. 271. MR0420610 (54 #8623b)

[May05] Peter J. May, *Operadic categories, A_∞-categories and n-categories*, 2005, talk given in Morelia, México, `http://www.math.uchicago.edu/~may/NCATS/PostMexico.pdf`.

[Por04] Timothy Porter, *S-categories, S-groupoids, Segal categories and quasi-categories*, preprint available as `arXiv:math/0401274 [math.AT]`, 2004.

[Qui67] Daniel G. Quillen, *Homotopical algebra*, Lecture Notes in Mathematics, No. 43, Springer-Verlag, Berlin-New York, 1967. MR0223432 (36 #6480)

[Qui69] Daniel Quillen, *On the formal group laws of unoriented and complex cobordism theory*, Bull. Amer. Math. Soc. **75** (1969), 1293–1298. MR0253350 (40 #6565)

[Ren06] Olivier Renaudin, *Théories homotopiques de Quillen combinatoires et dérivateurs de Grothendieck*, preprint available as `arXiv:math/0603339 [math.AT]`.

[Ren09] Olivier Renaudin, *Plongement de certaines théories homotopiques de Quillen dans les dérivateurs* (French, with French summary), J. Pure Appl. Algebra **213** (2009), no. 10, 1916–1935, DOI 10.1016/j.jpaa.2009.02.014. MR2526867 (2010c:18016)

[Rez01] Charles Rezk, *A model for the homotopy theory of homotopy theory*, Trans. Amer. Math. Soc. **353** (2001), no. 3, 973–1007 (electronic), DOI 10.1090/S0002-9947-00-02653-2. MR1804411 (2002a:55020)

[Rez05] Charles Rezk, *Toposes and homotopy toposes*, 2005.

[Rie14] Emily Riehl, *Categorical homotopy theory*, New Mathematical Monographs, vol. 24, Cambridge University Press, Cambridge, 2014. MR3221774

[RV13] Emily Riehl and Dominic Verity, *Homotopy coherent adjunctions and the formal theory of monads*, in preparation, 2013.

[Seg88] G. B. Segal, *The definition of conformal field theory*, Differential geometrical methods in theoretical physics (Como, 1987), NATO Adv. Sci. Inst. Ser. C Math. Phys. Sci., vol. 250, Kluwer Acad. Publ., Dordrecht, 1988, pp. 165–171. MR981378 (90d:58026)

[Seg04] Graeme Segal, *The definition of conformal field theory*, Topology, geometry and quantum field theory, London Math. Soc. Lecture Note Ser., vol. 308, Cambridge Univ. Press, Cambridge, 2004, pp. 421–577. MR2079383 (2005h:81334)

[Sei08] Paul Seidel, *Fukaya categories and Picard-Lefschetz theory*, Zurich Lectures in Advanced Mathematics, European Mathematical Society (EMS), Zürich, 2008. MR2441780 (2009f:53143)

[Sim96] Carlos Simspon, *Algebraic (geometric) n-stacks*, preprint available as `arXiv:alg-geom/9609014`, 1996.

[Sim99] Carlos Simpson, *A Giraud-type characterization of the simplicial categories associated to closed model categories as ∞-pretopoi*, preprint available as `arXiv:math/9903167 [math.AT]`.

[Sim12] Carlos Simpson, *Homotopy theory of higher categories*, New Mathematical Monographs, vol. 19, Cambridge University Press, Cambridge, 2012. MR2883823 (2012m:18019)

[SP09] Christopher John Schommer-Pries, *The classification of two-dimensional extended topological field theories*, ProQuest LLC, Ann Arbor, MI, 2009. Thesis (Ph.D.)–University of California, Berkeley. MR2713992

[Spi10] David I. Spivak, *Derived smooth manifolds*, Duke Math. J. **153** (2010), no. 1, 55–128, DOI 10.1215/00127094-2010-021. MR2641940 (2012a:57043)

[SS86] Stephen Schanuel and Ross Street, *The free adjunction* (English, with French summary), Cahiers Topologie Géom. Différentielle Catég. **27** (1986), no. 1, 81–83. MR845410 (87f:18005)

[Sta63] James Dillon Stasheff, *Homotopy associativity of H-spaces. I, II*, Trans. Amer. Math. Soc. 108 (1963), 275-292; ibid. **108** (1963), 293–312. MR0158400 (28 #1623)

[Ste67] N. E. Steenrod, *A convenient category of topological spaces*, Michigan Math. J. **14** (1967), 133–152. MR0210075 (35 #970)

[Str] Neil Strickland, *The category of Compactly Generated Weakly Hausdorff spaces*, notes from a course taught by the author, available at `http://neil-strickland.staff.shef.ac.uk/courses/homotopy/cgwh.pdf`.

[Tab07] Gonçalo Tabuada, *Homotopy theory of dg-categories*, Ph.D. thesis, Paris, 2007, available as `http://people.math.jussieu.fr/~keller/TabuadaThese.pdf`.

[Toë05] Bertrand Toën, *Vers une axiomatisation de la théorie des catégories supérieures* (French, with French summary), K-Theory **34** (2005), no. 3, 233–263, DOI 10.1007/s10977-005-4556-6. MR2182378 (2006m:55041)

[Toë07] Bertrand Toën, *The homotopy theory of dg-categories and derived Morita theory*, Invent. Math. **167** (2007), no. 3, 615–667, DOI 10.1007/s00222-006-0025-y. MR2276263 (2008a:18006)

[Toë09] Bertrand Toën, *Higher and derived stacks: a global overview*, Algebraic geometry—Seattle 2005. Part 1, Proc. Sympos. Pure Math., vol. 80, Amer. Math. Soc., Providence, RI, 2009, pp. 435–487, DOI 10.1090/pspum/080.1/2483943. MR2483943 (2010h:14021)

[Toë10] Bertrand Toën, *Lectures on dg-categories*, notes available from author's website as `http://www.math.univ-toulouse.fr/~toen/swisk.pdf`, 2010.

[Tri06] Todd Trimble, *Notes on Tetracategories*, `http://math.ucr.edu/home/baez/trimble/tetracategories.html`, October 2006.

[TV01] Bertrand Toën and Gabriele Vezzosi, *Algebraic geometry over model categories (a general approach to derived algebraic geometry)*, preprint available as `arXiv:math/0110109 [math.AG]`, 2001.

[TV02] Bertrand Toën and Gabriele Vezzosi, *Segal topoi and stacks over segal categories*, preprint available as `arXiv:math/0212330 [math.AG]`, 2002.

[TV04] Bertrand Toën and Gabriele Vezzosi, *From HAG to DAG: derived moduli stacks*, Axiomatic, enriched and motivic homotopy theory, NATO Sci. Ser. II Math. Phys. Chem., vol. 131, Kluwer Acad. Publ., Dordrecht, 2004, pp. 173–216, DOI 10.1007/978-94-007-0948-5_6. MR2061855 (2005f:14001)

[TV05] Bertrand Toën and Gabriele Vezzosi, *Homotopical algebraic geometry. I. Topos theory*, Adv. Math. **193** (2005), no. 2, 257–372, DOI 10.1016/j.aim.2004.05.004. MR2137288 (2007b:14038)

[TV08] Bertrand Toën and Gabriele Vezzosi, *Homotopical algebraic geometry. II. Geometric stacks and applications*, Mem. Amer. Math. Soc. **193** (2008), no. 902, x+224, DOI 10.1090/memo/0902. MR2394633 (2009h:14004)

[Vez10] Gabriele Vezzosi, *A note on the cotangent complex in derived algebraic geometry*, preprint available as `arXiv:1008.0601 [math.AG]`, 2010.

[Vis05] Angelo Vistoli, *Grothendieck topologies, fibered categories and descent theory*, Fundamental algebraic geometry, Math. Surveys Monogr., vol. 123, Amer. Math. Soc., Providence, RI, 2005, pp. 1–104. MR2223406

Department of Mathematics, Science Center, 1 Oxford Street, Harvard University, Cambridge Massachusetts 02138

E-mail address: `oantolin@math.harvard.edu`

Norm convergence in non-commutative central limit theorems: Combinatorial approach

Octavio Arizmendi and Carlos Vargas

ABSTRACT. In this note we study superconvergence, in the sense of Bercovici and Voiculescu, for the Boolean and monotone Central Limit Theorems. In these cases superconvergence is only achieved partially, namely, the endpoints of the supports of the normalized sums of bounded i.i.d. random variables converge to the endpoints of the support of the corresponding Gaussian distribution. The rate of convergence in both cases is $k^{-1/2}$. However, unlike the free case, uniform convergence of the densities does not hold.

Our method to show the convergence in norm is combinatorial and covers the free, Boolean and monotone cases uniformly.

1. Introduction

For a bounded random variable X with zero mean and unit variance, let us denote by $S_k^*(X) = (X_1 + X_2 + \cdots + X_k)/\sqrt{k}$ the normalized sum of k independent copies of X. The so-called Central Limit Theorem (CLT) states that $S_k^*(X)$ converges, as $k \to \infty$, to the standard Normal distribution $\mathcal{N}(0,1)$.

In the same way, in Voiculescu's Free Probability theory, the *free* Central Limit Theorem (see [**3**], [**18**]) states that the normalized sum of *free* copies of X converges weakly to the standard semicircle distribution with density

$$\frac{1}{2\pi}\sqrt{4-x^2}dx, \quad x \in [-2,2].$$

Bercovici and Voiculescu [**2**] showed that the convergence of the measures appearing in the *free* CLT is of much stronger kind than in the *classical* CLT.

THEOREM 1.1. *Let X be a centered, bounded random variable with unit variance, and let μ_k be the distribution of $S_k^{\boxplus}(X) = (X_1 + X_2 + \cdots + X_k)/\sqrt{k}$, which denotes the normalized sum of k free copies of X.*

(1) *There is some $N > 0$ such that, for all $k \geq N$, the distribution μ_k is absolutely continuous w.r.t. the Lebesgue measure.*
(2) *For $k \geq N$, the densities $d\mu_k/dx$ converge uniformly on $\mathbb{R}$ to the density of the semicircle law.*
(3) *If a_k and b_k are respectively, the lower and upper edges of the support of μ_k, then $a_k \to -2$ and $b_k \to 2$ as $k \to \infty$.*

2010 *Mathematics Subject Classification.* Primary 46L54, 46L53, 60E07.

Key words and phrases. Central limit theorem, free independence, boolean independence, monotone independence, superconvergence.

They referred to this phenomenon as *superconvergence*.

Other important notions of independence in non-commutative probability are the Boolean and monotone ones (See Section 2). Similar to the classical and free cases, the Boolean CLT states that the normalized sum of *Boolean*-independent copies of X converges weakly to the Bernoulli distribution, $\frac{1}{2}(\delta_{-1}+\delta_1)$, ([**16**]). On the other hand, for *monotone*-independent copies, the limiting distribution for the CLT is the Arcsine distribution, with density $(\pi\sqrt{2-x^2})^{-1}$ on the interval $[-\sqrt{2},\sqrt{2}]$ ([**10**]).

In this note we will show that Property (3) from Theorem 1.1 holds also for the Boolean and monotone CLT's with rate of convergence $k^{-1/2}$. However, one can find examples where (1) and (2) fail.

THEOREM 1.2. *Let $(X_n)_{n\geq 1}$ be Boolean independent, bounded, centered and i.d. random variables with unit variance. Let a_k and b_k be respectively, the lower and upper edges of the support of the random variable $S_k^{\uplus}(X)=(X_1+X_2+\cdots+X_k)/\sqrt{k}$. Then $a_k\to -1$, $b_k\to 1$ as $k\to\infty$. Moreover,*

$$\|S_k^{\uplus}(X)\|\leq 1+\|X_1\|/\sqrt{k}. \tag{1.1}$$

THEOREM 1.3. *Let $(X_n)_{n\geq 1}$ be monotone independent, bounded, centered and i.d. random variables with variance 1. Let a_k and b_k be respectively, the lower and upper edges of the support of the random variable $S_k^{\rhd}(X)=(X_1+X_2+\cdots+X_k)/\sqrt{k}$. Then $a_k\to -\sqrt{2}$, $b_k\to\sqrt{2}$ as $k\to\infty$. Moreover,*

$$\|S_k^{\rhd}(X)\|\leq \sqrt{2}+2\|X_1\|/\sqrt{k}. \tag{1.2}$$

Our combinatorial method to show Property (3) is uniform, in the sense that it can be adapted without much effort to the free, Boolean and monotone cases. We rely on the following observations for any of the non-classical convolutions $\circledast\in\{\boxplus,\uplus,\rhd\}$:

- For any bounded random variable X with mean 0 and variance 1, there exists a random variable Y with centered $\circledast$-Poisson distribution such that $\|S_k^{\circledast}(X)\|\leq\|S_k^{\circledast}(Y)\|$, for all $k\geq 1$.
- If Y has a centered $\circledast$-Poisson distribution, then $S_k^{\circledast}(Y)$ is also a (centered) $\circledast$-Poisson.
- The support of the centered $\circledast$-Poisson distribution is well understood.

The note is organized as follows. In Section 2 we introduce the main definitions concerning non-commutative probability spaces and non-classical independences, along with the corresponding partitions, cumulants and centered Poisson distributions. In Section 3 we reduce the problem to the case where X is the centered Poisson. In Section 4 the proof of the main theorems is given, along with some final remarks.

Acknowledgements. The authors express thanks to Dr. Monica Moreno for useful remarks. We thankfully acknowledge the careful reading of the referee and useful comments, which helped us improve the presentation of the paper.

2. Preliminaries

2.1. Non-Commutative probability spaces and notions of independence. A C^**-probability space* is a pair $(\mathcal{A},\tau)$, where $\mathcal{A}$ is a unital C^*-algebra and $\tau:\mathcal{A}\to\mathbb{C}$ is a positive unital linear functional. We will assume that τ is faithful.

The elements of $\mathcal{A}$ are called (non-commutative) random variables. An element $a \in \mathcal{A}$ such that $a = a^*$ is called self-adjoint.

The functional τ should be understood as the expectation in classical probability. For $a_1, \dots, a_k \in \mathcal{A}$, we will refer to the values of $\tau(a_{i_1} \cdots a_{i_n})$, $1 \le i_n \le k$, as the *mixed moments* of $a_1, \dots, a_k$.

For any self-adjoint element $a \in \mathcal{A}$ there exists a unique probability measure with compact support μ_a (its distribution) with the same moments as a, that is,

$$\int_{\mathbb{R}} x^k \mu_a(dx) = \tau(a^k), \quad \forall k \in \mathbb{N}.$$

Since a is faithful, if $a \in \mathcal{A}$ is self-adjoint, then its spectral radius, operator norm and $\max\{|t| : t \in Supp(\mu_a)\}$ coincide. We can calculate these by the formula

$$\|a\| = \lim_{n\to\infty} (\tau(a^{2n}))^{1/2n}. \tag{2.1}$$

Even if we know the individual distribution of two self-adjoint elements $a, b \in \mathcal{A}$, their joint distribution (mixed moments) can be quite arbitrary, unless some notion of independence is assumed to hold between a and b. Several classification works [**5, 11, 15**] have discussed the essential properties of classical independence which should be present in a non-commutative notion of independence. Under the most general assumptions, there are only five fundamental notions of independence: tensor (classical), free, Boolean, monotone and anti-monotone. Each type of independence can be thought as a rule for calculating mixed moments. Let $(A_n)_{n\ge1}$ be a sequence of subalgebras of $\mathcal{A}$.

We say that $(A_n)_{n\ge1}$ are *tensor-independent* iff, for any $k \ge 1$ and any k-tuple $a_1, \dots a_k \in \mathcal{A}$, such that $a_i \in A_{j(i)}$, $1 \le i \le k$,

$$\tau(a_1 a_2 \cdots a_k) = \prod_{s\in I} \tau\left(\prod_{j(i)=s} a_i\right), \tag{2.2}$$

where $I = \{j(i) | i \le k\} \subset \mathbb{N}$ and the a_i are multiplied in the same order as they appear in the left hand side.

The subalgebras $(A_n)_{n\ge1}$ are *Boolean-independent* iff

$$\tau(a_1 a_2 \cdots a_k) = \tau(a_1)\tau(a_2)\cdots\tau(a_k), \tag{2.3}$$

whenever $k \ge 1$, $a_1, \dots a_k \in \mathcal{A}$, are such that $a_i \in A_{j(i)}$, $1 \le i \le k$ and $j(i) \ne j(i+1)$.

If we write $\bar{a} := a - \tau(a)$ for $a \in \mathcal{A}$, we say that $(A_n)_{n\ge1}$ are *free* iff

$$\tau(\bar{a}_1 \bar{a}_2 \cdots \bar{a}_k) = 0, \tag{2.4}$$

whenever $k \ge 1$, $a_1, \dots a_k \in \mathcal{A}$, are such that $a_i \in A_{j(i)}$, $1 \le i \le k$ and $j(i) \ne j(i+1)$.

The linearly ordered subalgebras $A_1 < A_2 < \dots$ are *monotone-independent* iff the following rule holds for any $k \ge 1$ and any k-tuple $a_1, \dots a_k \in \mathcal{A}$, such that $a_i \in A_{j(i)}$, $1 \le i \le k$ and $j(i) \ne j(i+1)$:

$$\tau(a_1 \cdots a_{m-1} a_m a_{m+1} \cdots a_k) = \tau(a_1 \cdots a_{m-1} a_{m+1} \cdots a_k)\tau(a_m), \tag{2.5}$$

whenever $j(m) = \max\{j(s) | 1 \le s \le k\}$ (the *anti-monotone independence* is obtained by replacing max by min).

The monotone and anti-monotone are non-symmetric notions of independence: the fact that a, b are monotone-independent (iff b, a are anti-monotone independent) does not imply in general that b, a are monotone independent. Thus, a linear order of the subalgebras needs to be specified.

For any notion of independence, we say that the random variables $a_1, \ldots, a_n \in \mathcal{A}$ are independent if so are the (not necessarily unital) algebras $\langle a_1 \rangle, \ldots, \langle a_n \rangle$ ($\langle a_i \rangle$ is the algebra generated by a_i).

These rules allow to factorize a mixed moment into moments which involve only elements from the same subalgebra. For the tensor and Boolean case this can be seen directly, while for the free and monotone case an inductive argument is required.

As an example, let $a \in A_1$, $b \in A_2$, $c \in A_3$ and suppose that we want to compute $\tau(ab^2cbaca)$.

If A_1, A_2, A_3 are tensor independent, then

$$\tau(ab^2cbaca) = \tau(a^3)\tau(b^3)\tau(c^2)$$

If, instead, A_1, A_2, A_3 are Boolean independent, we get that

$$\tau(ab^2cbaca) = \tau(a)\tau(b^2)\tau(c)\tau(b)\tau(a)\tau(c)\tau(a).$$

If now A_1, A_2, A_3 are free, we get that

$$\tau((a-\tau(a))(b^2-\tau(b^2))(c-\tau(c))(b-\tau(b))(a-\tau(a))(c-\tau(c))(a-\tau(a))) = 0.$$

By the linearity of τ, we may then solve for $\tau(ab^2cbaca)$ in terms of mixed moments of lower order and proceed inductively. After cancellations we end up with the expression

$$\tau(ab^2cbaca) = \tau(a^3)\tau(b^3)\tau(c)^2 + \tau(a^2)\tau(a)\tau(b^2)\tau(b)\tau(c^2) - \tau(a^2)\tau(a)\tau(b^2)\tau(b)\tau(c)^2$$

For the case where $A_1 < A_2 < A_3$ are monotone-independent, we obtain that

$$\tau(ab^2cbaca) = \tau(ab^3aca)\tau(c) = \tau(ab^3a^2)(\tau(c))^2 = \tau(a^3)\tau(b^3)(\tau(c))^2.$$

Finally, if $A_1 < A_2 < A_3$ are anti-monotone-independent, we obtain that

$$\tau(ab^2cbaca) = \tau(b^2cbc)(\tau(a))^3 = \tau(c^2)\tau(b)\tau(b^2)(\tau(a))^3.$$

2.2. Transforms and convolutions. We denote by $\mathcal{M}$ the set of Borel probability measures on $\mathbb{R}$. Recall that the classical convolution of two probability measures μ_1, μ_2 on $\mathbb{R}$ is defined as the probability measure $\mu_1 * \mu_2$ on $\mathbb{R}$ such that $C_{\mu_1 * \mu_2}(t) = C_{\mu_1}(t) + C_{\mu_2}(t), t \in \mathbb{R}$, where $C_\mu(t) = \log \hat{\mu}(t)$, with $\hat{\mu}(t)$ the characteristic function of μ. The coefficients $c_n = c_n(\mu)$ in the series expansion

$$C_\mu(t) = \sum_{n=1}^{\infty} \frac{c_n}{n!} t^n.$$

are called (classical) cumulants.

Similar convolutions and related transforms exist for the free, Boolean and monotone theories.

2.2.1. *Free convolution.* Free convolution was defined in [**18**] for probability measures with compact support and later extended in [**9**] for the case of finite variance, and in [**4**] for the general unbounded case.

The upper half-plane and the lower half-plane are respectively denoted as $\mathbb{C}^+$ and $\mathbb{C}^-$. Let $G_\mu(z) = \int_{\mathbb{R}} \frac{\mu(dx)}{z-x}$ ($z \in \mathbb{C}^+$) be the Cauchy transform of $\mu \in \mathcal{M}$ and $F_\mu(z)$ its reciprocal $\frac{1}{G_\mu(z)}$.

It was proved in Bercovici and Voiculescu [4] that there are positive numbers η and M such that F_μ has a right inverse F_μ^{-1} defined on the region $\Gamma_{\eta,M} := \{z \in \mathbb{C}^+; |Re(z)| < \eta Im(z)\ |z| > M\}$.

The Voiculescu transform of μ is defined by $\phi_\mu(z) = F_\mu^{-1}(z) - z$, on any region of the form $\Gamma_{\eta,M}$ where F_μ^{-1} is defined; see [4]. The free cumulant transform or R-transform is a variant of ϕ_μ defined as $R_\mu(z) = z\phi_\mu(\frac{1}{z})$ for z in a domain $D_\mu \subset \mathbb{C}^-$ such that $1/z \in \Gamma_{\eta,M}$ where F_μ^{-1} is defined.

The *free additive convolution* of two probability measures μ_1, μ_2 on $\mathbb{R}$ is the probability measure $\mu_1 \boxplus \mu_2$ on $\mathbb{R}$ such that

$$\phi_{\mu_1 \boxplus \mu_2}(z) = \phi_{\mu_1}(z) + \phi_{\mu_2}(z), \quad \text{for } z \in \Gamma_{\eta_1,M_1} \cap \Gamma_{\eta_2,M_2}$$

or, equivalently,

$$R_{\mu_1 \boxplus \mu_2}(z) = R_{\mu_1}(z) + R_{\mu_2}(z), \quad \text{for } z \in D_{\mu_1} \cap D_{\mu_1}.$$

Free additive convolution corresponds to the sum of free random variables: $\mu_a \boxplus \mu_b = \mu_{a+b}$, for a and b free random variables. Free cumulants were introduced by Speicher [14] in his combinatorial approach to Free Probability. These are the coefficients $\kappa_n = \kappa_n(\mu)$ in the series expansion

$$R_\mu(z) = \sum_{n=1}^{\infty} \kappa_n z^n.$$

2.2.2. *Boolean convolution.* The *Boolean convolution* of two probability measures μ_1, μ_2 on $\mathbb{R}$ is defined as the probability measure $\mu_1 \uplus \mu_2$ on $\mathbb{R}$ such that the transform $K_\mu(z) = z - F_\mu(z)$ (usually called *self-energy*), satisfies

$$K_{\mu_1 \uplus \mu_2}(z) = K_{\mu_1}(z) + K_{\mu_2}(z), \ z \in \mathbb{C}^+$$

(see [16]). Boolean convolution corresponds to the sum of Boolean-independent random variables, that is, $\mu_a \uplus \mu_b = \mu_{a+b}$, for a and b Boolean-independent random variables. Boolean cumulants are defined as the coefficients $r_n = r_n(\mu)$ in the series

$$K_\mu(z) = \sum_{n=1}^{\infty} r_n z^{1-n}.$$

2.2.3. *Monotone convolution.* The monotone convolution was defined in [10]. The *monotone convolution* of two probability measures μ_1, μ_2 on $\mathbb{R}$ is defined as the probability measure $\mu_1 \rhd \mu_2$ on $\mathbb{R}$ such that

$$F_{\mu_1 \rhd \mu_2}(z) = F_{\mu_1}(F_{\mu_2}(z)), \ z \in \mathbb{C}^+,$$

and similarly, the anti-monotone convolution $\mu_1 \lhd \mu_2$ is defined as the probability measure on $\mathbb{R}$ such that $F_{\mu_1 \lhd \mu_2}(z) = F_{\mu_2}(F_{\mu_1}(z))$ for $z \in \mathbb{C}^+$,. Monotone (resp. anti-monotone) convolution corresponds to the sum of monotone (resp. anti-monotone) independent random variables: $\mu_a \rhd \mu_b = \mu_{a+b}$, for a and b monotone-independent random variables. Recently, Hasebe and Saigo [7] have defined the notion of monotone cumulants $(h_n)_{n\geq 1}$ which satisfy that $h_n(\mu^{\rhd k}) = kh_n(\mu)$.

2.3. Partitions and cumulants. The relation between the different cumulants and the moments $m_n = m_n(\mu_a) = \tau(a^n)$ of a random variable a can be better understood by considering different classes of partitions of the set $[n] := \{1, 2, \dots, n\}$.

A *partition* π is an equivalence relation on the set $[n]$, which is decomposed into equivalence classes $V_1, \dots, V_r$, called blocks. We write $\pi = \{V_1, \dots, V_r\}$ and $a \sim_\pi b$ for $a, b \in V_i$.

A partition π is called *non-crossing* iff $a \sim_\pi c, b \sim_\pi d \Rightarrow b \sim_\pi c$, for all $1 \leq a < b < c < d \leq n$.

A non-crossing partition is called *interval* partition if, $a \sim_\pi b \Rightarrow a \sim_\pi b-1$ for all $1 \leq a < b \leq n$. We denote by $\mathcal{P}(n) \supset NC(n) \supset \mathcal{I}(n)$ the sets of partitions, non-crossing partitions and interval partitions of $[n]$, respectively.

A *monotone* partition (π, λ) is a non-crossing partition, together with a linear order λ of the blocks $\{V_1, \dots, V_r\}$ of π, which respects the nesting structure of π. More precisely, if $1 \leq a < b < c \leq n$ are such that $a, c \in V_i$, $b \in V_j$ and $i \neq j$, then $\lambda(V_j) < \lambda(V_j)$.

For a single random variable $a \in \mathcal{A}$, its classical, free, Boolean and monotone *cumulants* $(c_n^a)_{n\geq 1}$, $(\kappa_n^a)_{n\geq 1}$, $(r_n^a)_{n\geq 1}$, $(h_n^a)_{n\geq 1}$, satisfy the moment-cumulant formulas

$$\tau(a^n) = \sum_{\pi \in \mathcal{P}(n)} c_\pi^a = \sum_{\pi \in NC(n)} \kappa_\pi^a = \sum_{\pi \in \mathcal{I}(n)} r_\pi^a = \sum_{(\pi,\lambda) \in \mathcal{M}(n)} \frac{h_\pi^a}{|\pi|!}, \tag{2.6}$$

where, for a sequence of complex numbers $(f_n)_{n\geq 1}$ and a partition $\pi = \{V_1, \dots, V_i\}$, we define $f_\pi := f_{|V_1|} \cdots f_{|V_i|}$ and $|\pi|$ is the number of blocks of the partition π.

The importance of cumulants is that they are additive with respect to sums of independent *identically distributed* random variables. In particular, we have

$$\begin{aligned} k^{n/2} c_n(S_k^*(X)) &= k c_n(X), \\ k^{n/2} \kappa_n(S_k^{\boxplus}(X)) &= k \kappa_n(X), \\ k^{n/2} r_n(S_k^{\uplus}(X)) &= k r_n(X), \\ k^{n/2} h_n(S_k^{\triangleright}(X)) &= k h_n(X). \end{aligned}$$

2.4. Centered Poisson distributions. The non-classical Poisson distributions $\pi^{\boxplus}(\lambda, \alpha)$, $\pi^{\uplus}(\lambda, \alpha)$ and $\pi^{\triangleright}(\lambda, \alpha)$ are of remarkable importance for each of the non-commutative notions of independence. Similar to the classical Poisson distribution $\pi^*(\lambda, \alpha)$, they arise as limit distributions for the corresponding laws of small numbers. Their cumulants are given by

$$c_n^{\pi^*(\lambda,\alpha)} = \kappa_n^{\pi^{\boxplus}(\lambda,\alpha)} = r_n^{\pi^{\uplus}(\lambda,\alpha)} = h_n^{\pi^{\triangleright}(\lambda,\alpha)} = \lambda \alpha^n \quad n \geq 1.$$

We will be specially interested in the supports of the *centered* Poisson distributions, namely, the distributions $\bar{\pi}^{\boxplus}(\lambda, \alpha)$, $\bar{\pi}^{\uplus}(\lambda, \alpha)$ and $\bar{\pi}^{\triangleright}(\lambda, \alpha)$, satisfying

$$\kappa_1^{\bar{\pi}^{\boxplus}(\lambda,\alpha)} = r_1^{\bar{\pi}^{\uplus}(\lambda,\alpha)} = h_1^{\bar{\pi}^{\triangleright}(\lambda,\alpha)} = 0,$$

and

$$\kappa_n^{\bar{\pi}^{\boxplus}(\lambda,\alpha)} = r_n^{\bar{\pi}^{\uplus}(\lambda,\alpha)} = h_n^{\bar{\pi}^{\triangleright}(\lambda,\alpha)} = \lambda \alpha^n \quad n \geq 2,$$

for $\lambda \geq 0$ and $\alpha \in \mathbb{R}$. It is important to note that for the Boolean and monotone cases, the centered Poisson is not just a shift of the non-centered Poisson.

The centered free Poisson distribution is obtained by applying a shift of $-\lambda\alpha$ to the usual (non-centered) free Poisson distribution (see **[13, 20]**). Hence

$$\bar{\pi}^{\boxplus}(\lambda, \alpha) = \begin{cases} (1-\lambda)\delta_{-\lambda\alpha} + \lambda \bar{\pi}_{\text{ac}}^{\boxplus}(\lambda, \alpha), & \text{if } 0 \leq \lambda \leq 1, \\ \bar{\pi}_{\text{ac}}^{\boxplus}(\lambda, \alpha), & \text{if } \lambda > 1, \end{cases}$$

where $\bar{\pi}^{\boxplus}_{\rm ac}(\lambda, \alpha)$ is the measure with density

$$\mathrm{d}\bar{\pi}^{\boxplus}_{\rm ac}(\lambda, \alpha)(t) = \frac{1}{2\pi\alpha(t+\alpha\lambda)}\sqrt{4\lambda\alpha^2 - (t-\alpha)^2}\mathrm{d}t,$$

which is supported on the interval

$$[\alpha(1-2\sqrt{\lambda}), \alpha(1+2\sqrt{\lambda})].$$

The centered Boolean Poisson is obtained as the Boolean convolution of a point mass at $-\alpha\lambda$ with the usual (non-centered) Boolean Poisson (see [**16**]). In the Boolean case, convolutions by a Dirac mass do not correspond to shifts on the measure. Nevertheless, the centered Boolean distribution can be easily calculated using its self-energy transform:

$$K_{\bar{\pi}^{\uplus}(\lambda,\alpha)}(z) = z\sum_{n=2}^{\infty}\frac{\lambda\alpha^n}{z^n} = \frac{\lambda\alpha^2}{z-\alpha}.$$

Thus

$$\begin{aligned} G(z)_{\bar{\pi}^{\uplus}(\lambda,\alpha)} &= (z - K_{\bar{\pi}^{\uplus}(\lambda,\alpha)}(z))^{-1} \\ &= \frac{z-\alpha}{z^2 - \alpha z - \lambda\alpha^2} \\ &= \frac{A}{z - \frac{\alpha - \sqrt{\alpha^2+4\lambda\alpha^2}}{2}} + \frac{B}{z - \frac{\alpha+\sqrt{\alpha^2+4\lambda\alpha^2}}{2}}, \end{aligned}$$

where

$$A = \frac{1}{2} + \frac{\alpha}{2\sqrt{\alpha^2+4\lambda\alpha^2}}, \quad B = \frac{1}{2} - \frac{\alpha}{2\sqrt{\alpha^2+4\lambda\alpha^2}}.$$

This means that the centered Boolean Poisson is concentrated on the points $\{\frac{1}{2}(\alpha \pm \sqrt{\alpha^2+4\lambda\alpha^2})\}$.

The centered monotone Poisson distribution (see [**10**]) $\bar{\pi}^{\rhd}(\lambda, \alpha)$ has density

$$d\bar{\pi}^{\rhd}_{\rm ac}(\lambda,\alpha)(x) = \frac{1}{\alpha\pi}\frac{\sqrt{2\lambda - (x/\alpha - 1)^2}}{1 + 2\lambda - (x/\alpha-1)^2}\mathbf{1}_{(\alpha(1-\sqrt{2\lambda}),\alpha(1+\sqrt{2\lambda}))}(x)$$

and singular part

$$\bar{\pi}^{\rhd}_{\rm sing}(\lambda, \alpha) = \frac{1}{1+2L}\delta_{\alpha(1-\sqrt{1+2\lambda})}.$$

In particular, we observe that the support of $\bar{\pi}^{\rhd}(\lambda,\alpha)$ is exactly

$$[(\alpha(1-\sqrt{2\lambda}), \alpha(1+\sqrt{2\lambda}))] \cup \{\alpha(1-\sqrt{1+2\lambda})\}.$$

3. Reduction to the Poisson case

We want to reduce the proof of our main theorems to the Poisson case, which can be easily controlled. For this we need some lemmas first.

LEMMA 3.1. *Let $(a_n)_{n\geq 1}$ and $(b_n)_{n\geq 1}$ be two families of Boolean independent identically distributed self-adjoint random variables in a C^*-probability space $(\mathcal{A}, \tau)$. If $|\tau(a_1^k)| \leq \tau(b_1^k)$ for all $k \geq 1$, then*

$$\|a_1 + \cdots + a_n\| \leq \|b_1 + \cdots + b_n\|.$$

The statement remains true if we replace Boolean independence by monotone independence.

PROOF. By equation (2.1), it suffices to show that $|\tau((a_1+\cdots+a_n)^k)| \leq \tau((b_1+\cdots+b_n)^k)$ for $k\geq 1$. Since

$$\left|\tau\left((a_1+a_2+\cdots+a_n)^k\right)\right| = \left|\sum_{1\leq j_1,\ldots,j_k\leq n}\tau\left(a_{j_1}a_{j_2}\ldots a_{j_k}\right)\right| \tag{3.1}$$

$$\leq \sum_{1\leq j_1,\ldots,j_k\leq n}\left|\tau\left(a_{j_1}a_{j_2}\ldots a_{j_k}\right)\right|, \tag{3.2}$$

it is enough to show that $|\tau(a_{j_1}a_{j_2}\ldots a_{j_k})| \leq \tau(b_{j_1}b_{j_2}\ldots b_{j_k})$ for each tuple $1\leq j_1,\ldots,j_k\leq n$. For this we only need to observe that $\tau(a_{j_1}\ldots a_{j_k})$ and $\tau(b_{j_1}\ldots b_{j_k})$ factorize in the same way, and by hypothesis each factor satisfies the desired inequality. The same proof works for the monotone case. □

LEMMA 3.2. *Let $a\in\mathcal{A}$ be a selfadjoint random variable in a C^*-probability space $(\mathcal{A},\tau)$, such that $\tau(a)=0$ and $\tau(a^2)=1$. Then $\tau(a^n)\leq\|a\|^{n-2}$.*

PROOF. For $n=1,2$ the statement is trivial and if $n\geq 3$

$$\tau(a^n)=\int_{\mathbb{R}}x^n d\mu_a(x)\leq\|a\|^{n-2}\int_{\mathbb{R}}x^2 d\mu_a(x)=\|a\|^{n-2}. \tag{3.3}$$

□

PROPOSITION 3.3. *Let $\circledast\in\{\uplus,\rhd\}$. For any bounded random variable X with mean 0 and variance 1, we have*

$$\|S_k^{\circledast}(X)\|\leq\|S_k^{\circledast}(Y)\|,\quad \textit{for all } k\geq 1,$$

where Y is the centered Poisson variable $Y\sim\bar{\pi}^{\circledast}((\|X\|)^{-2},\|X\|)$.

PROOF. By definition of Y, its nth $\circledast$-cumulant is $\|X\|^{n-2}$, for $n\geq 2$ and the first cumulant is zero. In particular, all the cumulants are non-negative. Hence, from the moment cumulant formulas (2.6) and Lemma 3.1, one sees that $\tau(Y^n)\geq\|X\|^{n-2}\geq\tau(X^n)$. The assertion follows by applying Lemma 3.2, for $(a_n)_{n\geq1}$ and $(b_n)_{n\geq1}$ $\circledast$-independent copies of X and Y, respectively. □

4. Proofs of the Main Theorems

We have now the tools to prove our main theorems.

PROOF OF THEOREMS 1.2 AND 1.3. Let $\circledast\in\{\uplus,\rhd\}$ and write $L=\|X\|$. Because of the Central limit theorems, we know that $S_k^{\circledast}(X)$ converges weakly to the corresponding Gaussian distribution. Hence

$$\limsup_{k\to\infty}a_k\leq\begin{cases}-1, & \text{if } \circledast=\uplus,\\ -\sqrt{2}, & \text{if } \circledast=\rhd,\end{cases}\qquad \liminf_{k\to\infty}b_k\geq\begin{cases}1, & \text{if } \circledast=\uplus,\\ \sqrt{2}, & \text{if } \circledast=\rhd.\end{cases}$$

Therefore, it is enough to show (1.1) and (1.2). By Proposition 3.3 we know that for $Y\sim\bar{\pi}^{\circledast}(L^{-2},L)$, we have

$$\|S_k^{\circledast}(X)\|\leq\|S_k^{\circledast}(Y)\|=\|\bar{\pi}_{\circledast}(kL^{-2},k^{-1/2}L)\|,$$

so it remains to check the support of such a Poisson for each case.

The measure $\bar{\pi}^{\uplus}(L^{-2}k,Lk^{-1/2})$ is supported on the points

$$\left\{\frac{1}{2}(Lk^{-1/2}\pm\sqrt{L^2k^{-1}+4})\right\},$$

so $\|S_k^{\uplus}(X)\| \leq \frac{1}{2}|Lk^{-1/2} \pm \sqrt{L^2k^{-1}+4}| \leq 1 + Lk^{-1/2} = 1 + \|X\|k^{-1/2}$.

The measure $\bar{\pi}^{\rhd}(L^{-2}k, Lk^{-1/2})$ is supported on

$$[Lk^{-1/2} - \sqrt{2}, Lk^{-1/2} + \sqrt{2}] \cup \{Lk^{-1/2} - \sqrt{L^2k^{-1}+2}\},$$

hence $\|S_k^{\rhd}(X)\| \leq \sqrt{2} + Lk^{-1/2} = \sqrt{2} + \|X\|k^{-1/2}$. □

REMARK 4.1. The proof of Theorems 1.2 and 1.3 applies also to the free case. One replaces Lemma 3.1 by the fact that

$$|\kappa_n^a| \leq \kappa_n^b \text{ for all } n \geq 1, \Rightarrow \|a\| \leq \|b\|, \tag{4.1}$$

which is easily shown by the use of (2.1) and the moment-cumulant formula. However, by working with cumulants (and not directly with moments) one needs to choose $L = 37\|X\|$. In the end we get that, since the measure $\bar{\pi}^{\boxplus}(L^{-2}k, Lk^{-1/2})$ is supported on $[Lk^{-1/2} - 2, Lk^{-1/2} + 2]$, then

$$\|S_k^{\boxplus}(X)\| \leq 2 + 37\|X\|k^{-1/2}.$$

REMARK 4.2. Convergence of the density is not achieved in the monotone or Boolean CLT's. This is clear for the Boolean Central Limit Theorem since the limiting distribution has two atoms. Now, for the monotone case, let μ be an atomic probability measure with exactly m different atoms and let ν be an atomic probability measure with exactly n different atoms. Then $\mu \rhd \nu$ consists of exactly mn different atoms (see [**6**]). In particular, $\mu^{\rhd k} = \mu \rhd \cdots \rhd \mu$ is a measure which consists of m^k atoms and has no density.

The successive composition of rational functions has been the subject of study for long in dynamical systems (see for example [**1**]). As an application of Theorem 1.3 we consider the behavior of zeros of the composition of rational functions. Similar results from the dynamical systems viewpoint can be found in [**17**].

EXAMPLE 4.3. Consider a rational function $F(z) = \frac{P(z)}{Q(z)}$ so that $P(z)$ and $Q(z)$ are degree n and $n-1$ polynomials, respectively, with leading coefficients equal to 1 and interlacing real roots $\{t_i\}$ and $\{s_j\}$, namely, $t_1 < s_1 < t_2 < \cdots < t_{n-1} < s_{n-1} < t_n$.

The reciprocal of $F(z) = 1/G(z)$, written in partial fractions,

$$G(z) = \frac{Q(z)}{P(z)} = \frac{a_1}{z - b_1} + \cdots + \frac{a_n}{z - b_n},$$

is the Cauchy transform $G_\mu(z)$ of the probability measure $\mu = \sum_i a_i \delta_{b_i}$ (see [**8**]). The mean $m = m_1(\mu) = \sum_i a_i b_i$ and second moment $\sigma^2 = m_2(\mu) = \sum_i a_i b_i^2$ will be assumed, without loss of generality, to be $m = 0$ and $\sigma^2 = 1$.

We are interested in the composition $F^{\circ n}(z) = F(F(\cdots F(z))$ which corresponds to the F transform of the measure $\mu^{\rhd n} = \mu \rhd \cdots \rhd \mu$. Denoting by $z_1(n)$ and $z_2(n)$ the smallest and largest real zeros of $F^{\circ n}$, Theorem 1.3 says that

$$\lim_{n\to\infty} z_1(n)/\sqrt{n} = -\sqrt{2}$$

and

$$\lim_{n\to\infty} z_2(n)/\sqrt{n} = \sqrt{2}.$$

References

[1] A. F. Beardon, *Iteration of rational functions*, Graduate Texts in Mathematics, vol. 132, Springer-Verlag, New York, 1991. Complex analytic dynamical systems. MR1128089 (92j:30026)

[2] H. Bercovici and D. Voiculescu, *Superconvergence to the central limit and failure of the Cramér theorem for free random variables*, Probab. Theory Related Fields **103** (1995), no. 2, 215–222, DOI 10.1007/BF01204215. MR1355057 (96k:46115)

[3] M. Bożejko, *On $\Lambda(p)$ sets with minimal constant in discrete noncommutative groups*, Proc. Amer. Math. Soc. **51** (1975), 407–412. MR0390658 (52 #11481)

[4] H. Bercovici and D. Voiculescu, *Free convolution of measures with unbounded support*, Indiana Univ. Math. J. **42** (1993), no. 3, 733–773, DOI 10.1512/iumj.1993.42.42033. MR1254116 (95c:46109)

[5] A. Ben Ghorbal and M. Schürmann, *Non-commutative notions of stochastic independence*, Math. Proc. Cambridge Philos. Soc. **133** (2002), no. 3, 531–561, DOI 10.1017/S0305004102006072. MR1919720 (2003k:46096)

[6] T. Hasebe, *Monotone convolution and monotone infinite divisibility from complex analytic viewpoint*, Infin. Dimens. Anal. Quantum Probab. Relat. Top. **13** (2010), no. 1, 111–131, DOI 10.1142/S0219025710003973. MR2646794 (2011i:60036)

[7] T. Hasebe and H. Saigo, *The monotone cumulants* (English, with English and French summaries), Ann. Inst. Henri Poincaré Probab. Stat. **47** (2011), no. 4, 1160–1170, DOI 10.1214/10-AIHP379. MR2884229 (2012j:46097)

[8] S. Kerov, *Interlacing measures*, Kirillov's seminar on representation theory, Amer. Math. Soc. Transl. Ser. 2, vol. 181, Amer. Math. Soc., Providence, RI, 1998, pp. 35–83. MR1618739 (99h:30034)

[9] H. Maassen, *Addition of freely independent random variables*, J. Funct. Anal. **106** (1992), no. 2, 409–438, DOI 10.1016/0022-1236(92)90055-N. MR1165862 (94g:46069)

[10] N. Muraki, *Monotonic independence, monotonic central limit theorem and monotonic law of small numbers*, Infin. Dimens. Anal. Quantum Probab. Relat. Top. **4** (2001), no. 1, 39–58, DOI 10.1142/S0219025701000339. MR1824472 (2002e:46076)

[11] N. Muraki, *The five independences as natural products*, Infin. Dimens. Anal. Quantum Probab. Relat. Top. **6** (2003), no. 3, 337–371, DOI 10.1142/S0219025703001365. MR2016316 (2005h:46093)

[12] A. Nica and R. Speicher, *Lectures on the combinatorics of free probability*, London Mathematical Society Lecture Note Series, vol. 335, Cambridge University Press, Cambridge, 2006. MR2266879 (2008k:46198)

[13] R. Speicher, *A new example of "independence" and "white noise"*, Probab. Theory Related Fields **84** (1990), no. 2, 141–159, DOI 10.1007/BF01197843. MR1030725 (90m:46116)

[14] R. Speicher, *Multiplicative functions on the lattice of noncrossing partitions and free convolution*, Math. Ann. **298** (1994), no. 4, 611–628, DOI 10.1007/BF01459754. MR1268597 (95h:05012)

[15] R. Speicher, *On universal products*, Free probability theory (Waterloo, ON, 1995), Fields Inst. Commun., vol. 12, Amer. Math. Soc., Providence, RI, 1997, pp. 257–266. MR1426844 (98c:46141)

[16] R. Speicher and R. Woroudi, *Boolean convolution*, Free probability theory (Waterloo, ON, 1995), Fields Inst. Commun., vol. 12, Amer. Math. Soc., Providence, RI, 1997, pp. 267–279. MR1426845 (98b:46084)

[17] B. O. Stratmann and M. Urbański, *The geometry of conformal measures for parabolic rational maps*, Math. Proc. Cambridge Philos. Soc. **128** (2000), no. 1, 141–156, DOI 10.1017/S0305004199003837. MR1724435 (2000i:37066)

[18] *Operator algebras and their connections with topology and ergodic theory*, Lecture Notes in Mathematics, vol. 1132, Springer-Verlag, Berlin, 1985. Edited by H. Araki, C. C. Moore, Ş. Strătilă and D. Voiculescu. MR799557 (86f:46001)

[19] D. Voiculescu, *Addition of certain noncommuting random variables*, J. Funct. Anal. **66** (1986), no. 3, 323–346, DOI 10.1016/0022-1236(86)90062-5. MR839105 (87j:46122)

[20] D. Voiculescu, *Limit laws for random matrices and free products*, Invent. Math. **104** (1991), no. 1, 201–220, DOI 10.1007/BF01245072. MR1094052 (92d:46163)

Department of Probability and Statistics, CIMAT, Aparatado Postal 402, Guanajuato, GTO, 36240, Mexico
E-mail address: octavius@cimat.mx

Department of Mathematical Structure Theory, Graz University of Technology. Steyrergasse 30 III, 8010, Graz, Austria
E-mail address: obieta@math.tugraz.at

Dynamical models for some torus homeomorphisms

P. Dávalos

ABSTRACT. Consider a homeomorphism of the torus $\mathbf{T}^2$ in the homotopy class of the identity. There is a topological invariant for f known as the *rotation set*, which is a compact convex subset of $H_1(\mathbf{T}^2, \mathbf{R}) \simeq \mathbf{R}^2$ and describes the homological direction and speed at which the orbits rotate on $\mathbf{T}^2$. In this paper we collect some results about the existence of dynamical models associated to this invariant.

1. Introduction

In **[21]** H. Poincaré defined the *rotation number* for circle homeomorphisms and proved it to be a topological invariant carrying dynamical information. For an orientation preserving homeomorphism $f : \mathbf{T}^1 \to \mathbf{T}^1$ and a lift $\tilde{f} : \mathbf{R} \to \mathbf{R}$ of f, he defined the rotation number of $\tilde{f}$ as the limit $\rho(\tilde{f}) := \lim_{n\to\infty}(\tilde{f}^n(x) - x)/n$, which he showed to exist and to be independent of $x \in \mathbf{R}$. From the topological point of view, the dynamics of f can be completely understood from the rotation number $\rho(\tilde{f})$. This number is rational if an only if there exist a periodic orbit for f, all the periodic orbits have the same period, and the complement of the set of periodic points of f is a union of periodic open intervals also with the same period. The rotation number $\rho(\tilde{f})$ is irrational if and only if f is semiconjugated to the rotation of $\mathbf{T}^1$ by the angle $2\pi\rho(\tilde{f})$. If the semiconjugacy is not an actual conjugacy, then it is injective on a Cantor set $K \subset \mathbf{T}^1$, and collapses the connected components of $\mathbf{T}^1 \setminus K$ to points. The set K is minimal for f, and the connected components of $\mathbf{T}^1 \setminus K$ are wandering open intervals.

Later, in **[20]** Misiurewicz and Ziemian generalized the concept of the rotation number for homeomorphisms of $\mathbf{T}^n$. For a homeomorphism $f : \mathbf{T}^n \to \mathbf{T}^n$ in the homotopy class of the identity, and a lift $\tilde{f} : \mathbf{R}^n \to \mathbf{R}^n$, the *rotation set* of $\tilde{f}$, denoted $\rho(\tilde{f})$, is defined as the set of accumulation points of sequences of the form

$$\left\{\frac{\tilde{f}^{n_i}(x_i) - x_i}{n_i}\right\}_{i\in\mathbf{N}}$$

with $x_i \in \mathbf{R}^2$ and $n_i \to \infty$ as $i \to \infty$. We will work with case that $n = 2$. In this case the rotation set $\rho(\tilde{f})$ is a compact convex subset of $\mathbf{R}^2$ **[20]** and it is also a topological invariant carrying dynamical information. For example, rational points

2010 *Mathematics Subject Classification.* Primary 37E45.
Key words and phrases. Torus homeomorphisms, rotation set.

of $\rho(\tilde{f})$ are related to periodic orbits of f [**9**, **10**] and extremal points of $\rho(\tilde{f})$ are in a sense related to ergodic measures of f [**20**].

Trying to emulate the case of $\mathbf{T}^1$, one could ask if there exist dynamical models associated to the rotation set. That is; to what extent, from the topological point of view, can one classify the dynamics of a torus homeomorphism from its rotation set? In this paper we gather some results related to this question, in the case that the rotation set has empty interior, namely, when it is either a singleton or a segment.

2. Notations and preliminaries

Consider a curve $\gamma : I \to \mathbf{T}^2$. We will denote its image $\mathrm{Im}(\gamma) \subset \mathbf{T}^2$ also by γ, and we will call it also a *curve*. We say that a closed curve $\gamma \subset \mathbf{T}^2$ is *essential* if it is not homotopically trivial, and we say that γ is *vertical* if it is homotopic to a vertical circle $\{x\} \times \mathbf{T}^1 \subset \mathbf{T}^2$. Similarily, we say that a topological annulus $A \subset \mathbf{T}^2$ is *essential* if it is not homotopic to a point, and we say that A is *essential and vertical* if it is homotopic to a vertical annulus $I \times \mathbf{T}^1$, with $I \subset \mathbf{T}^1$ an interval. A set $K \subset \mathbf{T}^2$ is said to be *annular* if $K = \cap_{n\geq 0} A_i$, where the A_i are topological compact annuli with $A_{i+1} \subset A_i$ such that $A_{i+1} \hookrightarrow A_i$ is a homotopy equivalence. If the annuli A_i are essential, the annular set K is called *essential*, and if the A_i are essential and vertical, the set K is called *essential and vertical.* We say that a set $K \subset \mathbf{T}^2$ is *fully essential* if its complement is a union of pairwise disjoint open topological discs.

A homeomorphism $f : \mathbf{T}^2 \to \mathbf{T}^2$ is called a *pseudo-rotation* if for a lift $\tilde{f}$ (and hence for any lift), the limit

$$\lim_{n\to\infty} \frac{\tilde{f}^n(x) - x}{n} \tag{2.1}$$

exists and is independent of x. The limit (2.1) is called the *rotation vector* of $\tilde{f}$, and in the case that the rotation vector is zero, we say that f and its lift $\tilde{f}$ are *irrotational.*

A pseudo-rotation f is said to have the *bounded mean motion property* if there is a lift $\tilde{f}$ with rotation vector v, such that the deviations

$$D(x, n) = |\tilde{f}^n(x) - x - nv|$$

are uniformly bounded in x and n.

In general, the limit (2.1) does not necessarily exist for every $x \in \mathbf{R}^2$, and we have the following definition.

DEFINITION 2.1 ([**20**]). The *rotation set of* $\tilde{f}$ is defined as

$$\rho(\tilde{f}) = \bigcap_{m=1}^{\infty} \mathrm{cl}\left(\bigcup_{n=m}^{\infty}\left\{\frac{\tilde{f}^n(x) - x}{n} : x \in \mathbf{R}^2\right\}\right) \subset \mathbf{R}^2.$$

If $\Lambda \subset \mathbf{T}^2$ is a compact f-invariant set, we define the *rotation set of* Λ as

$$\rho(\Lambda, \tilde{f}) = \bigcap_{m=1}^{\infty} \mathrm{cl}\left(\bigcup_{n=m}^{\infty}\left\{\frac{\tilde{f}^n(x) - x}{n} : x \in \pi^{-1}(\Lambda)\right\}\right) \subset \mathbf{R}^2,$$

where $\pi : \mathbf{R}^2 \to \mathbf{T}^2$ denotes the canonical projection.

Denote by $T_1, T_2 : \mathbf{R}^2 \to \mathbf{R}^2$ the integer translations $T_1(x,y) = (x+1,y)$, $T_2(x,y) = (x, y+1)$. It is easy to see that for integers n, m_1, m_2,

$$\rho(T_1^{m_1} T_2^{m_2} \tilde{f}^n) = n\rho(\tilde{f}) + (m_1, m_2).$$

Then, the rotation set of any other lift of f is an integer translate of $\rho(\tilde{f})$, and we can think of the 'rotation set of f' defined modulo $\mathbf{Z}^2$.

THEOREM 2.2 ([**20**]). *Let $f : \mathbf{T}^2 \to \mathbf{T}^2$ be a homeomorphism, and let $\tilde{f} : \mathbf{R}^2 \to \mathbf{R}^2$ be a lift of f. Then the rotation set set $\rho(\tilde{f})$ is compact and convex.*

Consider a homeomorphism $f : \mathbf{T}^2 \to \mathbf{T}^2$. We say that f has *disc-type dynamics* if there exists a topological open disc $U \subset \mathbf{T}^2$ such that $\mathbf{T}^2 \setminus U \subset \mathrm{Fix}(f)$. Now, f is said to have *weakly annular dynamics* if for any lift $\tilde{f}$ of f there is $M > 0$ and $v \in \mathbf{Z}^2 \setminus \{0\}$ such that

$$|\langle \tilde{f}^n(x) - x, v\rangle| \leq M$$

for all $x \in \mathbf{R}^2$ and $n \in \mathbf{Z}$. Finally, we say that f has *annular dynamics* if there exists an annular essential set set A which is periodic for f. In this case, if q is the period of A, to understand the dynamics of f one may study the dynamics of the restrictions of f^q to the open annulus $\mathbf{T}^2 \setminus A$, and to the annular set A.

If the dynamics of f is annular, it is easy to see that the rotation set of any lift is a (possibly degenerate) segment with rational slope containing rational points. If the rotation set of $\tilde{f}$ is a non-degenerate segment with rational slope containing rational points, we will say that $\tilde{f}$ has *annular rotation set* (cf. Question 4.2).

Consider an orientable surface M without boundary, and an isotopy $I : M \times [0,1] \to M$ from the identity to a homeomorphism f. For a point $x \in M$, denote $\gamma_x = I(x, \cdot)$. A fixed point p of f is called *contractible* if the closed path γ_p is not homotopically trivial. An oriented topological foliation $\mathcal{F}$ of M is said to be *transverse* to I if for any $x \in M$, the path γ_x is homotopic with fixed endpoints to a path which is positively transverse to $\mathcal{F}$. Having contractible fixed points is clearly an obstruction to the existence of a foliation of M transverse to I. The following theorem from [**5**] says that it is actually the only obstruction.

THEOREM 2.3. *If f has no contractible fixed points, then there exists a topological oriented foliation without singularities which is transverse to the isotopy I.*

In a similar way, if $\mathcal{F}$ is a foliation with singularities, we say that $\mathcal{F}$ is transverse to the isotopy I if for any $x \in M \setminus \mathrm{sing}(\mathcal{F})$, the path γ_x is homotopic with fixed endpoints to a path which is positively transverse to $\mathcal{F} \setminus \mathrm{sing}(\mathcal{F})$. The following result is a consequence of [**5**] and [**14**].

THEOREM 2.4. *If $\mathrm{Fix}(f)$ is totally disconnected, then there exists a compact set $X \subset \mathrm{Fix}(f)$, an oriented foliation $\mathcal{F}$ with singularities in X, and an isotopy I from the identity to f such that I fixes X, and $\mathcal{F}$ is transverse to I.*

3. Pseudorotations

The canonical model of a pseudo-rotation is a rigid rotation. There exist both *positive* and *negative* results, in the sense that certain hypotheses guarantee or not some similarity of the dynamics of a pseudo-rotation and the corresponding rigid rotation.

One way to relate a pseudo-rotation and a rotation is by means of a semiconjugacy. The question of whether such a semiconjugacy exists or not, has been systematically studied by T. Jäger. We start by stating a basic result of this kind. From now on, for $\rho \in \mathbf{R}^2$, R_ρ will denote the rotation $x \mapsto x + \rho \mod \mathbf{Z}^2$ on $\mathbf{T}^2$.

PROPOSITION 3.1. Let $f \in \text{Homeo}_0(\mathbf{T}^2)$ be a minimal pseudo-rotation with bounded mean motion and a totally irrational rotation vector $\rho \in \mathbf{R}^2$. Then f is semiconjugate to R_ρ.

In [**13**] it is proven a more general version of this proposition, for pseudorotations of $\mathbf{T}^n$, $n \geq 2$ (result which also deals with rotation sets that are not reduced to a point).

The proof of Proposition 3.1 consists in obtaining a semiconjugacy $h_i : \mathbf{T}^2 \to \mathbf{T}^1$ of the map f with the rotation in $\mathbf{T}^1$ by ρ_i, for $i = 1, 2$. Then, the semiconjugacy $h : \mathbf{T}^2 \to \mathbf{T}^2$ between f and R_ρ is defined as $h = (h_1, h_2)$. The h_i are defined as

$$h_i(z) = \sup_{n \in \mathbf{Z}}(\pi_i \circ F^n(z) - n\rho_i),$$

for $i = 1, 2$. Due to the bounded mean motion property, the h_i are well defined, and it is easy to check that $H(z) + \rho_i = H \circ F(z)$. The minimality of f is then used to prove continuity, and surjectivity is due to the minimality of the one dimensional rotation by ρ_i.

For conservative[1] pseudo-rotations, we have the following classification result. In such result, an important element is the concept of a *circloid*, which is a subset of $\mathbf{T}^2$ which is compact, connected, *essential* (not contained in any topological open disc), has a connected complement which contains an essential simple closed curve, and does not contain any strictly smaller subset with these properties.

THEOREM 3.2 ([**13**]). *Suppose $f \in \text{Homeo}_0(\mathbf{T}^2)$ is a conservative pseudo-rotation with rotation vector $\rho \in \mathbf{R}^2$ and bounded mean motion. Then the following hold:*

(1) *ρ is totally irrational if and only if f is semi-conjugate to R_ρ.*
(2) *ρ is neither totally irrational nor rational if and only if f has a periodic circloid.*
(3) *ρ is rational if and only if f has a periodic point.*

Item 3 of this theorem is a classical result of Franks [**8**]. For item (2), suppose that ρ is a horizontal vector $(a, 0)$, with a irrational. In this case Jäger proves that the bounded mean motion property is equivalent to having an actual horizontal 'barrier' for the dynamics. Precisely, he proves that for such ρ, there exists a horizontal f-invariant circloid if and only if f has the bounded mean motion property (a horizontal circloid is a circloid contained in an annulus homotopic to the horizontal annulus $\mathbf{T}^1 \times [0, 1/2]$). Then, the case for a general ρ that is neither totally irrational nor rational is then deduced from this.

In item (1), the 'if' part is elementary. The 'only if' part of this item is of principal interset in that article. One works with the lift $\hat{f}$ of f to $\mathbf{T}^1 \times \mathbf{R}$ with average vertical displacement ρ_2, and such that $|\pi_2(\hat{f}^n(z) - z) - n\rho_2| \leq c$ for all $n \in \mathbf{Z}$ and $z \in \mathbf{T}^1 \times \mathbf{R}$, where c is the bounded mean motion constant for f.

---

[1] By conservative, we mean that there exists an invariant probability measure of full topological support. By the Oxtoby-Ulam Theorem, we may actually assume that this measure is the Lebesgue measure on $\mathbf{T}^2$, but we will not make use of this fact.

For $r \in \mathbf{R}$, one defines the sets

$$A_r = \bigcup_{n \in \mathbf{Z}} \hat{f}^n(\mathbf{T}^1 \times \{r - n\rho_2\}).$$

By the mean motion property, the sets A_r are bounded in the vertical direction. A main step here carried out in [**13**] is to extract from each A_r a circloid C_r, in a way that the circloids are pairwise disjoint, and if $T : \mathbf{T}^1 \times \mathbf{R} \to \mathbf{T}^1 \times \mathbf{R}$ denotes the translation $T(x, y) = (x, y + 1)$, then

$$C_{r+1} = T(C_r), \tag{3.1}$$

$$\hat{f}(C_r) = C_{r+\rho_2}, \tag{3.2}$$

and

$$C_r \prec C_s \text{ if } r < s, \tag{3.3}$$

where the notation $C_r \prec A$ means that A is contained in the connected component of $(\mathbf{T}^1 \times \mathbf{R}) \setminus C_r$ that is unbounded from above and bounded from below.

Having shown this, one can construct a semiconjugacy $H_2 : \mathbf{T}^2 \to \mathbf{T}^1$ between f and the one dimensional rotation by ρ_2 in the following way. Set

$$H_2(z) = \sup\{r \in \mathbf{R} \,|\, C_r \prec z\}.$$

Using (3.1) and (3.2) it can be easily verified that

$$H_2 \circ T(z) = H(z) + 1, \tag{3.4}$$

and

$$H_2 \circ \hat{f}(z) = H(z) + \rho_2. \tag{3.5}$$

To see that H_2 is continuous, it is proved that for any open interval $(a, b) \subset \mathbf{R}$, the set $H_2^{-1}(a, b)$ is a union of open 'intervals' of the form $(C_r, C_s) \subset \mathbf{T}^1 \times \mathbf{R}$, $r < s$. Here, $(C_r, C_s) = \{z \in \mathbf{T}^1 \times \mathbf{R} \,|\, C_r \prec z \prec C_s\}$. By properties (3.4) and (3.5), H_2 projects then to a semiconjucacy h_2 between f and the irrational rotation $x \mapsto x + \rho_2$. In the same way, one constructs a semiconjugacy h_1 between f and the rotation $x \mapsto x + \rho_1$, and then $h = (h_1, h_2)$ gives a semiconjugacy between f and R_ρ on $\mathbf{T}^2$.

The conservative hypothesis in Theorem 3.2 is necessary, as the following proposition shows.

PROPOSITION 3.3 ([**13**]). *Given any totally irrational rotation vector $\rho \in \mathbf{R}^2$, there exists an irrational pseudo-rotation $f \in \mathrm{Homeo}_0(\mathbf{T}^2)$ which has rotation vector ρ and bounded mean motion, but which is not semi-conjugate to the irrational rotation R_ρ.*

One can easily see that the bounded mean motion property is also necessary in order to have a semiconjugacy between f a rotation R_ρ. Actually, the absence of this property allows to create exotic examples of pseudo-rotations, with dynamical properties far from the rigid rotations. An illustration of this phenomenon is given by a result of Koropecki and Tal, for the irrotational case. Before stating that result, we give a definition. Given a set $X \subset \mathbf{R}^2$, we say that X accumulates at infinity in the direction $v \in \mathbf{T}^1$ if there is a sequence of $x_n \in X$ such that $|x_n| \to \infty$ and $(x_n - x_0)/|x_n - x_0| \to v$ as $n \to \infty$. The boundary of X at infinity is defined as the set $\partial_\infty X$ consisting of all $v \in \mathbf{T}^1$ such that X accumulates in the direction v at infinity.

THEOREM 3.4 ([**18**]). *There exists a C^∞ area-preserving irrotational pseudo-rotation which is ergodic with respect to Lebesgue measure, and such that for almost every point $x \in \mathbf{R}^2$, the orbit of x accumulates in every direction at infinity, i. e.,*

$$\partial_\infty\{\tilde{f}^n(x) : n \in \mathbf{N}\} = \mathbf{T}^1.$$

For the case that the rotation vector ρ is any rational vector, an analogous statement is obtained for a power of f. The proof of Theorem 3.4 theorem uses the idea of embedding an open disc U in $\mathbf{T}^2$ in a way that U has full Lebesgue-measure and any lift of U to the universal covering accumulates in every direction at infinity. Then, one 'glues' in U a diffeomorphism of the open disc which is ergodic with respect to Lebesgue (or even isomorphic to a Bernoulli shift) and extends to the boundary of U as the identity. The construction of such a diffeomorphism of the unit disc can be done by slight modifications of results due to Katok [**15**].

More examples of pseudo-rotations without the bounded mean motion property and with dynamical properties distant from the rigid rotation, both from the topological and metrical point of view, can be found for example in [**12**], [**16**] and [**7**].

Koropecki and Tal have also found that the 'exotic' construction from Theorem 3.4 is to some extent the only way to create this behavior. That is, the existence of unbounded mean motion in many directions forces the existence of a 'large' (fully essential) set of fixed points. This is shown in the following result, which classifies conservative irrational pseudo-rotations.

THEOREM 3.5 ([**17**]). *Let $f \in \mathrm{Homeo}_0(\mathbf{T}^2)$ be a conservative irrotational pseudo-rotation, and let $\tilde{f}$ be its irrotational lift. Then one of the following holds (cf. Section* 1 *for definitions):*

(1) *f has disc-type dynamics, and* $\mathrm{Fix}(f)$ *is fully essential,*
(2) *every point in $\mathbf{R}^2$ has a bounded $\tilde{f}$-orbit,*
(3) *f has weakly annular dynamics.*

The proof of this theorem is done by contradiction. If the theorem does not hold, then $\mathrm{Fix}(f)$ is not fully essential, f is not annular, and there exists $x \in \mathbf{R}^2$ with unbounded $\tilde{f}$-orbit. If f is not annular, then one may show that $\mathrm{Fix}(f)$ is inessential, and in [**17**] it is showed that one may actually assume that $\mathrm{Fix}(f)$ is totally disconnected. A remarkable theorem proved in the same article (Theorem F) shows that the fact that f is conservative and irrotational implies that the irrotational lift $\tilde{f}$ is non-wandering. Now, the fact that $\tilde{f}$ is non-wandering allows Koropecki and Tal to prove that the Brouwer foliation of $\mathbf{R}^2$ given by Proposition 2.4 is actually a *gradient-like* foliation; that is, every leaf γ is such that $\alpha(\gamma) = \{p_1\}$ and $\omega(\gamma) = \{p_2\}$, for some $p_1, p_2 \in \tilde{X}$, $p_1 \neq p_2$. From there, using techniques from [**1**] and developing more machinery, it is worked to find a contradiction, which will prove the theorem.

4. Rotation sets which are segments

The only known examples of rotation sets which are segments are segments with rational slope containing rational points, and segments with irrational slope with one endpoint rational. If a conjecture by Franks and Misiurewicz is true [**11**], then these are the only possible examples of rotation sets which are non-degenerate segments.

Consider the simplest of such examples: a vertical segment of the form $\{0\} \times I$, containing $(0,0)$ in its interior. The canonical example of a homeomorphism with that rotation set is the twist-like map $(x,y) \mapsto (x, y+\sin(2\pi x))$, which clearly has annular dynamics. One would like to see to what extent this is a model for a torus homeomorphism with a rotation set of the form $\{0\} \times I$. A partial answer to this question is given by Theorem 4.1 below, which gives a qualitative description of the dynamics, in the case that the rotation set contains a rational vector which is not realized by a periodic orbit. It proves in particular that, in such case, the dynamics is indeed annular. Before stating the theorem, we recall some definitions. We say that a curve $\gamma \subset \mathbf{T}^2$ is *free forever* for f if $f^n(\gamma) \cap \gamma = \emptyset$ for all $n \in \mathbf{Z}$. Also, if $\gamma_1, \gamma_2 \subset \mathbf{T}^2$ are vertical and disjoint curves, using the covering $\mathbf{R} \times \mathbf{T}^1 \to \mathbf{T}^2$ one can define the colsed annulus $[\gamma_1, \gamma_2] \subset \mathbf{T}^2$ whose 'left' border component is γ_1 and whose 'right' border component is γ_2. By last, denote by $\Omega(\tilde{f})$ the *non-wandering set* of f, that is, the set of points $x \in \mathbf{T}^2$ such that for every neighborhood V of x, there is $n > 0$ such that $f^n(V) \cap V \neq \emptyset$.

THEOREM 4.1 ([6]). *Let f be a homeomorphism of $\mathbf{T}^2$ homotopic to the identity with a lift $\tilde{f} : \mathbf{R}^2 \to \mathbf{R}^2$ such that:*

- *$\rho(\tilde{f}) = \{0\} \times I$, where I is a non-degenerate segment containing 0 in its interior, and*
- *$(0,0)$ is not realized by a periodic point.*

Then, the dynamics of f is annular. Moreover, there exists a finite family $\{l_i\}_{i=0}^{r-1}$, $r \geq 2$, of curves in $\mathbf{T}^2$ which are simple, closed, vertical, and pairwise dijoint, and with the following properties. If

$$\Theta_i := \bigcap_{n \in \mathbf{Z}} f^n([l_i, l_{i+1}]) \quad \text{for } i \in \mathbf{Z}/r\mathbf{Z},$$

then,

(1) *at least one of the sets Θ_i is an annular, essential, f-invariant set which is a semi-attractor,*
(2) *the curves $l_0, l_1, \ldots, l_{r-1}$ are free forever for f,*
(3) *there is $\epsilon > 0$ such that $\rho(\Theta_i, \tilde{f})$ is contained either in $\{0\} \times (\epsilon, \infty)$, or in $\{0\} \times (-\infty, -\epsilon)$, and*
(4) *$\Omega(f) \subset \cup\Theta_i$, (see Fig. 1).*

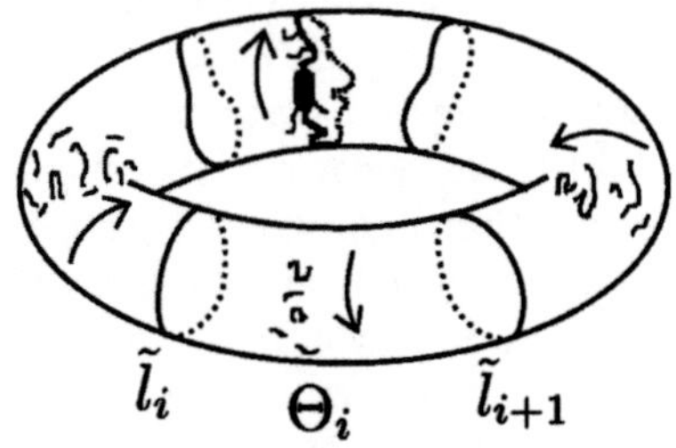

FIGURE 1. Illustration for Theorem 4.1. At least one of the sets Θ_i must be annular and essential.

This theorem tells us that, if $(0,0) \in \rho(\tilde{f})$ is not realized by a periodic orbit, then there is a 'filtration' for the dynamics of f, given by the curves l_i. At each level of the filtration, the maximal invariant set Θ_i rotates either upwards or downwards, and one of these maximal invariant sets is a 'vertical barrier' for the dynamics. Also, in [**6**] is deduced from this an analogous theorem for the case that the rotation set is a general segment with rational slope and containing rationals.

In Theorem 4.1, the hypothesis that $(0,0)$ is not realized by a periodic orbit is equivalent to the fact that the lift $\tilde{f}$ has no fixed points (this is Frank's Lemma [**8**]), which in turn is equivalent to the fact that f has no contractible fixed points. Therefore, applying Theorem 2.3 one easily obtains a foliation $\mathcal{F}$ of $\mathbf{T}^2$ such that the lift $\ell \subset \mathbf{R}^2$ of any leaf is a Brouwer curve for $\tilde{f}$ (that is, $\tilde{f}(\ell) \cap \ell = \tilde{f}^{-1}(\ell) \cap \ell = \emptyset$). Then, in [**6**] it is proved that there exists a finite family of leaves l_i of $\mathcal{F}$ that are compact, essential and vertical, and such that the maximal invariant sets Θ_i for f between them rotate either upwards or downwards. This is done using the fact that $\mathcal{F}$ is positively transverse to the isotopy I from Id to f, with techniques similar to those in [**4**] and with the use of Atkinson's Lemma [**2**] from ergodic theory.

The main and new part in theorem 4.1 is the existence of an annular essential f-invariant set. This is done in the following way. Let $\ell \subset \mathbf{R}^2$ be a lift of some of the curves $l_i \subset \mathbf{T}^2$. Then, ℓ is a free curve for $\tilde{f}$. By contradiction, suppose that none of the sets Θ_i is an essential set. Then, one may easily see that

$$\tilde{f}^{n_0}(\ell) \cap T_1(\ell) \neq \emptyset,$$

for some $n_0 \in \mathbf{Z}$ (assume without loss of generality that $n_0 > 0$). The main work in [**6**] consists of proving that the fact that $\tilde{f}^{n_0}(\ell) \cap T_1(\ell) \neq \emptyset$ implies actually that there exists $x \in \mathbf{R}^2$ such that

$$\lim_{n \to \infty} \frac{\mathrm{pr}_1(\tilde{f}^n(x) - x)}{n} = \infty,$$

which yields the contradiction $\max \mathrm{pr}_1(\rho(\tilde{f})) > 0$. Hence, one of the sets Θ_i must be an annular essential set. The proof of the other items of Theorem 4.1 follows from this.

For the general case of a homeomorphism with an annular rotation set, the following question is still open.

QUESTION 4.2. If a torus homeomorphism f has an annular rotation set, then is the dynamics of f annular?

Progress in this direction has been made by Bortolatto an Tal [**3**], showing that the answer to this question is affirmative in the case that Lebesgue measure is ergodic and has a non-rational rotation vector (that is, Lebesgue-almost every point in $\mathbf{T}^2$ has the same well defined non-rational rotation vector). More recently, Guelman, Koropecki and Tal have shown that the answer to Question 4.2 is also affirmative if one assumes only that Lebesgue measure is preserved [**19**].

Now, for the case that the rotation set is a segment with irrational slope, as we mentioned above, the only known examples have a rational endpoint. Such an example can be given in the following way. Let $v \in \mathbf{R}^2$ be a vector with irrational slope, and let χ denote the constant vector field $\chi \equiv v$ in $\mathbf{T}^2$. For $p \in \mathbf{T}^2$, let $\psi : \mathbf{T}^2 \to \mathbf{R}$ be a continuous function such that $\psi \geq 0$ and $\psi(x) = 0$ if and only if $x = p$. Now, let $f : \mathbf{T}^2 \to \mathbf{T}^2$ be the time-1 map of the flow given by the vector field $\psi\chi$. We have that $\mathrm{Fix}(f) = \{p\}$, and that the future orbit of every point passes arbitrarily close of p. Let $\tilde{f} : \mathbf{R}^2 \to \mathbf{R}^2$ be the lift of f wich fixes the lifts of p. If ψ is chosen adequately (not too close to zero), one can prove that there are points in $\mathbf{R}^2$ with non-zero rotation vector for $\tilde{f}$, and then the rotation set is a segment with irrational slope (equal to the slope of v) and with $(0,0)$ as an endpoint.

In such example, one can of course replace the point p by any totally disconnected closed set X, or even 'explode' the set X and some flux lines, remaining with examples with the same rotation set and which are semi-conjugate to the example constructed above. Another way to construct an example with such a rotation set, in the conservative setting, may be found in **[22]**. These are the only two mechanisms known to the author to construct a homeomorphism whose rotation set has irrational slope and a rational endpoint.

References

[1] S. Addas-Zanata and F. A. Tal, *Homeomorphisms of the annulus with a transitive lift*, Math. Z. **267** (2011), no. 3-4, 971–980, DOI 10.1007/s00209-009-0657-x. MR2776069 (2012c:37083)

[2] G. Atkinson, *Recurrence of co-cycles and random walks*, J. London Math. Soc. (2) **13** (1976), no. 3, 486–488. MR0419727 (54 #7745)

[3] R. B. Bortolatto and F. A. Tal, *Ergodicity and annular homeomorphisms of the torus*, Qual. Theory Dyn. Syst. **12** (2013), no. 2, 377–391, DOI 10.1007/s12346-012-0095-8. MR3101267

[4] P. Le Calvez, *Propriétés dynamiques des difféomorphismes de l'anneau et du tore* (French, with English and French summaries), Astérisque **204** (1991), 131. MR1183304 (94d:58092)

[5] P. Le Calvez, *Une version feuilletée équivariante du théorème de translation de Brouwer* (French, with English summary), Publ. Math. Inst. Hautes Études Sci. **102** (2005), 1–98, DOI 10.1007/s10240-005-0034-1. MR2217051 (2007m:37100)

[6] P. Dávalos, *On torus homeomorphisms whose rotation set is an interval*, Math. Z. **275** (2013), no. 3-4, 1005–1045, DOI 10.1007/s00209-013-1168-3. MR3127045

[7] B. R. Fayad, *Weak mixing for reparameterized linear flows on the torus*, Ergodic Theory Dynam. Systems **22** (2002), no. 1, 187–201, DOI 10.1017/S0143385702000081. MR1889570 (2003a:37012)

[8] J. Franks, *Recurrence and fixed points of surface homeomorphisms*, Ergodic Theory Dynam. Systems **8*** (1988), no. Charles Conley Memorial Issue, 99–107, DOI 10.1017/S0143385700009366. MR967632 (90d:58124)

[9] J. Franks, *Realizing rotation vectors for torus homeomorphisms*, Trans. Amer. Math. Soc. **311** (1989), no. 1, 107–115, DOI 10.2307/2001018. MR958891 (89k:58239)

[10] J. Franks, *The rotation set and periodic points for torus homeomorphisms*, Dynamical systems and chaos, Vol. 1 (Hachioji, 1994), World Sci. Publ., River Edge, NJ, 1995, pp. 41–48. MR1479903 (98j:58092)

[11] J. Franks and M. Misiurewicz, *Rotation sets of toral flows*, Proc. Amer. Math. Soc. **109** (1990), no. 1, 243–249, DOI 10.2307/2048385. MR1021217 (90i:58091)

[12] T. Jäger, *The concept of bounded mean motion for toral homeomorphisms*, Dyn. Syst. **24** (2009), no. 3, 277–297, DOI 10.1080/14689360802641800. MR2561442 (2010j:37067)

[13] T. Jäger, *Linearization of conservative toral homeomorphisms*, Invent. Math. **176** (2009), no. 3, 601–616, DOI 10.1007/s00222-008-0171-5. MR2501297 (2011e:37095)

[14] O. Jaulent, *Existence d'un feuilletage positivement transverse à un homéomorphisme de surface*, eprint: arXiv:1206.0213.

[15] A. Katok, *Bernoulli diffeomorphisms on surfaces*, Ann. of Math. (2) **110** (1979), no. 3, 529–547, DOI 10.2307/1971237. MR554383 (81a:28015)

[16] A. Kocsard and A. Koropecki, *A mixing-like property and inexistence of invariant foliations for minimal diffeomorphisms of the 2-torus*, Proc. Amer. Math. Soc. **137** (2009), no. 10, 3379–3386, DOI 10.1090/S0002-9939-09-09903-1. MR2515407 (2010i:37092)

[17] A. Koropecki and F. A. Tal, *Bounded and unbounded behavior for area preserving rational pseudorotaions*, eprint: arXiv:1207.5573v3.

[18] A. Koropecki and F. A. Tal, *Area-preserving irrotational diffeomorphisms of the torus with sublinear diffusion*, Proc. Amer. Math. Soc. **142** (2014), no. 10, 3483–3490, DOI 10.1090/S0002-9939-2014-12062-4. MR3238423

[19] N. Guelman A. Koropecki and Fabio A. Tal, *A caracterization of annularity for area preserving toral homeomorphisms*, eprint arXiv:1211.5044v1 (2012).

[20] M. Misiurewicz and K. Ziemian, *Rotation sets for maps of tori*, J. London Math. Soc. (2) **40** (1989), no. 3, 490–506, DOI 10.1112/jlms/s2-40.3.490. MR1053617 (91f:58052)

[21] H. Poincaré, *Oeuvres complètes*, Gauthier-Villars, Paris, 1952.

[22] F. A. Tal and S. Addas-Zanata, *On periodic points of area preserving torus homeomorphisms*, Far East J. Dyn. Syst. **9** (2007), no. 3, 371–378. MR2371838 (2008m:37071)

Instituto Tecnológico y de Estudios Superiores de Occidente, Periférico Sur Manuel Gómez Morín 8585, C.P. 45604, Tlaquepaque, Jalisco, México

E-mail address: pablod@iteso.mx

A review of No Free Lunch Theorems for search

Edgar A. Duéñez-Guzmán and Marte A. Ramírez-Ortegón

Abstract. Black-box search techniques are used as general purpose optimizers and are purported to robustly solve computationally complex problems. The No-free-lunch theory (NFL) aims to understand the limitations of black-box search. As its name implies, there are typically tradeoffs when designing an algorithm: in order to perform better in a particular scenario, it needs to perform worse in another one. Disagreements over the ultimate consequences of the theory have persisted over almost two decades, with some researchers insisting that NFL results have limited applicability, and others refining the extant results to address such limitations. Regardless of their immediate applicability, NFL theorems help us understand the underlying symmetries and structure of search algorithms. Here we review the main NFL theorems and discuss their application to real world problem solving.

Solutions to problems
are easy to find:
the problem's a great
contribution.
What's truly an art
is to wring from your mind
a problem to fit
a solution.

–Piet Hein

1. Introduction

Since the 1960s, search algorithms inspired by physical and biological processes have been a popular approach to general purpose optimization [**3, 4, 8, 11, 12, 14, 17, 23, 28**]. These algorithms are especially attractive to researchers working on computationally complex problems. They are frequently implemented as black-box search; that is, they are applied to optimization and search problems without exploiting *a priori* information about the objective function. Such algorithms acquire information about the objective function only by evaluating it. Many of these black-box search algorithms have been regarded as robust optimization techniques [**12–14, 21**].

2010 *Mathematics Subject Classification.* Primary 68Q99.
Key words and phrases. Benchmarks, black-box search, minimax, no free lunch, optimization.

The *No free lunch* (NFL) theory aims to understand the fundamental limitations of algorithms. As the name implies, there are typically tradeoffs when designing an algorithm: an algorithm that performs better in one scenario performs worse in another. NFL theorems [**34, 35**] were conceived to investigate whether different search algorithms could differ in robustness. The original NFL theorem states that the graph of an algorithm's performance against all objective functions has the same expected, maximum, and minimum as any other algorithm's performance graph. In fact, all performance graphs are equal, up to a permutation. NFL-like results have been proposed in diverse areas including machine learning [**7, 36, 38**], multi-objective optimization [**5**], discrete Laplace operators [**29**], induction and combinatorial problems [**37**], Walsh analysis [**31**], noise prediction [**20**], and others. In this review, we concentrate on NFL theorems for search.

NFL theorems have attracted considerable controversy due to their potential implications for popular black-box search techniques such as evolutionary algorithms, tabu search, simulated annealing and other nature-inspired algorithms [**1, 7, 9, 15, 18**]. Auger & Teytaud argued that NFL does not apply to functions over continuous domains [**1**]. The apparent limitations discussed in [**1, 2**] arise from the probabilistic framework within which original NFL theorems were expressed. However, the set-theoretic NFL theorems of Rowe, *et al.* [**24**] obviate these limitations by dispensing with probability altogether.

Other criticisms of NFL, however, are still debated. Droste, *et al.* [**9**] observed that the likelihood of facing a particular optimization problem is often not uniform. They argued that weighted average performance is, thus, the most meaningful metric to summarize aggregate optimization behavior. The original NFL assumes a uniform distribution, and was later extended to nonuniform distributions [**35**]. However, the nonuniformities considered by Wolpert & Macready [**35**] do not address all the limitations pointed out in [**9**]. As an attempt to alleviate the requirement for uniform average performance, and the equal consideration of all target functions, Igel & Toussaint [**16**] proposed a Non-Uniform NFL, which still failed to address all limitations mentioned in [**9**]. The Non-Uniform NFL was later proven by Rowe, *et al.* [**24**] to be a corollary of the Sharpened NFL by Schumacher, *et al.* [**26, 27**].

Similarly, Domingos [**7**] pointed out that instead of measuring performance in all scenarios, real world problem-solvers are employed in special, restricted contexts. Progressively more specific NFL theorems have been proposed that hold if the set of functions considered, the objective function distribution, or the relationship between functions and algorithms has special properties; permutation closure in the case of the Sharpened NFL [**26, 27**] and focused sets in the case of Focused NFL [**32**]. In practice, the assumption of permutation closure is rarely satisfied [**15**]; however, the generality of focused sets remains unclear.

The pitfalls for NFL applicability noted above stem from the traditional focus in characterizing the classes of functions over which some or all algorithms have the same performance. Researchers of black-box search, instead, typically establish a benchmark, and evaluate multiple algorithms over it. Performance over the benchmark is hoped to be indicative of performance over a related group of target problems. In response to this, Whitley a & Rowe [**32**] changed the focus of NFL to arbitrary benchmarks, and Duéñez-Guzmán & Vose [**10**] have subsequently begun to develop this perspective. These results apply to collections of functions that need not be permutation closed, nor focused.

Importantly, the cautionary observations mentioned above do not invalidate NFL results, but instead warn against their misapplication. An example for which a particular NFL theorem does not apply indicates a potential opening for future research.

In this article we introduce the main theorems of NFL, focusing on their interpretation and implications for optimization problems. To facilitate reading, we exclude all proofs; the interested reader is invited to look at the original sources for these. We will first discuss deterministic algorithms and later examine stochastic algorithms. For each, we review the theoretical background and major NFL theorems. We conclude by discussing general implications of NFL for black-box search, and potential future directions of NFL research.

Part 1. Deterministic Algorithms

2. Theoretical background

The original No Free Lunch (NFL) theorems were expressed for functions of finite domain and codomain and in the language of probability [**34, 35**]. However, Auger & Teytaud [**1, 2**] showed that probability is inadequate to affirm unconstrained NFL results in continuous cases. An alternative set-theoretic expression which eliminates the limitations of probability on NFL theorems has been employed by several researchers [**10, 24–27, 32**]. In this article, we focus on this set-theoretic expression of NFL.

Before we can enunciate the NFL theorems, we need to formalize some concepts that are commonly used in NFL theorems. These concepts include traces, algorithms, black-box search and permutations of functions.

Start by fixing two (arbitrary) finite sets $\mathcal{X}$ and $\mathcal{Y}$. A target function (or simply, a function) $f : \mathcal{X} \to \mathcal{Y}$ is an element of the space of all functions $\mathcal{Y}^{\mathcal{X}}$ with domain $\mathcal{X}$ and codomain $\mathcal{Y}$. A set of target functions will be called a *benchmark*. Let y_i denote $f(x_i)$. A function itself can be considered as a set of pairs (x_i, y_i) with the condition that $x_i = x_j$ implies $y_i = y_j$. A *sequence* $S = \langle s_0 \dots s_i \dots \rangle$ can be considered as a function mapping i to s_i. As an example, the space of all sequences of elements of $\mathcal{Y}$ of length 3 is given by $\mathcal{Y}^3$, where the exponent 3 in this case is interpreted as a set containing three elements ($\{0, 1, 2\}$). A sequence of values from $\mathcal{Y}$ will be called a *performance vector*.

Of central importance to NFL is the concept of a *trace*. A *trace* T corresponding to f is a finite sequence of elements from f,

$$T = \langle (x_0, y_0), \dots \rangle \,,$$

where the x-components are unique. A trace represents a particular search history of the space $\mathcal{X}$ and its corresponding target values in $\mathcal{Y}$. For instance, one can interpret trace $\langle (x_0, y_0), (x_1, y_1), (x_2, y_2) \rangle$ as a process visiting first x_0, evaluating f on x_0 to obtain y_0, the process later visits x_1 obtaining y_1, and finally visits x_2 obtaining y_2.

To refer to the sub-components of a trace the following notation is used

$$\begin{aligned} T_x &= \langle x_0, \dots \rangle \quad \text{sequence of } x\text{-components} \\ T_y &= \langle y_0, \dots \rangle \quad \text{sequence of } y\text{-components} \end{aligned}$$

The performance vector associated with T is T_y.

For a sequence S, let S^* be the range of S (without order). In particular, $T^* \subset f$. Trace T is *total* if $T_x^* = \mathcal{X}$ (equivalently, if $T^* = f$). A trace that is not total is *partial.*

Let $\mathcal{T}(f)$ be the set of all partial traces corresponding to f, and let the set of all partial traces be

$$\mathcal{T} = \bigcup_{f \in \mathcal{Y}^{\mathcal{X}}} \mathcal{T}(f)$$

A *search operator* is a function $g : \mathcal{T} \to \mathcal{X}$ such that $g(T) \notin T_x^*$. A *non-repeating/non-revisiting deterministic black box search algorithm* $\mathcal{A}$ corresponds to a search operator g, and will be referred to simply as an *algorithm.* Algorithm $\mathcal{A}$ applied to function f is denoted by $\mathcal{A}_f$, and maps traces to traces

$$\mathcal{A}_f(T) = \begin{cases} T \,|\, \langle (g(T), f \circ g(T)) \rangle & \text{if } T \in \mathcal{T}(f) \\ T & \text{otherwise} \end{cases}$$

where $|$ denotes concatenation.

This formal definition of an algorithm corresponds to the intuitive notion of a program running on a particular target problem as follows. Algorithm $\mathcal{A}$ runs on function f by beginning with the empty trace $\emptyset$, and repeatedly applying $\mathcal{A}_f$. At each step, the search operator of $\mathcal{A}$ produces the new point in the search space $\mathcal{X}$ that the algorithm will visit, and it makes this decision based on the complete history of which points the algorithm has already visited. Of course a particular search operator (and thus, an algorithm) might only use a subset of this information, or none of it at all.

Notice that this formal definition of an algorithm abstracts many details of its implementation. For instance, algorithmic complexity and running time are ignored. The restriction to deterministic algorithms is weaker that it might seem if we consider that the implementation of many stochastic algorithms, in fact, uses pseudo-random number generators that are deterministic. A pseudo-stochastic algorithm will behave differently if and only if the random seed used is different. Therefore, we simply consider the combination of the algorithm and a particular seed as a new deterministic algorithm. In Part 2, we will relax the deterministic constrain and discuss NFL results for truly stochastic algorithms.

In addition, many real-world algorithms do revisit points in the search space $\mathcal{X}$. However, such repeating algorithms produce the same trace as a non-repeating algorithm by ignoring repeats in the trace (equivalently, taking as output of the search operator the first point in T_x not previously visited). NFL theory typically ignores the case of algorithms that never visit the whole of the search space, regardless of running time. Note that such algorithms can have no better performance than alternative implementations of the same algorithm where, when caught in such a loop, simply restarts at a previously unvisited point.

Following [**27**], denote by $\mathcal{A}(f)$ the total trace produced (by running $\mathcal{A}$ on f to convergence). Note that $\mathcal{A}(f)^* = f$. Algorithms $\mathcal{A}$ and $\mathcal{A}'$ are regarded as equal if and only if $\mathcal{A}(f) = \mathcal{A}'(f)$ for all f. Let $\tilde{\mathcal{A}}(f)$ denote the performance vector $\mathcal{A}(f)_y$ associated with running to completion algorithm $\mathcal{A}$ on function f.

Permutations of the search space are central in NFL theory, and are used to define group actions on functions and algorithms [**22, 26**]. Symmetries of the space of functions or algorithms are expressed as permutations of the search space. Let us denote by $\mathcal{X}!$ the set of permutations (bijections) of $\mathcal{X}$. Given a permutation (i.e. a bijection) $\sigma : \mathcal{X} \to \mathcal{X}$, the *permutation of f by σ* (that is, the action of σ on the function f) is a new function denoted σf and defined by

$$(\sigma f)(x) = f(\sigma^{-1}(x)).$$

Thus, a permutation σ of $\mathcal{X}$ may also be considered as a permutation of the space of functions $\mathcal{Y}^{\mathcal{X}}$ (i.e. an element of $\mathcal{Y}^{\mathcal{X}}!$) via $f \mapsto \sigma f$. Intuitively, the function σf behaves exactly as f but over a permutation of $\mathcal{X}$. For example, if $\mathcal{X} = \{0, 1, 2\}$, $\mathcal{Y} = \{0, 1\}$, $\sigma(x) = (x + 1) \mod (3)$ and

$$f(x) = \begin{cases} 0 & \text{if } x = 0 \\ 1 & \text{if } x = 1 \\ 0 & \text{if } x = 2 \end{cases}, \quad \text{then} \quad (\sigma f)(x) = \begin{cases} 0 & \text{if } x = 0 \\ 0 & \text{if } x = 1 \\ 1 & \text{if } x = 2 \end{cases}.$$

One can also define the action of a permutation on algorithms. Given $\sigma \in \mathcal{X}!$, define the corresponding function σ_x which maps traces to traces by

$$\sigma_x(\langle (x_0, y_0), \ldots \rangle) = \langle (\sigma(x_0), y_0), \ldots \rangle.$$

Intuitively, σ_x operates on the x-components of a trace by applying σ to each of them while leaving the y-components unchanged. The *permutation $\sigma\mathcal{A}$ of $\mathcal{A}$ by σ* is the algorithm corresponding to search operator σg defined by $(\sigma g)(T) = \sigma^{-1}(g(\sigma_x(T)))$, where g is the search operator of $\mathcal{A}$.

Actions of permutations on algorithms and functions are intimately intertwined. The following theorem shows precisely how they correspond to one another, and was first proved in [**26**].

THEOREM 1. *(Duality) For any algorithm $\mathcal{A}$, permutation $\sigma \in \mathcal{X}!$, and function $f \in \mathcal{Y}^{\mathcal{X}}$,*

$$\sigma_x(\mathcal{A}(\sigma f)) = \sigma\mathcal{A}(f)$$

This theorem shows that the trace of a permuted algorithm ($\sigma\mathcal{A}$) is the same as the trace of the algorithm over the permuted function (σf) after the permutation has also been applied to the search space of the trace (σ_x). The strength of this result is perhaps best shown by projecting both sides of the equation to their y-components. This yields $\tilde{\mathcal{A}}(\sigma f) = \widetilde{\sigma\mathcal{A}}(f)$ where on the left hand side the permutation σ is acting on function f (it is regarded as an element of $\mathcal{Y}^{\mathcal{X}}!$), and on the right-hand side the permutation is acting on algorithm $\mathcal{A}$.

The map $\tilde{\mathcal{A}}$ is a bijection between functions and performance vectors. Theorem 1 implies that the performance vector of running $\sigma\mathcal{A}$ on function f is the same as the performance vector of algorithm $\mathcal{A}$ ran on the permuted function σf.

A set of functions $\mathcal{F} \subset \mathcal{Y}^{\mathcal{X}}$ is *closed with respect to a set $S \subset \mathcal{X}!$* of permutations if and only if

$$\mathcal{F} = \{\sigma f : f \in \mathcal{F}, \sigma \in S\}$$

that is, every permutation σ in S, when applied to a function f in $\mathcal{F}$, yields another (possibly the same) function also in $\mathcal{F}$. For example, if $\mathcal{X} = \{0, 1, 2\}$, $\mathcal{Y} = \{0, 1\}$, $\sigma_i(x) = (x + i) \mod (3)$ and

$$f_j(x) = \begin{cases} 1 & \text{if } x = j \\ 0 & \text{otherwise} \end{cases},$$

then $\mathcal{F} = \{f_0, f_1, f_2\}$ is closed with respect to $S = \{\sigma_1, \sigma_2\}$, and any benchmark is closed with respect to σ_0, since σ_0 is the identity.

The set $\mathcal{F}$ is *permutation closed* if and only if it is closed with respect to $\mathcal{X}!$ (all permutations of $\mathcal{X}$).

3. No Free Lunch theorems

The performance of algorithm $\mathcal{A}$ over a given function f is, in principle, a function of the whole trace $\mathcal{A}(f)$. In practice, performance is independent on the order of points visited in the search space ($\mathcal{X}$). Rather, it depends only on the values (in $\mathcal{Y}$) the function takes at those points. This leads to the definition of a performance measure as given in [**24**]:

DEFINITION 2. A *performance measure with respect to* $\mathcal{F} \subset \mathcal{Y}^{\mathcal{X}}$ is any function m_F defined over the collection of all search algorithms such that $m_{\mathcal{F}}(\mathcal{A})$ is a function of the multiset $\mathcal{A}(\mathcal{F}) = \{\!\{\tilde{\mathcal{A}}(f) : f \in \mathcal{F}\}\!\}$.

Note that for finite domains, one can think of a peformance measure as a function mapping performance vectors (of size $|\mathcal{X}|$) to real values. This specialized definition of performance measures leads naturally to the original NFL theorem.

THEOREM 3. *(Original NFL) Every algorithm $\mathcal{A}$ has the same average performance over the set of all target functions $f \in \mathcal{Y}^{\mathcal{X}}$.*

This theorem, originally enunciated by Wolpert & Macready [**34**], is in fact a corollary of Theorem 1.

When the multi-set of performance vectors is the same for two algorithms $\mathcal{A}$ and $\mathcal{B}$ ($\mathcal{A}(\mathcal{F}) = \mathcal{B}(\mathcal{F})$), then, regardless of the performance measure m both algorithms necessarily have the same average performance. It should be noted that this is a sufficient, but not necessary, condition for equal average performance. However, the NFL literature preferentially uses this definition of equal performance for it allows ignoring the role of performance measures. In the remainder of this review we might simply say *equal performance* to refer to this notion of equal average performance.

While Theorem 3 shows that any two algorithms have equal performance over all target functions, it does not characterize all sets of functions for which this is true. This characterization is the purpose of the Sharpened NFL by Schumacher [**26, 27**].

THEOREM 4. *(Sharpened NFL) Let $\mathcal{F} \subset \mathcal{Y}^{\mathcal{X}}$.*

$$\forall \mathcal{A}, \mathcal{B} \text{ algorithms. } \mathcal{A}(\mathcal{F}) = \mathcal{B}(\mathcal{F})$$

if and only if $\mathcal{F}$ is permutation closed.

This characterization shows that equal performance of all possible algorithms over a benchmark is equivalent to the benchmark having all possible symmetries (i.e. being permutation closed). For example, all algorithms have equal performance over the benchmark $\mathcal{F} = \{f_0, f_1, f_2\}$ defined at the end of Section 2.

The Sharpened NFL has also been extended to arbitrary sets $\mathcal{X}$ and $\mathcal{Y}$ (e.g., countable, continuous, etc.) [**24**] with only the additional requirement that algorithms be *efficient.*

DEFINITION 5. Search algorithm $\mathcal{A}$ is efficient if the domain of $\mathcal{A}(f)$ is equal to $I(\lfloor \mathcal{X} \rfloor)$ for all $f \in \mathcal{Y}^{\mathcal{X}}$, where $I(\alpha)$ denote the set of ordinal numbers less than α, and $\lfloor S \rfloor$ denotes the smallest ordinal α such that $\bar{\alpha} = |S|$.

Efficient algorithms are, intuitively, ones that, when applied to any function, cover any point in the search space $\mathcal{X}$ in at most $|\mathcal{X}|$ steps (which is trivially true for finite $\mathcal{X}$). One example of a non-efficient algorithm is given in [**24**]:

> Let $\mathcal{X}$ be the positive integers and consider $\mathcal{A}$ which explores $1, 3, 5, \ldots$ before moving on to $2, 4, 6, \ldots$ unless $f(1) = 1$ in which case $\mathcal{A}$ enumerates $1, 2, 3, \ldots$

Notice that there are functions for which this algorithm spends an infinite amount of steps and still has not visited 2.

A word of caution: When the cardinality of $\mathcal{X}$ is larger than countable, efficiency of an algorithm is not synonymous with the algorithm visiting every point of the search space $\mathcal{X}$ in finite time.

The requirement of permutation closure in Theorem 4 has been seen as a potential drawback [**15, 16**], since the number of benchmarks that are permutation closed is

$$2^{\binom{|\mathcal{X}|+|\mathcal{Y}|-1}{|\mathcal{X}|}}$$

which is small when compared to the number of possible benchmarks $2^{|\mathcal{Y}^{\mathcal{X}}|}$.

Igel & Toussaint [**16**] attempted to generalize Theorem 4 in such a way that the average performance is weighed over an special probability distribution over a benchmark. The so-called *Non-Uniform Sharpened NFL* theorem is, however, a corollary of the Sharpened NFL [**24**].

To alleviate the need for permutation closed benchmarks, Whitley and Rowe [**32**] shifted from all possible algorithms to only a subset of algorithms having equal performance.

Benchmark $\mathcal{F}$ is *focused* with respect to a set $\mathfrak{A}$ of algorithms, if and only if $\mathcal{A}(\mathcal{F})$ is independent of $\mathcal{A} \in \mathfrak{A}$. In other words, all algorithms in $\mathfrak{A}$ have the same set of performance vectors over the benchmark. Notice the parallel between the notion of $\mathcal{A}(\mathcal{F})$ being independent of $\mathcal{A}$ for all algorithms used in the Sharpened NFL and this notion of independence for a *particular* set of algorithms $\mathfrak{A}$ in focused benchmarks. In fact, we can rephrase Theorem 4 as:

> Benchmark $\mathcal{F}$ is focused with respect to the set of all algorithms if and only if it is permutation closed.

As we will see later, the concept of a focused benchmark is tied to a group of permutations in an analogous way to the concept of permutation closure of benchmarks.

For each pair of algorithms $\mathcal{A}$ and $\mathcal{B}$, $\tilde{\mathcal{A}}^{-1}\tilde{\mathcal{B}}$ induces a permutation of $\mathcal{Y}^{\mathcal{X}}$. This happens because for any algorithm $\mathcal{A}$, $\tilde{\mathcal{A}}$ is a bijection between the set of all target functions and the set of all performance vectors. These permutations induced by pairs of algorithms need not form a group, but one may consider the group they generate. This is precisely what the following definition does.

DEFINITION 6. For any set $\mathfrak{A}$ of algorithms, let $\mathcal{G}_{\mathfrak{A}}$ be the subgroup of $\mathcal{Y}^{\mathcal{X}}!$ generated by

$$\{\tilde{\mathcal{A}}^{-1}\tilde{\mathcal{B}} \,:\, \mathcal{A}, \mathcal{B} \in \mathfrak{A}\}$$

This leads to the Focused NFL theorem.

THEOREM 7. *(Focused NFL) Benchmark $\mathcal{F}$ is focused with respect to a set $\mathfrak{A}$ of algorithms if and only if it is closed with respect to $\mathcal{G}_{\mathfrak{A}}$.*

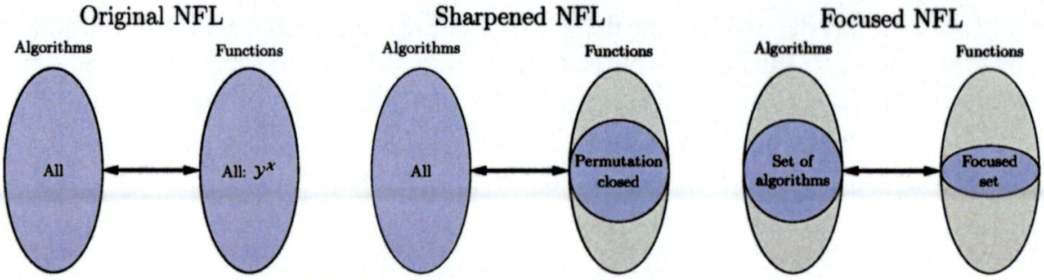

FIGURE 3.1. Schematic representation of the progressive specificity of NFL theorems. The Original NFL was concerned with all algorithms having equal performance over all functions. The Sharpened NFL strengthens the requirement from all functions to permutation closed benchmarks. Finally, Focused NFL deals with subsets of algorithms having equal performance over focused benchmarks.

The original Focused NFL theorem was expressed as an implication in [**32**]. Here we opt for its most general expression as given in [**10**].

The implications of Focused NFL merit some discussion. It characterizes all possible benchmarks over which arbitrary collections of algorithms must necessarily have equal performance. In principle, one could imagine starting with a benchmark $\mathcal{F}$ and an algorithm $\mathcal{A}$ (as is customary in performance evaluation) and ask if there exists any other algorithm $\mathcal{B}$ with the same performance as $\mathcal{A}$ (i.e. if $\mathcal{A}(\mathcal{F}) = \mathcal{B}(\mathcal{F})$). As it turns out, such a $\mathcal{B}$ exists if and only if there is some non-trivial permutation under which $\mathcal{F}$ is closed. Focused NFL tells us that, in general, we cannot do better.

Figure 3.1 depicts the transition from considering all algorithms and all target functions in the Original NFL, to permutation closed benchmarks in Sharpened NFL, and finally to focused sets of functions for particular sets of algorithms in Focused NFL.

In real world applications, algorithms are seldom run until they have exhausted the search space. For instance, the search space might be vast, and the practical reason for using a black-box search algorithms is precisely to avoid enumerating all possible solutions.

Consider now algorithms restricted to $\gamma \in \mathbb{N}$ steps. The projection π_γ of a sequence $x = \langle x_0, x_1, \ldots \rangle$ to their first γ elements is simply given by

$$\pi_\gamma(x) = \langle x_0, \ldots, x_{\gamma-1} \rangle.$$

Following [**10**], let us abbreviate $\pi_\gamma \circ \tilde{\mathcal{A}}$ by $\tilde{\mathcal{A}}_\gamma$. In this way, applying algorithm $\mathcal{A}$ to function f for γ steps yields the performance vector $\tilde{\mathcal{A}}_\gamma(f)$. Notice that, even though $\tilde{\mathcal{A}}$ is a bijection between functions and performance vectors, $\tilde{\mathcal{A}}_\gamma$ is no-longer 1–1. Because of this, we will consider multisets of performance vectors instead of sets. A multiset extends concept of a set by allowing elements to appear more than once and is denoted by double brackets $\{\!\{\ \}\!\}$. We extend π_γ to multisets by

$$\pi_\gamma S = \{\!\{ \pi_\gamma(s) : s \in S \}\!\}.$$

A benchmark $\mathcal{F}$ is *focused* with respect to a set $\mathfrak{A}$ of algorithms and integer γ if and only if the multiset

$$\tilde{\mathcal{A}}_\gamma(\mathcal{F}) = \{\!\{ \tilde{\mathcal{A}}_\gamma(f) \,:\, f \in \mathcal{F} \}\!\}$$

is independent of $\mathcal{A} \in \mathfrak{A}$. Similarly to the case of Focused NFL above, we define the set of permutations $G_{\mathcal{A}} \subset \mathcal{G}_{\mathfrak{A}}$ defined for $\mathcal{A} \in \mathfrak{A}$ by

$$G_{\mathcal{A}} = \{\tilde{\mathcal{A}}^{-1}\tilde{\mathcal{B}} : \mathcal{B} \in \mathfrak{A}\}$$

The following theorem, due to Whitley, *et al.* **[32]** (and later refined in **[10]**) generalizes the Focused NFL to a number of steps.

THEOREM 8. *(γ-step Focused NFL) The benchmark $\mathcal{F}$ is focused with respect to a collection $\mathfrak{A}$ of algorithms and integer γ if and only if for some $\mathcal{A} \in \mathfrak{A}$ and all $\alpha \in G_{\mathcal{A}}$,*

$$\tilde{\mathcal{A}}_\gamma(\mathcal{F}) = \tilde{\mathcal{A}}_\gamma(\alpha\mathcal{F})$$

Moreover, if the above holds for some $\mathcal{A} \in \mathfrak{A}$ then it holds for all $\mathcal{A} \in \mathfrak{A}$.

In contrast to Focused NFL, in the γ-step Focused NFL (γFNFL) theorem above, $G_{\mathcal{A}}$ is not necessarily the group $\mathcal{G}_{\mathfrak{A}}$. Indeed, in general it is not even a group. Closure over $G_{\mathcal{A}}$ is enough to guarantee equal performance when restricted to γ steps, whereas closure with respect to the full group $\mathcal{G}_{\mathcal{A}}$ was necessary in Theorem 7.

Focused benchmarks, characterized in Theorem 8, exhibit some interesting cyclic properties. Given algorithm $\mathcal{A}$ and integer $0 \leq \gamma \leq |\mathcal{X}|$, define the equivalence relation $\equiv$ on $\mathcal{Y}^{\mathcal{X}}$ by

$$f \equiv f' \iff \tilde{\mathcal{A}}_\gamma(f) = \tilde{\mathcal{A}}_\gamma(f')$$

that is, two functions are considered equivalent at γ steps under algorithm $\mathcal{A}$, if they produce the exact same performance vector.

DEFINITION 9. A benchmark $\mathcal{F}$ is *cyclic* with respect to algorithm $\mathcal{A}$, integer $0 \leq \gamma \leq |\mathcal{X}|$, and permutation $\alpha \in \mathcal{Y}^{\mathcal{X}}!$ if and only if for some positive integer ℓ,

$$\mathcal{F} = \{f_0, \ldots, f_{\ell-1}\} \text{ where } \alpha\,(f_{i \bmod \ell}) \equiv f_{i+1 \bmod \ell}$$

This leads to the following decomposition theorem by Duéñez-Guzmán & Vose **[10]**: every focused benchmark can be decomposed into a disjoint union of cyclic benchmarks.

THEOREM 10. *Benchmark $\mathcal{F}$ is focused with respect to a set $\mathfrak{A}$ of algorithms and $0 \leq \gamma \leq |\mathcal{X}|$ if and only if for some $\mathcal{A} \in \mathfrak{A}$ and all $\alpha \in G_{\mathcal{A}}$,*

$$\mathcal{F} = \bigcup_{j>0} \mathcal{F}_j$$

where the union above is disjoint, and each $\mathcal{F}_j$ (which may depend on $\mathfrak{A}$, $\mathcal{A}$, γ, and α) is cyclic with respect to $\mathcal{A}$, γ, and α.

While the decomposition is not (necessarily) unique, it is a complete characterization of focused benchmarks, like in γFNFL.

Part 2. Stochastic Algorithms

4. Theoretical background

In the previous section, we considered exclusively deterministic algorithms. Here, we show the extension of NFL theory to truly stochastic algorithms (sometimes called *randomized algorithms* **[10]**). A *stochastic algorithm* is simply a probability distribution over the space of deterministic algorithms. In any particular run on a target function f, an stochastic algorithm must necessarily produce a trace that is compatible with f. Therefore, a particular run of an stochastic algorithm on function f will produce exactly the same trace as some deterministic algorithm$\mathcal{A}$ (i.e. it behaves indistinguishably from $\mathcal{A}$).

Formally, a *stochastic algorithm* is described by a probability vector μ indexed over the set $\mathfrak{D}$ of all deterministic algorithms. We will use the probability vector μ itself to refer to the stochastic algorithm. We index the components of μ with deterministic algorithms $\mathcal{A}$. Thus, the $\mathcal{A}$-th component $\mu_{\mathcal{A}}$ is the probability the stochastic algorithm behaves like $\mathcal{A}$. In procedural terms, stochastic algorithm μ runs on f by choosing $\mathcal{A}$ with probability $\mu_{\mathcal{A}}$ and then applying algorithm $\mathcal{A}$ to f.

Stochastic algorithms are identified with elements of the simplex $\Lambda_{\mathfrak{D}}$ defined by

$$\Lambda_{\mathfrak{D}} = \{x \in \mathbb{R}^{|\mathfrak{D}|} : i \in \mathfrak{D},\ x_i \geq 0, \sum_i x_i = 1\}.$$

Note that the collection of stochastic algorithms $\Lambda_{\mathfrak{D}}$ contains the set of (deterministic) algorithms at the corners of the simplex. The *support* $\mathcal{S}_\mu$ of random algorithm μ is the set of algorithms $\mathcal{A}$ for which $\mu_{\mathcal{A}} > 0$. A deterministic algorithm μ has a unique nonzero component $\mu_{\mathcal{A}} = 1$ (we will call such support, *trivial*). In this case, we use $\mathcal{A}$ indistinguishably from μ.

Like before, we can define the action of a permutation on stochastic algorithms. Given $\sigma \in \mathcal{X}!$ and $\mu \in \Lambda_{\mathfrak{D}}$, the stochastic algorithm $\sigma\mu$ is defined by

$$(\sigma\mu)_{\mathcal{A}} = \mu_{\sigma^{-1}\mathcal{A}}$$

Intuitively, the stochastic algorithm $\sigma\mu$ applied to f, chooses algorithm $\mathcal{A}$ with probability $\mu_{\mathcal{A}}$ and then runs $\sigma\mathcal{A}$ on f. Random algorithms μ and μ' are *equivalent*, denoted by $\mu \equiv \mu'$, if and only if for all functions $f \in \mathcal{Y}^{\mathcal{X}}$ and every trace T,

$$\text{Prob}\,(\mu(f) = T) = \text{Prob}\,(\mu'(f) = T)$$

where the total trace $\mu(f)$ is the result of applying μ to f is equal to $\mathcal{A}(f)$ with probability $\mu_{\mathcal{A}}$. In other words, two stochastic algorithms are equivalent if the probability of generating a particular trace from a particular function is the same for both.

A performance measure m (simply called a measure) is extended to stochastic algorithms in the natural way: the performance of μ on f as measured by m is

$$m(\mu, f) = \sum_{\mathcal{A}} m(\tilde{\mathcal{A}}(f))\, \mu_{\mathcal{A}}$$

Measure m has, therefore, two interpretations. In one, it assigns real values to performance vectors, whereas in the other, it measures the expected performance of μ on f , aggregated and weighted over all deterministic algorithms. With the performance of the stochastic algorithm μ over a function defined, the expected performance of stochastic algorithm μ over benchmark $\mathcal{F}$ is

$$\mathcal{E}(\mu, \mathcal{F}) = \frac{1}{|\mathcal{F}|} \sum_{f \in \mathcal{F}} m(\mu, f)$$

and is referred to as *expected average performance*. Notice that $\mathcal{E}(\cdot, \cdot)$ depends on the measure m. An important observation is that the expected average performance is a linear function of algorithms.

PROPOSITION 11. *(Linearity) $\mathcal{E}(\cdot, \cdot)$ is linear in its first argument.*

The Duality Theorem (1) can be extended to stochastic algorithms [**10**]. This theorem shows the link between permutations of stochastic algorithms and permutations of functions in the same way the deterministic Duality Theorem did.

THEOREM 12. *(Expected Duality) For every measure m, stochastic algorithm μ, benchmark $\mathcal{F}$, and permutation $\sigma \in \mathcal{X}!$,*

$$\mathcal{E}(\sigma\mu, \mathcal{F}) = \mathcal{E}(\mu, \sigma\mathcal{F})$$

Intuitively, this theorem states that applying a permuted stochastic algorithm to a benchmark is the same, in the sense of expected performance, as applying the stochastic algorithm to a permuted benchmark.

5. No free lunch theorems

The restriction of traditional NFL theorems to deterministic algorithms might appear a shortcoming. However, even if one insists on considering truly stochastic algorithms, Proposition 11 has some immediate implications. A linear function over a convex set such as $\Lambda_{\mathfrak{D}}$ must necessarily attain its maximum and minimum at the boundary (in particular, since $\Lambda_{\mathfrak{D}}$ is a simplex, it attains them at a corner). With this, it is possible to prove that regardless of the measure m and benchmark $\mathcal{F}$ used, a non-trivial (i.e. non deterministic) stochastic algorithm will have the same performance as infinitely many other stochastic algorithms. Moreover, if m is not constant with respect to $\mathcal{F}$ and the support of μ, then there necessarily exist stochastic algorithms with larger and smaller average expected performance than μ. This is expressed formally in the following proposition [**10**].

PROPOSITION 13. *For every measure m, stochastic algorithm μ, and benchmark $\mathcal{F}$, if m is non-constant with respect to $\mathcal{F}$ and S_μ, then there exist infinitely many μ' and μ'' with support S_μ such that*

$$\mathcal{E}(\mu', \mathcal{F}) < \mathcal{E}(\mu, \mathcal{F}) < \mathcal{E}(\mu'', \mathcal{F})$$

If m is constant with respect to $\mathcal{F}$ and S_μ, then $S_\eta \subset S_\mu \Longrightarrow \mathcal{E}(\eta, \mathcal{F}) = \mathcal{E}(\mu, \mathcal{F})$.

Given benchmark $\mathcal{F}$, the group $H(\mathcal{F})$ of *benchmark symmetries* is

$$H(\mathcal{F}) = \{\sigma \in \mathcal{X}! : \sigma\mathcal{F} = \mathcal{F}\}.$$

Notice that the set of benchmark symmetries is intrinsic to the benchmark $\mathcal{F}$. As a consequence of Expected Duality, for any permutation σ in $H(\mathcal{F})$ and any stochastic algorithm μ, the expected performance of μ is the same as that of $\sigma\mu$. The following theorem by Rowe & Vose [**25**] expresses this formally.

THEOREM 14. *Given a benchmark $\mathcal{F}$, the set of algorithms $H(\mathcal{F})\mu$ has the same expected average performance over benchmark $\mathcal{F}$. Moreover if $\mathcal{F}$ is permutation closed, $\mathcal{E}(\mu, \mathcal{F}) = \mathcal{E}(\mu', \mathcal{F})$ for any stochastic algorithm μ'.*

In the theorem above, $H(\mathcal{F})\mu$ denotes the set of all the permutations of μ given by elements of $H(\mathcal{F})$ (the group of benchmark symmetries.) If benchmark $\mathcal{F}$ is permutation closed, them $H(\mathcal{F})$ is the group of all permutations $\mathcal{X}!$, and thus all stochastic algorithms have the same performance.

Benchmark symmetries were generalized in [**25**] to depend upon the measure m as well as the benchmark $\mathcal{F}$, and a result similar to the Focused NFL emerges which, likewise, depends upon m. Given benchmark $\mathcal{F}$, the collection $H_m(\mathcal{F})$ of *benchmark invariants* is

$$H_m(\mathcal{F}) = \{\sigma \in \mathcal{X}! : \forall \mathcal{A} . \mathcal{E}(\sigma\mathcal{A}, \mathcal{F}) = \mathcal{E}(\mathcal{A}, \mathcal{F})\}.$$

The set of benchmark invariants is intrinsic to the benchmark and the measure. The following theorem by Duéñez-Guzmán & Vose [**10**] is analogous to the previous theorem, but for benchmark invariants instead of for benchmark symmetries.

THEOREM 15. *The set of benchmark invariants $H_m(\mathcal{F})$ is a group (under composition) for every measure m and benchmark $\mathcal{F}$. For every stochastic algorithm μ, the stochastic algorithms in $H_m(\mathcal{F})\mu$ all have the same expected average performance over $\mathcal{F}$.*

This theorem can be specialized to algorithms run for γ steps. Since benchmark invariants depend on the measure m, we can consider a measure that is restricted to only the first γ components of a performance vector. Formally, let m_γ be $m \circ \pi_\gamma$ for some $0 \leq \gamma \leq |\mathcal{X}|$ and some measure m. For every stochastic algorithm μ, the γ-step expected average performance of the collection of algorithms $H_{m_\gamma}(\mathcal{F})\mu$ is identical over benchmark $\mathcal{F}$.

The sets of benchmark symmetries and benchmark invariants are important because they capture the inherent symmetries of a benchmark, in the sense that algorithms transformed by the permutations in them have the same average performance. The following theorem ties together the concept of benchmark invariants with the familiar concept of permutation closure of deterministic NFL theorems [**10**].

THEOREM 16. *For every measure m and benchmark $\mathcal{F}$,*

$$\sigma \in H_m(\mathcal{F}) \implies H_m(\sigma\mathcal{F}) = H_m(\mathcal{F})$$

By the Focused NFL theorem, we know that given a collection of algorithms $\mathfrak{A}$, $\mathcal{G}_{\mathfrak{A}}$ is a group such that if benchmark $\mathcal{F}$ is focused with respect to $\mathfrak{A}$, then $\mathcal{F}$ is closed with respect to the permutations in $\mathcal{G}_{\mathfrak{A}}$. Moreover, the γ-step Focused NFL theorem simplifies to the normal Focused NFL when $\gamma = |\mathcal{X}|$. In contrast, it is not necessarily true that $\mathcal{F}$ is closed with respect to the permutations in $H_m(\mathcal{F})$, even when $\gamma = |\mathcal{X}|$.

Theorem 16 tells us that for any $\sigma \in H_m(\mathcal{F})$, benchmark $\sigma\mathcal{F}$ has the same benchmark invariant as $\mathcal{F}$. This means that while the benchmark itself is not preserved under the action of $H_m(\mathcal{F})$, the benchmark invariant itself is preserved under the action of $H_m(\mathcal{F})$.

We conclude this section with two results that deal with matching the performance of a particular algorithm on a benchmark. It turns out that, for every measure, the performance of any stochastic algorithm on any benchmark can be matched in a non-trivial way.

PROPOSITION 17. *For every measure m, stochastic algorithm μ, and benchmark $\mathcal{F}$, there exist $\mathcal{F}'$ and μ' such that $\mathcal{E}(\mu, \mathcal{F}) = \mathcal{E}(\mu', \mathcal{F}')$ where $\mathcal{F}' \neq \mathcal{F}$ or μ' may be chosen arbitrarily.*

One might hope for a variant of Proposition 17 asserting that for every m, μ, and $\mathcal{F}$, there exists $\mu' \neq \mu$ for which $\mathcal{E}(\mu, \mathcal{F}) = \mathcal{E}(\mu', \mathcal{F})$ when μ is nontrivial. However, this is not the case in general. The following example was given in [**10**]:

If $\mathcal{X} = \mathcal{Y} = \{0, 1\}$, then there are only two possible deterministic algorithms $\mathfrak{D} = \{\mathcal{A}, \mathcal{B}\}$ where

$$\begin{aligned} \mathcal{A}(f) &= \langle (0, f(0)), (1, f(1)) \rangle \\ \mathcal{B}(f) &= \langle (1, f(1)), (0, f(0)) \rangle \end{aligned}$$

If benchmark $\mathcal{F}$ contains only the identity function, and the performance measure is

$$m(\langle y_0, y_1 \rangle) = 2y_0 + y_1 ,$$

it then follows that $H_m(\mathcal{F})$ is the trivial group (it contains only the identity). The example above demonstrates that the following cannot be improved.

PROPOSITION 18. *Let* $|\mathfrak{D}| > 2$. *For every measure* m, *benchmark* $\mathcal{F}$, *and nontrivial stochastic algorithm* μ, *there exist infinitely many* $\mu' \neq \mu$ *such that* $\mathcal{E}(\mu, \mathcal{F}) = \mathcal{E}(\mu', \mathcal{F})$.

These two results highlights how the performance of stochastic algorithms allows for even more symmetries than deterministic ones. There is no restriction of permutation closure with respect to any set or group of permutations.

6. Discussion

Several progressively more general No Free Lunch theorems have been proposed [**10, 16, 24, 27, 32, 35**]. All of these theorems improve our understanding of the applicability and limitations of black-box optimization algorithms. Understanding these limitations is relevant in two different ways. First, there is a need for a more meaningful way of comparing algorithm performance than simply analyzing the performance over a benchmark set of functions. Second, they provide a way to clearly define when algorithms are truly better than others.

The traditional approach of NFL theorems, discussed in Part 1, is the characterization of the sets of functions for which a particular set of algorithms have equal performance. For instance, the original NFL considered all target functions. Schumacher, *et al.* [**27**] later found that the set of all functions is not required to guarantee equal performance, showing that permutation closure is the relevant property. In a similar way, the result proposed by Whitley & Rowe [**32**] was sharpened and extended in [**10**] to characterize all focused sets of functions.

While the original formulation of the NFL theorem was concerned with all algorithms having the same average performance over a set of functions, later the Focused NFL [**32**] restricted its study on fixing a set of algorithms and finding sets of functions over which the algorithms in question had equal performance. In [**10**] a different approach to NFL results was also explored. Instead of trying to find the set of functions for which a set of algorithms perform equally on average, it is possible to ask under which conditions, given a set of functions and an algorithm, one can find other algorithms or sets of functions that have equal performance. Section 5 gives the main results in this direction. This result addresses, at least partially, one of the long standing criticisms to NFL theorems, namely that they only apply to benchmarks with restrictive properties.

The extensions of NFL to stochastic algorithms in Part 2 show that regardless of the benchmark function and random algorithm under consideration, there is always at least another random algorithm with the same performance as the original algorithm over that benchmark. Moreover, under arguably general conditions, given any algorithm, there are (infinitely many) other algorithms that perform better and others that perform worse. In Section 5, we discussed that the performance of an algorithm on a particular benchmark can be matched by some other algorithm on some other benchmark. Also, the maximal performance of a stochastic algorithm on a benchmark is attainable, but typically at a vertex of the simplex, thus reducing to a deterministic algorithm. Furthermore, this maximization of performance on the benchmark offers no clear indication of performance on other functions. Indeed, by the Sharpened NFL theorem, there exists another deterministic algorithm that performs at least as well in some other benchmark. The concept of benchmark invariants given at the end of Section 4, is analogous for stochastic algorithms to the concept of permutation closure for deterministic algorithms.

Future directions. There is a large amount of research done in developing and empirically improving heuristic optimization algorithms. In contrast, little effort has been devoted to understanding their theoretical limitations and overall behavior. Virtually nobody argues against the practical importance and success of heuristics. However, it is crucial to balance the empirical progress with theoretical understanding. Despite claims that everything on NFL has already been said [**30**], new and more powerful NFL theorems are still being found.

The applicability of NFL results to classes of functions considered of "real-world importance" is frequently questioned. While some progress has been recently achieved in this regard (see Section 5), there is still ample room for research in this area. Droste, *et al.* [**9**] describe what they consider realistic scenarios for black-box search. Their observations present an opportunity for further NFL research, and they can be summarized as pertaining to three broad categories: modality, separability and compressibility.

The *modality* of a function is how many local maxima (or minima) the function has. A unimodal function has a unique maximum (minimum), while a multimodal function has many. Modality is, therefore, generally used as a proxy for optimization difficulty, an observation that is empirically supported. Some efforts to understand the relationship between black-box search and modality have been published [**6**, **33**], but more research is needed.

Separability means that the optimum of a target function can be obtained by independently optimizing each of its variables. Separability is considered to indicate ease of optimization. The extent of the relevance of the class of separable functions to NFL theorems is currently unknown.

Compressibility refers to the possibility of describing a target function in a compact form instead of having to enumerate all its (x, y) pairs. Intuitively, highly structured functions can be represented succinctly, like $f(x) = x^2$, while completely unstructured ones require the enumeration of all (x, y) pairs. Formally, a function is compressible if its Kolmogorov length [**19**] is shorter than its plain representation length. Interestingly, Schumacher, *et al.* [**27**] pointed out that permutation closed benchmarks can be highly compressible, and their argument extends to focused benchmarks. Therefore, not all compressible functions violate the assumptions of NFL, but it is not known to what extent they are relevant for NFL theory.

There are, however, many other classes of functions that researchers can consider realistic. Perhaps the most difficult obstacle in finding NFL results for problems of "real-world importance" is not mathematical at all: researchers do not agree on what precisely "real-world importance" means. Without a precise definition, there is little hope for the formulation of a mathematical result.

From a purely theoretical perspective, there are still several open questions in NFL. A characterization of benchmark symmetries $H(\mathcal{F})$ and benchmark invariants $H_m(\mathcal{F})$ would be desirable. As of yet, little has been said about stochastic algorithms limited to a number of steps (but see Theorem 15). Compared to properties of benchmarks, little is known about the properties of search operators (the workhorses of algorithms). The particular way an algorithm explores the search space is precisely what differentiates it from others. More research is needed on the symmetries underlying the space of algorithms (beyond the Duality theorems), and on identifying the properties of search operators that are relevant for performance; especially for performance over a "real-world" class of target functions.

At its inception, NFL theory caused some stir in the optimization community. Some people saw it as highligting a fundamental flaw in black-box search algorithms, while other saw it as merely a mathematical curiosity. The ensuing debate lasted several years, and many researchers settled on a view of NFL as a modest cautionary tale: there are some classes of problems for which blind search is useless, but if you steer away from those, you are fine. However, there is, perhaps, a broader lesson here. Developing a branch of knowledge exclusively resting on empirical assessments is a risky enterprise. Whether NFL applies to a given set of functions and algorithms or not, critical thinking about algorithms and their performance assessment may enable us to avoid pitfalls. Showcasing an algorithm's properties on a benchmark might be appropriate sometimes, but if we are to ever truly understand what makes an algorithm *better*, we need to understand what *better* means mathematically.

Acknowledgments

The authors would like to thank Suzanne Sadedin for her very helpful comments on an early version of this manuscript.

References

[1] A. Auger and O. Teytaud, *Continuous lunches are free!*, Proceedings of the 9th annual conference on Genetic and Evolutionary Computation, GECCO-2007 (2007), 916–922.

[2] Anne Auger and Olivier Teytaud, *Continuous lunches are free plus the design of optimal optimization algorithms*, Algorithmica **57** (2010), no. 1, 121–146, DOI 10.1007/s00453-008-9244-5. MR2581081 (2011j:68061)

[3] T. Back and H. P. Schwefel, *Evolutionary computation: an overview*, Evolutionary Computation, 1996., Proceedings of IEEE International Conference on, IEEE, May 1996, pp. 20–29.

[4] Hans-Georg Beyer and Hans-Paul Schwefel, *Evolution strategies—a comprehensive introduction*, Nat. Comput. **1** (2002), no. 1, 3–52, DOI 10.1023/A:1015059928466. MR1907492 (2003d:68213)

[5] D.W. Corne and J.D. Knowles, *No free lunch and free leftovers theorems for multiobjective optimisation problems*, Evolutionary Multi-Criterion Optimization, Lecture Notes in Computer Science, vol. 2632/2003, Springer Berlin / Heidelberg, 2003, p. 66.

[6] Martin Dietzfelbinger, Jonathan E. Rowe, Ingo Wegener, and Philipp Woelfel, *Precision, local search and unimodal functions*, Algorithmica **59** (2011), no. 3, 301–322, DOI 10.1007/s00453-009-9352-x. MR2770461 (2012b:68304)

[7] P. Domingos, *How to get a free lunch: A simple cost model for machine learning applications*, Proc. AAAI98/ICML98, Workshop on the Methodology of Applying Machine Learning (1998), 1–7.

[8] M. Dorigo, V. Maniezzo, and A. Colorni, *Ant system: optimization by a colony of cooperating agents*, Systems, Man, and Cybernetics, Part B: Cybernetics, IEEE Transactions on **26** (1996), no. 1, 29–41.

[9] S. Droste, T. Jansen, and I. Wegener, *Perhaps not a free lunch but at least a free appetizer*, Proceedings of the First Genetic and Evolutionary Computation Conference, GECCO-1999 (1999), 833–839.

[10] EA Duéñez Guzmán and MD Vose, *No free lunch and benchmarks*, Evolutionary Computation **In press** (2012).

[11] Lawrence J. Fogel, *Intelligence through simulated evolution: forty years of evolutionary programming*, John Wiley & Sons, Inc., New York, NY, USA, 1999.

[12] D. Goldberg, *Genetic algorithm in search, optimization and machine learning*, Addison-Wesley Pub. Co. ISBN: 0201157675, 1989.

[13] Randy L. Haupt and Sue Ellen Haupt, *Practical genetic algorithms*, 2nd ed., Wiley-Interscience [John Wiley & Sons], Hoboken, NJ, 2004. With 1 CD-ROM (Windows). MR2059939

[14] John H. Holland, *Adaptation in natural and artificial systems*, University of Michigan Press, Ann Arbor, Mich., 1975. An introductory analysis with applications to biology, control, and artificial intelligence. MR0441393 (55 #14256)

[15] C. Igel and M. Toussaint, *On classes of functions for which no free lunch results hold*, 2001.

[16] Christian Igel and Marc Toussaint, *A no-free-lunch theorem for non-uniform distributions of target functions*, J. Math. Model. Algorithms **3** (2004), no. 4, 313–322, DOI 10.1023/B:JMMA.0000049381.24625.f7. MR2231453 (2006m:90140)

[17] S. Kirkpatrick, C. D. Gelatt Jr., and M. P. Vecchi, *Optimization by simulated annealing*, Science **220** (1983), no. 4598, 671–680, DOI 10.1126/science.220.4598.671. MR702485 (85f:90091)

[18] Gary J. Koehler, *Conditions that obviate the no-free-lunch theorems for optimization*, INFORMS Journal on Computing **19** (2007), no. 2, 273–279.

[19] Ming Li and Paul Vitányi, *An introduction to Kolmogorov complexity and its applications*, 3rd ed., Texts in Computer Science, Springer, New York, 2008. MR2494387 (2010c:68058)

[20] Malik Magdon-Ismail, *No free lunch for noise prediction*, Neural Computation **12** (2000), no. 3, 547–564.

[21] M. Mitchell, *An introduction to genetic algorithms*, third ed., Bradford, 1998.

[22] Nicholas J. Radcliffe and Patrick D. Surry, *Fundamental limitations on search algorithms: evolutionary computing in perspective*, Computer science today, Lecture Notes in Comput. Sci., vol. 1000, Springer, Berlin, 1995, pp. 275–291, DOI 10.1007/BFb0015249. MR1389581 (97b:68205)

[23] Ingo Rechenberg, *Evolutionsstrategie–optimierung technisher systeme nach prinzipien der biologischen evolution*, Werkstatt Bionik und Evolutionstechnik, vol. 1, Friedrich Frommann Verlag (Günther Holzboog KG), Stuttgart, 1994.

[24] J. Rowe, M.D. Vose, and A.H. Wright, *Reinterpreting no free lunch*, Evolutionary Computation **17** (2009), 117–129.

[25] J.E. Rowe and M.D. Vose, *Unbiased black box search algorithms*, Proceedings of the 13th annual conference on Genetic and evolutionary computation (Natalio Krasnogor, ed.), ACM New York, NY, USA, 2011, pp. 2035–2042.

[26] C. Schumacher, *Black box search - framework and methods.*, PhD thesis, The University of Tennesee, Knoxville (2000).

[27] C. Schumacher, M.D. Vose, and D. Whitley, *The no free lunch and problem description length*, Proceedings of the 3rd annual conference on Genetic and evolutionary computation, GECCO-2001 (2001), 565–570.

[28] Rainer Storn and Kenneth Price, *Differential evolution—a simple and efficient heuristic for global optimization over continuous spaces*, J. Global Optim. **11** (1997), no. 4, 341–359, DOI 10.1023/A:1008202821328. MR1479553 (98e:90172)

[29] M. Wardetzky, S. Mathur, F. Kalberer, and E. Grinspun, *Discrete laplace operators: No free lunch*, Eurographics Symposium on Geometry Processing (A. Belyaev and M. Garland, eds.), 2007.

[30] Ingo Wegener, *Computational complexity and ec.*, Tutorial at GECCO **04** (2004).

[31] D. Whitley, *Functions as permutations: Regarding no free lunch, walsh analysis and summary statistics*, Parallel Problem Solving from Nature PPSN VI (Marc Schoenauer, Kalyanmoy Deb, Günther Rudolph, Xin Yao, Evelyne Lutton, JuanJulian Merelo, and Hans-Paul Schwefel, eds.), Lecture Notes in Computer Science, vol. 1917, Springer Berlin Heidelberg, 2000, pp. 169–178.

[32] D. Whitley and J. Rowe, *Focused no free lunch theorems*, Proceedings of the 10th annual conference on Genetic and evolutionary computation, GECCO-2008 (2008), 811–818.
[33] Darrell Whitley and Jonathan Rowe, *Subthreshold-seeking local search*, Theoret. Comput. Sci. **361** (2006), no. 1, 2–17, DOI 10.1016/j.tcs.2006.04.008. MR2254222 (2007c:90121)
[34] D. Wolpert and M. Macready, *No free lunch theorems for search*, Technical Report **SFI-TR-95-02-010** (1995).
[35] D. Wolpert and M. Macready, *No free lunch theorems for optimization*, IEEE Transactions on Evolutionary Computation **1(1)** (1997), 67–82.
[36] David H. Wolpert, *The supervised learning no-free-lunch theorems*, In Proc. 6th Online World Conference on Soft Computing in Industrial Applications, 2001, pp. 25–42.
[37] J.R. Woodward and J.R. Neil, *No free lunch, program induction and combinatorial problems*, Genetic Programming, Lecture Notes in Computer Science, vol. 2610/2003, Springer Berlin / Heidelberg, 2003, pp. 287–313.
[38] Huan Xu, C. Caramanis, and S. Mannor, *Sparse algorithms are not stable: A no-free-lunch theorem*, Pattern Analysis and Machine Intelligence, IEEE Transactions on **34** (2012), no. 1, 187–193.

Google Inc., 1600 Amphitheatre Parkway, Mountain View, California 94043
E-mail address: `duenez@google.com`

Google Inc., 1600 Amphitheatre Parkway, Mountain View, California 94043
E-mail address: `darkmars@google.com`

On triangulations, quivers with potentials and mutations

Daniel Labardini-Fragoso

ABSTRACT. In this survey article we give a brief account of constructions and results concerning the quivers with potentials associated to triangulations of surfaces with marked points. Besides the fact that the mutations of these quivers with potentials are compatible with the flips of triangulations, we mention some recent results on the representation type of Jacobian algebras and the uniqueness of non-degenerate potentials. We also mention how the quivers with potentials associated to triangulations give rise to CY2 and CY3 triangulated categories that have turned out to be useful in the subject of stability conditions and in theoretical physics.

CONTENTS

Introduction

Around 11 years ago, Fomin-Zelevinsky defined cluster algebras *in an attempt to create an algebraic framework for dual canonical bases and total positivity in semisimple groups* (cf. **[19]**, from whose abstract the emphasized line was taken). Since then, cluster algebras have been found to possess interactions with a wide variety of areas, like Poisson geometry **[25, 26]**, integrable systems **[20]**, Teichmüller theory **[16–18]**, Lie theory **[24]**, representation theory of associative algebras **[3, 4, 11, 12, 39, 40]**, hyperbolic 3-manifolds **[38]**, commutative and non-commutative algebraic geometry **[6]**, mirror symmetry **[32]**, KP solitons **[29–31]**, and even with string theory in Physics **[1, 2]**.

Fundamental in the definition of cluster algebras is the notion of *quiver mutation*, which is a combinatorial operation on quivers. In a representation-theoretic

2010 *Mathematics Subject Classification.* Primary 16S99, 13F60, 05E99.
Key words and phrases. Mutation, quiver with potential, triangulation.

approach to cluster algebras, Derksen-Weyman-Zelevinsky developed in **[11]** a *mutation theory of quivers with potentials*, which lifts quiver mutation from the combinatorial to the algebraic level. A quiver with potential (*QP* for short) is a pair consisting of a quiver Q and a potential S on Q, that is, a linear combination of cycles of Q. The mutation theory of quivers with potentials ultimately leads to the notion of mutation of representations of QPs, thus providing a representation-theoretic interpretation for the combinatorial operation of quiver mutations.

On the other hand, a class of cluster algebras arising from triangulations of Riemann surfaces was introduced and systematically studied in **[17]** by Fomin-Shapiro-Thurston. These authors show that the elementary operation of *flip* of arcs in triangulations can be interpreted as the operation of mutation inside the corresponding cluster algebra. In particular, they show that every triangulation τ of a Riemann surface with marked points has a naturally associated quiver $Q(\tau)$, and that the flip of triangulations is reflected in the quiver level as quiver mutation.

In this survey article we describe a construction from **[33]** that associates a potential $S(\tau)$ to each triangulation τ, in such a way that the operation of flip is reflected at the level of QPs as the mutation of Derksen-Weyman-Zelevinsky. Then we state some results regarding the finite-dimensionality and the representation type of the Jacobian algebras of the QPs $(Q(\tau), S(\tau))$, as well as the uniqueness of non-degenerate potentials for the quivers $Q(\tau)$. Finally, we mention a couple of applications that the QPs $(Q(\tau), S(\tau))$ have had in the subject of stability conditions and in theoretical physics.

The paper is divided in five sections. In Section 1, after recalling some elementary facts concerning (complete) path algebras of quivers (Subsection 1.1), we describe the combinatorial operation of quiver mutation (Subsection 1.2), then we give a quick overview of mutations of quivers with potentials (Subsection 1.3), and close the section with a brief reminder of the setup of surfaces with marked points, their triangulations and flips of triangulations (Subsection 1.4).

In Section 2, we quickly say how to attach a quiver $Q(\tau)$ to each triangulation τ of a surface with marked points and state the compatibility between flips and quiver mutations (Subsection 2.1). In Subsection 2.2 we lift the story to the level of QPs, that is, we describe a way to associate a QP $(Q(\tau), S(\tau))$ to each triangulation τ, and state the compatibility between flips and QP-mutations.

In Section 3 we state results regarding the finite-dimensionality and the representation type of the Jacobian algebras of the QPs $(Q(\tau), S(\tau))$. It turns out that the Jacobian algebras $\mathcal{P}(Q(\tau), S(\tau))$ are always finite-dimensional and tame. Also, the tori with exactly one marked point are the only surfaces whose triangulations have quivers that admit a non-degenerate potential with wild Jacobian algebra. Moreover, if Q is a quiver admitting a non-degenerate potential with tame Jacobian algebra, then Q is either the quiver of a triangulation, or mutation-equivalent to one of nine exceptional quivers.

In section 4 we mention some results on the uniqueness of non-degenerate potentials for the quivers $Q(\tau)$. It turns out that most quivers arising from surfaces admit exactly one non-degenerate potential up to right-equivalence.

Finally, in Section 5 we indicate how QPs give rise to triangulated Calabi-Yau categories, and mention how, via such categories, the QPs $(Q(\tau), S(\tau))$ have had applications in the subject of stability conditions and in theoretical physics.

1. Three operations

This section is devoted to the description of three operations, namely, quiver mutations (Subsection 1.2), mutations of quivers with potentials (Subsection 1.3), and flips of triangulations (Subsection 1.4). These operations are the basis for the objects of study of this paper.

In our first subsection we give the basic background that is necessary to define quiver mutations and mutations of quivers with potentials.

1.1. Quivers and path algebras.

Recall that a *quiver* is a finite graph with oriented edges, that is, a quadruple $Q = (Q_0, Q_1, t, h)$ consisting of a finite set of *vertices* Q_0, a finite set of *arrows*, and a pair of functions $t, h : Q_1 \to Q_0$ that determine the *tail* $t(\alpha)$ and the head $h(\alpha)$ of any given arrow $\alpha \in Q_1$. We write $\alpha : j \to i$ to indicate that $t(\alpha) = j$ and $h(\alpha) = i$. We will always assume that the quivers we work with are *loop-free*, that is, that no arrow α satisfies $h(\alpha) = t(\alpha)$.

A *path of length* $\ell > 0$ on Q is a sequence $a = \alpha_1\alpha_2\dots\alpha_\ell$ of arrows with $t(\alpha_j) = h(\alpha_{j+1})$ for $j = 1, \dots, \ell-1$. We set $h(a) = h(\alpha_1)$ and $t(a) = t(\alpha_\ell)$. Positive-length paths are composed as functions, that is, if $a = \alpha_1\dots\alpha_\ell$ and $b = \beta_1\dots\beta_{\ell'}$ are paths with $h(b) = t(a)$, then the concatenation ab is defined as the path $\alpha_1\dots\alpha_\ell\beta_1\dots\beta_{\ell'}$, which starts at $t(ab) = t(\beta_{\ell'})$ and ends at $h(ab) = h(\alpha_1)$.

For each vertex $i \in Q_0$ we formally introduce a *length-0 path* e_i. By A^ℓ we denote the $\mathbb{C}$-vector space with basis the set of paths of length $\ell \geq 0$. We use the notations $R = A^0$ and $A = A^1$. Note that R is the vector space with basis the set of length-0 paths, hence has dimension equal to the cardinality of Q_0, while A is the vector space with basis the set of arrows of Q. If we define $e_ie_j = \delta_{i,j}e_i$, where $\delta_{i,j}$ is the *Kronecker delta*, R becomes a commutative $\mathbb{C}$-algebra. Furthermore, if we define $e_i\alpha = \delta_{i,h(\alpha)}\alpha$ and $\alpha e_i = \delta_{i,t(\alpha)}\alpha$, then A, and actually every A^ℓ with $\ell > 0$, becomes an R-R-bimodule.

DEFINITION 1.1. The *path algebra* of Q is the $\mathbb{C}$-vector space $R\langle Q\rangle$ consisting of all finite linear combinations of paths in Q, that is,

$$R\langle Q\rangle = \bigoplus_{\ell=0}^{\infty} A^\ell. \tag{1.1}$$

The *complete path algebra* of Q is the $\mathbb{C}$-vector space $R\langle\langle Q\rangle\rangle$ consisting of all possibly infinite linear combinations of paths in Q, that is,

$$R\langle\langle Q\rangle\rangle = \prod_{\ell=0}^{\infty} A^\ell. \tag{1.2}$$

Both $R\langle Q\rangle$ and $R\langle\langle Q\rangle\rangle$ have their multiplications induced by concatenation of paths (the product of two paths is their concatenation if they can be concatenated, and 0 if they cannot be concatenated). In terms of the homogeneous components in the decomposition (1.2) (resp. (1.1)), the multiplication of two elements of $R\langle\langle Q\rangle\rangle$ (resp. $R\langle Q\rangle$) resembles the multiplication of formal power series (resp. polynomials). That is, if we have $u = \sum_{\ell\geq 0} u^{(\ell)}$ and $v = \sum_{\ell\geq 0} v^{(\ell)}$, with $u^{(\ell)}, v^{(\ell)} \in A^\ell$ for every $\ell \geq 0$, then

$$uv = \sum_{\ell\geq 0}\sum_{\ell_1+\ell_2=\ell} u^{(\ell_1)}v^{(\ell_2)}.$$

(In this equality, the right-hand side is a well-defined element of $R\langle\langle Q\rangle\rangle$: for $\ell \geq 0$ fixed, we have $\sum_{\ell_1+\ell_2=\ell} u^{(\ell_1)}v^{(\ell_2)} = \sum_{k=0}^{\ell} u^{(k)}v^{(k-\ell)}$.)

Note that $R\langle Q\rangle$ is a $\mathbb{C}$-subalgebra of $R\langle\langle Q\rangle\rangle$. Actually, $R\langle Q\rangle$ is dense in $R\langle\langle Q\rangle\rangle$ under the $\mathfrak{m}$-*adic topology* of $R\langle\langle Q\rangle\rangle$. The fundamental system of open neighborhoods of this topology around 0 is given by the powers of the two-sided ideal $\mathfrak{m}$ of $R\langle\langle Q\rangle\rangle$ generated by the arrows of Q.

We are ready to describe the three operations this survey article is about: quiver mutations, mutations of quivers with potentials, and flips of surface triangulations.

1.2. Quiver mutations.

DEFINITION 1.2. Let Q be a quiver. An *ℓ-cycle* on Q is a path $\alpha_1\alpha_2\ldots\alpha_\ell$, with $\ell > 0$, such that $h(\alpha_1) = t(\alpha_\ell)$. A quiver is *2-acyclic* if it does not have 2-cycles.

Central in the definition of cluster algebras is the notion of *quiver mutation.* This is a combinatorial operation on 2-acyclic quivers that can be described as an elementary 3-step procedure as follows. Start with a 2-acyclic quiver Q and a vertex i of Q.

(Step 1) Every time we have an arrow $\alpha : j \to i$ and an arrow $\beta : i \to k$ in Q, add an arrow $[\beta\alpha] : j \to k$;

(Step 2) replace each arrow γ incident to i with an arrow γ^* going in the opposite direction;

(Step 3) delete 2-cycles one by one (2-cycles may have been created when applying Step 1).

The result is a 2-acyclic quiver $\mu_i(Q)$, called the *mutation of Q with respect to i.* See Figure 1 for an example.

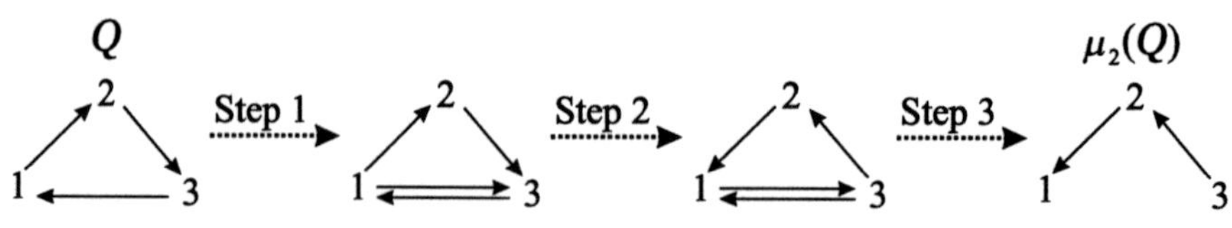

FIGURE 1. The three steps of quiver mutation. Here we are applying μ_2.

1.3. Mutations of quivers with potentials.

In a representation-theoretic approach to cluster algebras, Derksen-Weyman-Zelevinsky developed in [**11**] a *mutation theory of quivers with potentials*, which lifts quiver mutation from the combinatorial to the algebraic level.

DEFINITION 1.3. Let Q be a quiver. An element S of $R\langle\langle Q\rangle\rangle$ is called a *potential* if it is a possibly infinite linear combination of cycles of Q, with the property that no two different cycles appearing in S with non-zero coefficient can be obtained from each other by rotation. If S is a potential on Q, we say that the pair (Q, S) is a *quiver with potential*, or simply a *QP*.

Quiver mutation is lifted to the algebraic level of QPs by providing lifts of the three steps described in Subsection 1.2. Among the three steps, the one that turns out to be the hardest to lift is Step 3: one needs an algebraic procedure to delete

2-cycles algebraically. The procedure is provided by a technical result (Theorem 1.6 below) that requires some preparation.

DEFINITION 1.4. Let Q and Q' be quivers with the same vertex set $Q_0 = Q'_0$.

(1) Two potentials S and S' on Q are *cyclically-equivalent* if $S - S'$ lies in the closure of the vector subspace of $R\langle\langle Q\rangle\rangle$ spanned by all the elements of the form $\alpha_1 \dots \alpha_\ell - \alpha_2 \dots \alpha_\ell \alpha_1$ with $\alpha_1 \dots \alpha_\ell$ a cycle of positive length.
(2) We say that two QPs (Q,S) and (Q',S') are *right-equivalent* if there exists a *right-equivalence* between them, that is, a $\mathbb{C}$-algebra isomorphism $\varphi : R\langle\langle Q\rangle\rangle \to R\langle\langle Q'\rangle\rangle$ satisfying $\varphi(e_i) = e_i$ for all $i \in Q_0 = Q'_0$, and such that $\varphi(S)$ is cyclically-equivalent to S'.
(3) For each arrow $\alpha \in Q_1$ and each cycle $\alpha_1 \dots \alpha_\ell$ in Q we define the *cyclic derivative*

$$\partial_\alpha(\alpha_1 \dots \alpha_\ell) = \sum_{k=1}^{\ell} \delta_{\alpha,\alpha_k} \alpha_{k+1} \dots \alpha_\ell \alpha_1 \dots \alpha_{k-1},$$

and extend ∂_α by linearity and continuity so that $\partial_\alpha(S)$ is defined for every potential S.
(4) The *Jacobian ideal* $J(S)$ is the topological closure of the two-sided ideal of $R\langle\langle Q\rangle\rangle$ generated by $\{\partial_\alpha(S) \mid \alpha \in Q_1\}$, and the *Jacobian algebra* $\mathcal{P}(Q,S)$ is the quotient algebra $R\langle\langle Q\rangle\rangle/J(S)$.
(5) A QP (Q,S) is *trivial* if $S \in A^2$ and $\{\partial_\alpha(S) \mid \alpha \in Q_1\}$ spans A as a $\mathbb{C}$-vector space.
(6) A QP (Q,S) is *reduced* if the degree-2 component of S is 0, that is, if the expression of S involves no 2-cycles.
(7) The *direct sum* $Q \oplus Q'$ is the quiver whose vertex set is $Q_0 = Q'_0$ and whose arrow set is the disjoint union $Q_1 \sqcup Q'_1$, with the tail and head functions defined in the obvious way.
(8) The *direct sum* of two QPs (Q,S) and (Q',S') is the QP $(Q,S)\oplus(Q',S') = (Q \oplus Q', S + S')$.

If S and S' are potentials on the same quiver Q, and the QPs (Q,S) and (Q,S') are right-equivalent, we will say, in a slight abuse of nomenclature, that S and S' are right-equivalent.

PROPOSITION 1.5. **[11]** *If $\varphi : R\langle\langle Q\rangle\rangle \to R\langle\langle Q'\rangle\rangle$ is a right-equivalence between (Q,S) and (Q',S'), then φ sends $J(S)$ onto $J(S')$ and therefore induces an algebra isomorphism $\mathcal{P}(Q,S) \to \mathcal{P}(Q',S')$.*

THEOREM 1.6. **[11]** *For every QP (Q,S) there exist a trivial QP $(Q_{\mathrm{triv}}, S_{\mathrm{triv}})$ and a reduced QP $(Q_{\mathrm{red}}, S_{\mathrm{red}})$ such that (Q,S) is right-equivalent to the direct sum $(Q_{\mathrm{triv}}, S_{\mathrm{triv}}) \oplus (Q_{\mathrm{red}}, S_{\mathrm{red}})$. The right-equivalence class of each of the QPs $(Q_{\mathrm{triv}}, S_{\mathrm{triv}})$ and $(Q_{\mathrm{red}}, S_{\mathrm{red}})$ is determined by the right-equivalence class of (Q,S).*

ON THE PROOF. The proof of the second statement (uniqueness of $(Q_{\mathrm{triv}}, S_{\mathrm{triv}})$ and $(Q_{\mathrm{red}}, S_{\mathrm{red}})$ up to right-equivalence) is rather long and relies on a series of technical (albeit elementary) preliminary results. One of the problems is that, for an arbitrary right-equivalence $\varphi : R\langle\langle Q' \oplus C'\rangle\rangle \to R\langle\langle Q'' \oplus C''\rangle\rangle$ between $(Q',W')\oplus(C',T')$ and $(Q'',W'')\oplus(C'',T'')$, with $(Q',W'), (Q'',W'')$ reduced QPs and $(C',T'), (C'',T'')$ trivial QPs, it is not necessarily true that φ restricts to an isomorphism $R\langle\langle Q'\rangle\rangle \to R\langle\langle Q''\rangle\rangle$, which means that the restriction of φ to $R\langle\langle Q'\rangle\rangle$ is not necessarily a right-equivalence between (Q', W') and (Q'', W'').

Let us discuss Derksen-Weyman-Zelevinsky's proof of existence of $(Q_{\text{triv}}, S_{\text{triv}})$ and $(Q_{\text{red}}, S_{\text{red}})$. Using basic linear algebra, it is easy to show that there exists an R-R-bimodule automorphism f of A such that the algebra automorphism ψ of $R\langle\langle Q\rangle\rangle$ induced by f sends S to a potential cyclically equivalent to

$$W = \sum_{k=1}^{N} a_N b_N + W^{(\geq 3)} \tag{1.3}$$

for some $N \geq 0$, some set $\{a_1, b_1, \dots, a_N, b_N\}$ of $2N$ distinct arrows of Q such that each $a_k b_k$ is a 2-cycle, and some potential $W^{(\geq 3)} \in \mathfrak{m}^3$. Up to cyclical equivalence one can actually assume that

$$W = \left(\sum_{k=1}^{N} a_N b_N + a_k u_k + b_k v_k\right) + W' \tag{1.4}$$

for some u_k and v_k belonging to $\mathfrak{m}^\ell$ for some $\ell \geq 2$, and some potential $W' \in \mathfrak{m}^3$ that does not involve any of the arrows $a_1, b_1, \dots, a_N, b_N$. If $u_k = v_k = 0$ for all $k = 1, \dots, N$, then (Q, W) is already reduced and the existence of $(Q_{\text{triv}}, S_{\text{triv}})$ and $(Q_{\text{red}}, S_{\text{red}})$ is established. Otherwise, one uses (1.4) to define an algebra automorphism f of $R\langle\langle Q\rangle\rangle$ by setting

$$f(a_k) = a_k - v_k, \quad f(b_k) = b_k - u_k, \quad \text{for } k = 1, \dots, N,$$

and $f(c) = c$ for every arrow $c \in Q_1 \setminus \{a_1, b_1, \dots, a_N, b_N\}$. A little algebraic manipulation shows that $f(W)$ is cyclically equivalent to a potential of the form (1.4), but with the corresponding factors u_k and v_k belonging to a power $\mathfrak{m}^{\ell'}$ with $\ell' > \ell$. This property and the fact that the *depth* of f is at least $\ell - 1$ allow a recursive construction of a sequence $(f_n)_{n>0}$ of algebra automorphisms of $R\langle\langle Q\rangle\rangle$ such that the limit $\varphi = \lim_{n\to\infty} f_n f_{n-1} \cdots f_2 f_1$ is a well-defined algebra automorphism of $R\langle\langle Q\rangle\rangle$ that sends W to a potential cyclically-equivalent to a potential $\widetilde{W}$ of the form (1.4) with all factors u_k and v_k equal to zero[1]. Setting Q_{red} (resp. Q_{triv}) to be the subquiver of Q whose arrow set is $Q_1 \setminus \{a_1, b_1, \dots, a_N, b_N\}$ (resp. $\{a_1, b_1, \dots, a_N, b_N\}$), and $S_{\text{red}} = \widetilde{W} - \sum_{k=1}^{N} a_k b_k$ (resp. $S_{\text{triv}} = \sum_{k=1}^{N} a_k b_k$), we see that $(Q_{\text{red}}, S_{\text{red}})$ and $(Q_{\text{triv}}, S_{\text{triv}})$ are a reduced and a trivial QP, respectively, and the composition $\varphi\psi$ is a right-equivalence between (Q, S) and $(Q_{\text{red}}, S_{\text{red}}) \oplus (Q_{\text{triv}}, S_{\text{triv}})$. □

Definition 1.7. In the situation of Theorem 1.6, the QPs $(Q_{\text{red}}, S_{\text{red}})$ and $(Q_{\text{triv}}, S_{\text{triv}})$ are called, respectively, the *reduced part* and the *trivial part* of (Q, S).

We now turn to the definition of mutation of a QP. Let (Q, S) be a QP on the vertex set Q_0 and let $i \in Q_0$. Assume that Q is 2-acyclic. Let $\widetilde{\mu}_i(Q)$ be the quiver obtained right after applying the first two steps of quiver mutation, but before applying Step 3. Replacing S if necessary with a cyclically equivalent potential, we assume that every cycle appearing in the expression of S starts at a vertex different from i. Then we define $[S]$ to be the potential on $\widetilde{\mu}_i(Q)$ obtained from S by replacing each length-2 path $\beta\alpha$ of Q such that $h(\alpha) = i = t(\beta)$, with the arrow $[\beta\alpha]$ of $\widetilde{\mu}_i(Q)$. Also, we define $\Delta_i(Q) = \sum \alpha^* \beta^* [\beta\alpha]$, where the sum runs over all length-2 paths

[1] This is the reason why we work with the complete path algebra $R\langle\langle Q\rangle\rangle$ rather than with the path algebra $R\langle Q\rangle$: the only way to ensure convergence of this limit process for arbitrary (Q, S) is to work with the complete path algebra.

$\beta\alpha$ of Q such that $h(\alpha) = i = t(\beta)$. Finally, we set $\widetilde{\mu}_i(S) = [S] + \Delta_i(Q)$, which clearly is a potential on $\widetilde{\mu}_i(Q)$.

DEFINITION 1.8. **[11]** Under the assumptions and notation just stated, we define the *mutation* $\mu_i(Q, S)$ of (Q, S) with respect to i to be the reduced part of the QP $\widetilde{\mu}_i(Q, S) = (\widetilde{\mu}_i(Q), \widetilde{\mu}_i(S))$.

EXAMPLE 1.9. Consider the QP (Q, S), where Q is the quiver

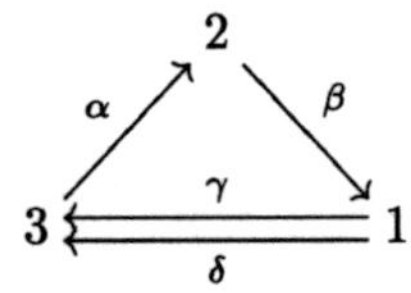

and $S = \gamma\beta\alpha$. Then $\widetilde{\mu}_2(Q)$ is the quiver

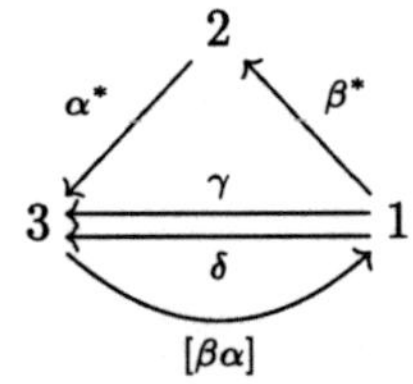

and $\widetilde{\mu}_2(S) = \gamma[\beta\alpha] + \alpha^*\beta^*[\beta\alpha]$. The rule $\gamma \mapsto \gamma - \alpha^*\beta^*$ extends to an algebra automorphism φ of $R\langle\langle\widetilde{\mu}_2(Q)\rangle\rangle$ that serves as a right-equivalence between $(\widetilde{\mu}_2(Q), \widetilde{\mu}_2(S))$ and $(Q', S') \oplus (C, T)$, where the quivers Q' and C are

Q' :
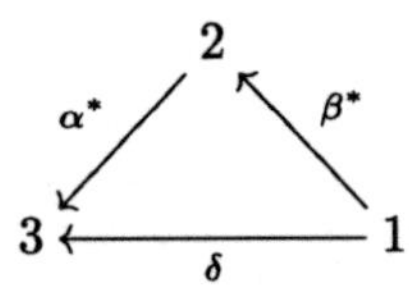

C :
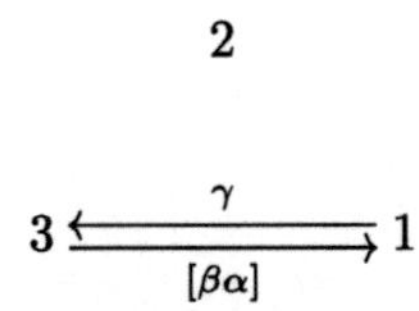

and $S' = 0$, $T = \gamma[\beta\alpha]$. Since (Q', S') is reduced and (C, T) is trivial, this implies that $(Q', S') = \mu_2(Q, S)$, ie, that (Q', S') is the mutation of (Q, S) with respect to 2.

"Unfortunately", given a QP (Q, S) with Q 2-acyclic, the underlying quiver of the mutated QP $\mu_i(Q, S)$ is not necessarily 2-acyclic, its 2-acyclicity depends heavily on the potential S.

DEFINITION 1.10. **[11]** A QP (Q, S) is *non-degenerate* if Q is 2-acyclic and the underlying quiver of the QP obtained after any possible sequence of QP-mutations is 2-acyclic.

EXAMPLE 1.11. Consider the quiver Q from Example 1.9, then $(Q, 0)$ is a degenerate QP. Indeed, the underlying quiver of the QP $\mu_2(Q, 0)$ is not 2-acyclic.

EXAMPLE 1.12. The QP (Q, S) from example 1.9 is non-degenerate. This follows from the fact that the quiver underlying $\mu_2(Q, S)$ is acyclic, which implies that $\mu_2(Q, S)$ is non-degenerate (cf. **[11]**). Alternatively, the non-degeneracy of (Q, S) can be seen as a consequence of Theorem 2.6 below, for (Q, S) is the QP associated to a triangulation of a surface (cf. Definition 2.2 below).

THEOREM 1.13. **[11]** *The following hold if* $\mathbb{C}$ *is the ground field:*

(1) *Mutations of QPs are well defined up to right-equivalence.*
(2) *Mutations of QPs are involutive up to right-equivalence.*
(3) *Every 2-acyclic quiver Q admits a non-degenerate potential on it.*
(4) *Finite-dimensionality of Jacobian algebras is invariant under QP-mutations.*

REMARK 1.14. Theorem 1.13 still holds for ground fields different from $\mathbb{C}$. Indeed, parts (1), (2) and (4) hold over any ground field, while part (3) holds over any uncountable ground field.

1.4. Flips of triangulations.

DEFINITION 1.15. A *surface with marked points*, or simply a *surface*, is a pair $(\Sigma, \mathbb{M})$, where Σ is a compact connected oriented Riemann surface with (possibly empty) boundary, and $\mathbb{M}$ is a non-empty finite subset of Σ containing at least one point from each connected component of the boundary of Σ. We refer to the elements of $\mathbb{M}$ as *marked points*. The marked points that lie in the interior of Σ are called *punctures*, and the set of punctures of $(\Sigma, \mathbb{M})$ is denoted $\mathbb{P}$.

We think of $\mathbb{M}$ as a prescribed set of vertices for triangulations of Σ. More formally:

DEFINITION 1.16. Let $(\Sigma, \mathbb{M})$ be a surface with marked points.

(1) An *arc* on $(\Sigma, \mathbb{M})$, is a curve i on Σ such that:
   - the endpoints of i belong to $\mathbb{M}$;
   - i does not intersect itself, except that its endpoints may coincide;
   - the points in i that are not endpoints do not belong to $\mathbb{M}$ nor to the boundary of Σ;
   - i does not cut out an unpunctured monogon nor an unpunctured digon.

(2) Two arcs i_1 and i_2 are *isotopic relative to* $\mathbb{M}$ if there exists a continuous function $H : [0,1] \times \Sigma \to \Sigma$ such that
   (a) $H(0,x) = x$ for all $x \in \Sigma$;
   (b) $H(1,i_1) = i_2$;
   (c) $H(t,m) = m$ for all $t \in I$ and all $m \in \mathbb{M}$;
   (d) for every $t \in I$, the function $H_t : \Sigma \to \Sigma$ given by $x \mapsto H(t,x)$ is a homeomorphism.

   Arcs will be considered up to isotopy relative to $\mathbb{M}$, parametrization, and orientation.

(3) Two arcs are *compatible* if there are arcs in their respective isotopy classes that, except possibly for their endpoints, do not intersect.

(4) An *ideal triangulation* of $(\Sigma, \mathbb{M})$ is any maximal collection τ of pairwise compatible arcs.

REMARK 1.17. The adjective *ideal* in the term *ideal triangulation* comes from the connection with Teichmüller theory, see [**18**].

If τ is any ideal triangulation, then, replacing if necessary each arc in τ with an isotopic one (relative to $\mathbb{M}$), one can assume that any two arcs in τ intersect at most at their endpoints. This fact is less trivial than it may seem at first glance, see for example [**21**].

Any ideal triangulation τ of $(\Sigma, \mathbb{M})$ splits Σ into *triangles*. If $(\Sigma, \mathbb{M})$ has punctures, some of the triangles of τ may be *self-folded*, see Figure 2. A self-folded triangle always contains a *folded side.* If $i \in \tau$ is an arc which is not the folded side

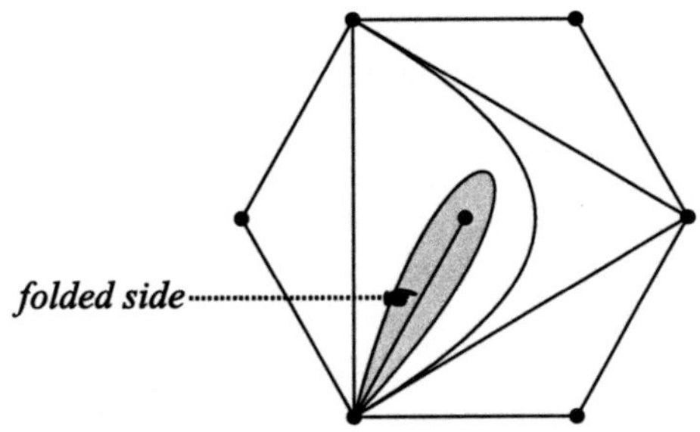

FIGURE 2. Ideal triangulation with a self-folded triangle (the self-folded triangle has been highlighted).

of a self-folded triangle, then there exists a unique arc $j \neq i$ on $(\Sigma, \mathbb{M})$, such that $\sigma = (\tau \setminus \{i\}) \cup \{j\}$ is an ideal triangulation of $(\Sigma, \mathbb{M})$. We say that σ is obtained from τ by the *flip* of the arc i. See Figure 3. Intuitively speaking, flipping an arc of a triangulation corresponds to the operation that replaces a diagonal of a square with the other diagonal.

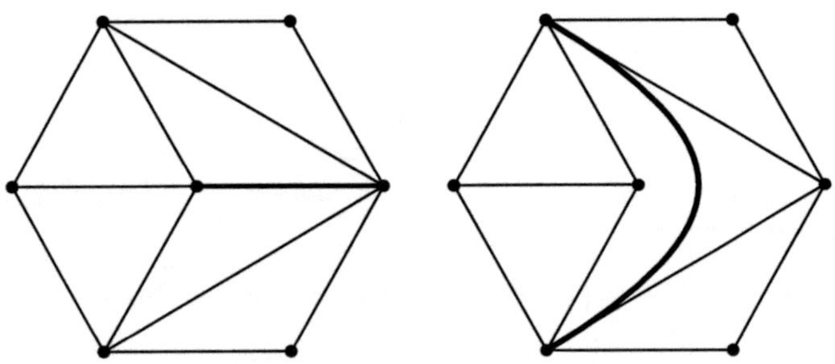

FIGURE 3. Two ideal triangulations related by a flip, the arcs involved in the flip have been drawn bolder than the rest.

Note that with the notion of triangulation we have thus far (that of ideal triangulation), it is not possible to flip folded sides of self-folded triangles. In order to be able to flip these, Fomin-Shapiro-Thurston introduced in **[17]** the concept of *tagged triangulation*, a notion of triangulation which is more general than the notion of ideal triangulation we have defined above. The combinatorics of flips of tagged triangulations becomes rather subtle at some points, but the following does hold:

THEOREM 1.18. *Let $(\Sigma, \mathbb{M})$ be a surface with marked points.*

(1) **[17]** *If τ is a tagged triangulation of $(\Sigma, \mathbb{M})$, and i is a tagged arc belonging to τ, then there exists a unique tagged arc $j \neq i$ on $(\Sigma, \mathbb{M})$ such that $\sigma = (\tau \setminus \{i\}) \cup \{j\}$ is a tagged triangulation of $(\Sigma, \mathbb{M})$. In other words, every tagged arc in a tagged triangulation can be flipped.*
(2) **[37]** *Every two ideal triangulations can be connected by a sequence of flips of ideal triangulations.*

(3) **[17]** *If* $(\Sigma, \mathbb{M})$ *is not a closed surface with exactly one puncture, then every two tagged triangulations of* $(\Sigma, \mathbb{M})$ *can be connected by a sequence of flips.*
(4) **[17]** *Every quiver* $Q(\tau)$ *with* τ *a tagged triangulation is isomorphic to a quiver of the form* $Q(\sigma)$ *with* σ *an ideal triangulation.*

2. The quiver with potential of a triangulation

The aim of this section is to give an overview of the definition and flip/mutation dynamics of the QPs associated in **[33]**, **[9]** and **[34]** to the triangulations of a surface with marked points $(\Sigma, \mathbb{M})$. To this end, we first describe the quivers associated to "nice" triangulations of $(\Sigma, \mathbb{M})$ (Subsection 2.1). The construction of the corresponding QPs is laid out in Subsection 2.2.

As a manner of warning: in the forthcoming exposition of constructions we have restricted our attention to "nice" triangulations (those satisfing (2.1) below), although it is possible to carry them out in full generality. From the cluster algebra perspective, it is necessary to consider the triangulations that are not "nice", and not only that, in order to obtain the fully general picture, it is necessary to introduce the concept of *tagged triangulation*, a notion of triangulation more general than the one considered here. We have chosen to leave tagged triangulations out of our exposition only in order to avoid going into some intricate technicalities. The reader is referred to **[17]** for the definition and combinatorics of these objects, and to **[34]** for the corresponding QP counterpart.

2.1. The quiver of a triangulation. Every ideal triangulation τ has a quiver $Q(\tau)$ associated in a natural way. This was first observed by Fock-Goncharov **[16]**, Fomin-Shapiro-Thurston **[17]** and Gekhtman-Shapiro-Vainshtein **[26]**. Let us describe $Q(\tau)$ under the assumption that

(2.1) every puncture of $(\Sigma, \mathbb{M})$ is incident to at least three arcs of τ.

The vertices of $Q(\tau)$ are the arcs of τ, the arrows are drawn in the clockwise direction within each triangle of τ. See Figure 4.

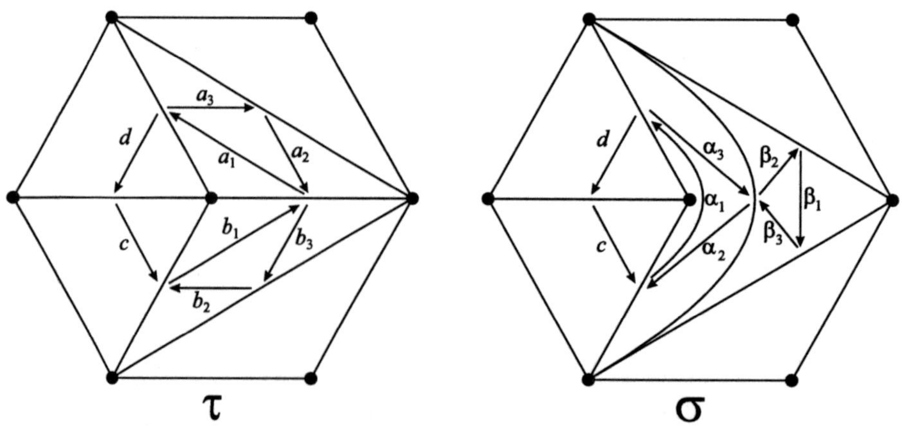

FIGURE 4. Two ideal triangulations related by a flip, with their associated quivers drawn on the surface.

If an ideal triangulation τ is such that

(2.2) there are punctures incident to less than three arcs of τ,

or more generally, if τ is a tagged triangulation, then the definition of $Q(\tau)$ is slightly more involved, but we stress that all triangulations, including the tagged ones, have naturally associated quivers. Let us also remark that the definition of the quivers of tagged triangulations is due to Fomin-Shapiro-Thurston alone.

THEOREM 2.1. [**16, 17, 26**] *Let τ and σ be ideal triangulations of $(\Sigma, \mathbb{M})$. If σ is obtained from τ by the flip of an arc i, then $Q(\sigma) = \mu_i(Q(\tau))$. That is, if two ideal triangulations are related by a flip, then their associated quivers are related by the corresponding quiver mutation.*

Thus, for example, the two quivers drawn in Figure 4 are related by quiver mutation. Fomin-Shapiro-Thurston have shown that Theorem 2.1 is valid in the more general setting of tagged triangulations.

2.2. The potential of a triangulation. We know that every triangulation has a quiver associated to it, and we know that flips of triangulations are compatible with mutations of quivers, in the sense that if two triangulations are related by a flip, then their associated quivers are related by the corresponding quiver mutation. Could this story be lifted to the level of QPs? To try and answer this question, we first notice that for each ideal triangulation τ of $(\Sigma, \mathbb{M})$ satisfying (2.1), the quiver $Q(\tau)$ possesses two obvious types of cycles:

- 3-cycles arising from triangles $\triangle$ of τ, and
- simple cycles (that is, without repeated arrows) surrounding the punctures.

To avoid redundancies, in the following definition we consider cycles up to cyclical, that is, we take only one cycle per cyclical equivalence class of cycles.

DEFINITION 2.2. [**33**] Let τ be a triangulation of $(\Sigma, \mathbb{M})$ satisfying (2.1). The potential $S(\tau)$ associated to τ is the potential on $Q(\tau)$ that results from adding all the 3-cycles that arise from triangles of τ and all the simple cycles that surround the punctures of $(\Sigma, \mathbb{M})$.

REMARK 2.3.
(1) In the case when $(\Sigma, \mathbb{M})$ is a surface without punctures and non-empty boundary, the potentials $S(\tau)$ were found independently by Assem-Brüstle-Charbonneau-Plamondon in [**5**].
(2) In [**33**], the definition of $S(\tau)$ was given for every ideal triangulation τ, including those satisfying (2.2).

EXAMPLE 2.4. The potentials associated to the ideal triangulations τ and σ shown in Figure 4 are $S(\tau) = a_1a_2a_3 + b_1b_2b_3 + a_1b_1cd$ and $S(\sigma) = \alpha_1\alpha_2\alpha_3 + \beta_1\beta_2\beta_3 + \alpha_1cd$. The QPs $(Q(\tau), S(\tau))$ and $(Q(\sigma), S(\sigma))$ turn out to be related by QP-mutation. This can be checked directly, or seen as a consequence of the following theorem.

THEOREM 2.5. [**33**] *Let τ and σ be ideal triangulations of a surface with marked points $(\Sigma, \mathbb{M})$.*

(1) *If τ and σ are related by the flip of an arc i, then the QPs $\mu_i(Q(\tau), S(\tau))$ and $(Q(\tau), S(\tau))$ are right-equivalent. In other words, $(Q(\tau), S(\tau))$ and $(Q(\sigma), S(\sigma))$ are related by the QP-mutation μ_i.*

(2) *If the boundary of* Σ *is not empty, then all QPs* $(Q(\tau), S(\tau))$ *associated to the ideal triangulations of* $(\Sigma, \mathbb{M})$ *are non-degenerate.*

ON THE PROOF. Note that if τ and σ are ideal triangulations related by the flip of an arc $i \in \tau$, then i cannot be the folded side of a self-folded triangle of τ.

Fomin-Shapiro-Thurston introduce three different kinds of *puzzle pieces* –planar unpunctured triangles, planar punctured digons triangulated in a very specific way, and planar twice-punctured monogons also triangulated in a specific way. These puzzle pieces come in handy because any given ideal triangulation of an arbitrary surface $(\Sigma, \mathbb{M})$ can be obtained by gluing some set of puzzle pieces.

Given a *puzzle-piece decomposition* of an ideal triangulation τ, any flip of an arc of τ occurs either inside a puzzle piece, or at the arc shared by two puzzle pieces that are glued together. These possibilities for a flip comprise a basic list of cases to be considered in the proof of Part (1) of Theorem 2.5, although a couple of slight subtleties have to be considered. The first subtlety concerns the fact that a puzzle-piece decomposition gives not only a gluing of pairs of sides of puzzle pieces, but in many cases also an identification of different vertices of puzzle pieces as the same marked point in $(\Sigma, \mathbb{M})$. Such identification affects how the potential $S(\tau)$ looks like locally, in that the local configuration of $S(\tau)$ around a pair of glued puzzle pieces is not determined only by the gluing of the side(s) shared by the puzzle pieces, but depends also on how the vertices of these are identified to obtain τ. The second subtlety concerns the fact that some gluing of puzzle pieces yield some 2-cycles that "are not seen" by the quiver $Q(\tau)$.

The two mentioned subtleties make it necessary to refine the basic list of cases to be checked for the proof of Part (1) of Theorem 2.5. After such refinement, one has a finite (albeit larger) list of cases. Note that, by what we said in the first paragraph, none of these cases involves the flip of a folded side. In the cases that are indeed considered, one reads the QPs $(Q(\tau), S(\tau))$ and $(Q(\sigma), S(\sigma))$ on the one hand, and the mutation $\mu_i(Q(\tau), S(\tau))$ on the other. Then an explicit right-equivalence between $\mu_i(Q(\tau), S(\tau))$ and $(Q(\tau), S(\tau))$ is given.

The proof of part (2) is done by induction on the number of punctures of $(\Sigma, \mathbb{M})$, using the non-emptiness of the boundary in an essential way. Given an unpunctured surface $(\Sigma, \mathbb{M}_0)$ with non-empty boundary, it is straightforward to see that any ideal triangulation τ_0 of $(\Sigma, \mathbb{M}_0)$ has the property that its QP $(Q(\tau_0), S(\tau_0))$ is *rigid.* Then we add punctures to $(\Sigma, \mathbb{M}_0)$ one by one, thus obtaining an n-punctured surface $(\Sigma, \mathbb{M}_n) = (\Sigma, \mathbb{M}_0 \cup \{p_1, \dots, p_n\})$ for each $n \geq 1$. Every time we add a puncture we also complete the ideal triangulation τ_{n-1} of $(\Sigma, \mathbb{M}_{n-1})$ to an ideal triangulation τ_n of $(\Sigma, \mathbb{M}_n)$ in a very specific. We then show the rigidity of $(Q(\tau_n), S(\tau_n))$ under the assumption that $(Q(\tau_{n-1}), S(\tau_{n-1}))$ is rigid. By Part (1) of Theorem 2.5, this establishes the rigidity of the QP associated to any ideal triangulation of $(\Sigma, \mathbb{M}_n)$, since rigidity is preserved by QP-mutations thanks to a result of Derksen-Weyman-Zelevinsky. And since another result of Derksen-Weyman-Zelevinsky shows that rigid QPs are non-degenerate, Part (2) of Theorem 2.5 follows. □

Potentials corresponding to arbitrary tagged triangulations were not even mentioned in [**33**], despite the fact that it is possible to read an 'obvious' potential from any tagged triangulation. The reason is that, at that moment, a "tagged version" of Part (1) of Theorem 2.5 had not been established: it was not clear whether the 'obvious' potentials would have the property that arbitrary tagged triangulations related by a flip would have QPs related by the corresponding QP-mutation.

Among other things, a "tagged version" of Part (1) of Theorem 2.5 would allow us to deduce a fact that is not an immediate consequence of Theorem 2.5, namely, that the QPs $(Q(\tau), S(\tau))$ associated to triangulations of surfaces with empty boundary are non-degenerate.

Potentials for arbitrary tagged triangulations were defined in **[9]** under the assumption that the underlying surface Σ has non-empty boundary; but even with this assumption, the corresponding "tagged version" of Theorem 2.5 was not proved for all flips of tagged triangulations, but only for some of them. Indeed, it was proved in **[9]** that for every two tagged triangulations τ and σ of a surface with non-empty boundary, there exists a sequence $(\tau = \tau_0, \tau_1, \tau_2, \ldots, \tau_{n-1}, \tau_n = \sigma)$ of tagged triangulations with the property that each τ_ℓ is obtained from $\tau_{\ell-1}$ by the flip of a tagged arc $i_\ell \in \tau_{\ell-1}$ and the QP $\mu_{i_\ell}(Q(\tau), S(\tau))$ is right-equivalent to the QP $(Q(\tau), S(\tau))$. However, for two arbitrary tagged triangulations τ and σ related by the flip of a tagged arc i, it was not proved that the sequence of flips just described can always be taken to be the sequence $(\tau = \tau_0, \tau_1 = \sigma)$.

The referred assumption on the boundary was removed in **[34]**, where potentials were defined for all tagged triangulations of surfaces, including both the surfaces with boundary and the surfaces without boundary, and the "tagged version" of Theorem 2.5 was proved for all flips of tagged triangulations:

THEOREM 2.6. *Let $(\Sigma, \mathbb{M})$ be a surface with marked points. Suppose $(\Sigma, \mathbb{M})$ is not a sphere with less than five punctures. If τ and σ are tagged triangulations of $(\Sigma, \mathbb{M})$ related by the flip of a tagged arc i, then*

1. **[34]** *the QPs $\mu_i(Q(\tau), S(\tau))$ and $(Q(\sigma), S(\sigma))$ are right-equivalent if $(\Sigma, \mathbb{M})$ is not a sphere with exactly five punctures;*
2. **[23]** *the QP $\mu_i(Q(\tau), S(\tau))$ is right-equivalent to $(Q(\sigma), \lambda S(\sigma))$ for some non-zero scalar λ if $(\Sigma, \mathbb{M})$ is a sphere with exactly five punctures.*

Consequently, all QPs $(Q(\tau), S(\tau))$ associated to the tagged triangulations of $(\Sigma, \mathbb{M})$ are non-degenerate.

ON THE PROOF. The proofs of (1) and (2) differ at a crucial point (note that (2) is weaker than (1)). We only sketch the proof of (1). Every tagged triangulation τ gives rise to a function $\epsilon_\tau : \mathbb{P} \to \{-1, 1\}$, called *weak signature of* τ, that takes the value -1 at a puncture p if and only if at least two notches of tagged arcs in τ are incident to p.

It is easy to deduce from Theorem 2.5 that if τ and σ are tagged triangulations whose weak signatures ϵ_τ and ϵ_σ are equal, and τ and σ are related by the flip of a tagged arc i, then the QPs $\mu_i(Q(\tau), S(\tau))$ and $(Q(\sigma), S(\sigma))$ are right-equivalent. This reduces the proof of Theorem 2.6 to the case of tagged triangulations τ and σ related by a flip, but with different weak signatures ϵ_τ and ϵ_σ. A moment of reflection shows that in this case ϵ_τ and ϵ_σ differ at exactly one puncture, say p.

One can assume, without loss of generality, that $\varepsilon_\tau(p) = 1 = -\varepsilon_\sigma(p)$ (this is because every flip is an involution and every QP-mutation is an involution up to right-equivalence). Applying a combinatorial procedure of "deletion of notches", one can further assume that τ is an ideal triangulation. The proof of Theorem 2.6 is hence reduced to showing that if σ is the tagged triangulation obtained by flipping the folded side i of an arbitrary self-folded triangle of an ideal triangulation τ, then $\mu_i(Q(\tau), S(\tau))$ is right-equivalent to $(Q(\sigma), S(\sigma))$. The proof of this last implication is rather involved and relies crucially on **[34**, Theorem 6.1**]**, a technical result that guarantees the existence of a right-equivalence between $(Q(\tau), S(\tau))$ and $(Q(\tau), W(\tau))$ for certain specific potential $W(\tau)$ that, on the other hand, has the property that $(Q(\tau), W(\tau))$ can be easily seen to be right-equivalent to $\mu_i(Q(\sigma), S(\sigma))$. □

REMARK 2.7. The crucial right-equivalence in the proof of Theorem 2.6, namely, the right-equivalence between $(Q(\tau), S(\tau))$ and $(Q(\tau), W(\tau))$ above, is not exhibited explicitly: for one triangulation it is constructed as the limit of certain algebra automorphisms of $R\langle\langle Q(\tau)\rangle\rangle$, and for the other triangulations it is only shown to exist. This is very unlike the proof of Part (1) of Theorem 2.5, where, despite the division into cases, all right-equivalences are defined explicitly.

REMARK 2.8. The sphere with four punctures has been dealt with in **[22]** and **[23]**. We stress the fact that, in order to obtain a non-degenerate potential under Definition 2.2, it is strictly necessary to multiply exactly one of the cycles around punctures by a scalar $\lambda \in \mathbb{C} \setminus \{0, 1\}$ (the rest of the cycles are still taken as are, that is, multiplied by 1).

3. Dimension and representation type of Jacobian algebras

From the perspective of representation theory of associative algebras, there are several natural questions one can ask regarding the Jacobian algebras of the QPs $(Q(\tau), S(\tau))$. For example: are they finite-dimensional? are they tame/wild?

THEOREM 3.1. **[33]** *Let* $(\Sigma, \mathbb{M})$ *be a surface with non-empty boundary (and any number of punctures). Then for any ideal triangulation* τ *of* $(\Sigma, \mathbb{M})$*, the Jacobian algebra* $\mathcal{P}(Q(\tau), S(\tau))$ *has finite dimension over* $\mathbb{C}$.

THEOREM 3.2. **[36]** *Let* $(\Sigma, \mathbb{M})$ *be a surface with empty boundary. Suppose that* $(\Sigma, \mathbb{M})$ *is not a sphere with less than 5 punctures. Then for any ideal triangulation* τ *of* $(\Sigma, \mathbb{M})$*, the Jacobian algebra* $\mathcal{P}(Q(\tau), S(\tau))$ *has finite dimension over* $\mathbb{C}$.

REMARK 3.3.
(1) Theorem 3.1 can be either proved independently of Theorem 3.2, or deduced from it via *restriction of QPs*.
(2) In the case of unpunctured surfaces with non-empty boundary, Theorem 3.1 was proved by Assem-Brüstle-Charbonneau-Plamondon in **[5]** independently of **[33]** and **[36]**.
(3) For polygons with at most one puncture, the finite-dimensionality of the Jacobian algebras $\mathcal{P}(Q(\tau), S(\tau))$ was already known to Caldero-Chapoton-Schiffler **[7]** and Schiffler **[41]**, although in the referred papers the authors did not work with complete path algebras or potentials.
(4) Theorem 3.2 is due to Ladkani. In the case of spheres with at least five punctures, it was proved independently by Trepode–Valdivieso-Díaz in **[43]**.
(5) For the sphere with exactly four punctures, the finite-dimensionality of Jacobian algebras of non-degenerate potentials follows from **[22]** (where these algebras are shown to be tubular) and **[23]**. See Remark 2.8 above.

Let us turn to the problem of whether the Jacobian algebras $\mathcal{P}(Q(\tau), S(\tau))$ are tame or wild.

DEFINITION 3.4. Let Λ be a finite-dimensional associative $\mathbb{C}$-algebra.

(1) We say that Λ is *tame* if for each dimension vector $\mathbf{d}$ there are finitely many Λ-$\mathbb{C}[X]$-bimodules $M_1, \dots, M_t$, free of finite rank as right $\mathbb{C}[X]$-modules, such that every indecomposable Λ-module N with $\dim(N) = \mathbf{d}$ is isomorphic to a Λ-module of the form

$$M_i \otimes_{\mathbb{C}[X]} (\mathbb{C}[X]/(X - \lambda))$$

for some $1 \leq i \leq t$ and some $\lambda \in \mathbb{C}$.

(2) We say that Λ is *wild* if there is a Λ-$\mathbb{C}\langle X, Y\rangle$-bimodule M, free of finite rank as right $\mathbb{C}\langle X, Y\rangle$-module, such that the exact functor

$$M \otimes_{\mathbb{C}\langle X,Y\rangle} - \colon \operatorname{mod}(\mathbb{C}\langle X, Y\rangle) \to \operatorname{mod}(\Lambda)$$

sends indecomposable modules to indecomposable ones and pairwise non-isomorphic modules to pairwise non-isomorphic ones. Here, $\mathbb{C}\langle X, Y\rangle$ denotes the free $\mathbb{C}$-algebra in two (non-commuting) generators X and Y.

A famous result of Drozd asserts that any given finite-dimensional $\mathbb{C}$-algebra is either tame or wild, and not simultaneously tame and wild. This is Drozd's *tame/wild dichotomy theorem.*

In the case when $(\Sigma, \mathbb{M})$ is a surface with non-empty boundary and without punctures, Assem-Brüstle-Charbonneau-Plamondon have shown in **[5]** that $\mathcal{P}(Q(\tau), S(\tau))$ is a *gentle algebra*, and this implies its tameness, for gentle algebras are well-known to be tame. More generally, we have:

THEOREM 3.5. **[23]**

(1) *For any QP (Q, S), if $\mathcal{P}(Q, S)$ is tame, then $\mathcal{P}(\mu_i(Q, S))$ is tame as well.*
(2) *Any surface with marked points has a triangulation τ such that the Jacobian algebra $\mathcal{P}(Q(\tau), S(\tau))$ is tame.*
(3) *Consequently, for every surface $(\Sigma, \mathbb{M})$ and every tagged triangulation τ of $(\Sigma, \mathbb{M})$, the Jacobian algebra $\mathcal{P}(Q(\tau), S(\tau))$ is tame.*

ON THE PROOF. The proof of Part (1) of Theorem 3.5 presents a challenge: Quotients of finite-rank free $\mathbb{C}[X]$-modules are not always free $\mathbb{C}[X]$-modules, not every inclusion of $\mathbb{C}[X]$-modules is a section, and not every surjective morphism of $\mathbb{C}[X]$-modules is a retraction. So, one cannot define the mutation of $\mathcal{P}(Q, S)$-$\mathbb{C}[X]$-bimodules in the "obvious" way, since the mutation process involves taking certain cokernel, as well as a section and a retraction. What Geiss-LF-Schröer end up doing in **[23]** is the following: For each localization R of $\mathbb{C}[X]$, they define the mutation of a $\mathcal{P}(Q, S)$-R-bimodule M (assumed to be a finite-rank free right R-module), as a $\mathcal{P}(\mu_i(Q, S))$-R'-bimodule M' for some other localization R' of $\mathbb{C}[X]$ that depends on M (as a right R'-module, M' turns out to be finite-rank free). With such definition, to show Part (1) of Theorem 3.5 they then use a characterization of tameness given by Dowbor-Skowroński **[13]** in terms of localizations of the polynomial ring $\mathbb{C}[X]$.

If $(\Sigma, \mathbb{M})$ is a surface with non-empty boundary (and any number of punctures), different from a (punctured) monogon, then there exists a triangulation τ of $(\Sigma, \mathbb{M})$ such that the Jacobian algebra $\mathcal{P}(Q(\tau), S(\tau))$ is a *clannish algebra*. In the case of (punctured) monogons, one can always find a triangulation such that the Jacobian algebra is a *deformation* of a *skewed-gentle algebra*. If, on the other hand, the boundary of $(\Sigma, \mathbb{M})$ is empty, then there exists a triangulation τ of $(\Sigma, \mathbb{M})$ such that the Jacobian algebra $\mathcal{P}(Q(\tau), S(\tau))$ is a deformation of a gentle algebra. In each of the three situations just described, it is easy to find an explicit triangulation τ with the stated property.

Part (3) is an immediate consequence of Parts (1) and (2). □

Theorem 3.5 tells us that for every QP of the form $(Q(\tau), S(\tau))$ the associated Jacobian algebra is tame. Something stronger is true: except for a couple of surfaces, for any non-degenerate potential on the quiver of a triangulation the Jacobian algebra is tame. To be precise:

THEOREM 3.6. **[23]** *Let $(\Sigma, \mathbb{M})$ be a surface with marked points and let τ be a triangulation of $(\Sigma, \mathbb{M})$.*

(1) *If $(\Sigma, \mathbb{M})$ is not a torus with exactly one marked point, then for any non-degenerate potential S on $Q(\tau)$, the Jacobian algebra $\mathcal{P}(Q(\tau), S)$ is tame.*
(2) *If $(\Sigma, \mathbb{M})$ is a torus with exactly one marked point (hence Σ has either empty boundary or exactly one boundary component), then $Q(\tau)$ admits a non-degenerate potential W such that $\mathcal{P}(Q(\tau), W)$ is a wild algebra.*

ON THE PROOF. To prove (1), in **[23]** we show that if $(\Sigma, \mathbb{M})$ is not a torus with exactly one marked point, nor a sphere with less than 6 punctures, nor an annulus with exactly two marked points, then $(\Sigma, \mathbb{M})$ admits a triangulation σ such that every puncture is incident to at least 4 arcs of σ, and moreover, the quiver $Q(\sigma)$ does not have double arrows. These properties of σ imply that

- any non-degenerate potential on $Q(\sigma)$ is right-equivalent to $S' + S''$ for some potential S'' involving only cycles of length at least 4, where S' is the sum of all 3-cycles of $Q(\sigma)$;
- the Jacobian algebra $\mathcal{P}(Q(\sigma), S')$ is gentle (albeit possibly infinite-dimensional).
- (each *truncation* of) the Jacobian algebra $\mathcal{P}(Q(\sigma), S' + S'')$ is a deformation of (the corresponding truncation) of $\mathcal{P}(Q(\sigma), S')$.

Using a theorem of Crawley-Boevey **[10]**, this allows to deduce the tameness of $\mathcal{P}(Q(\sigma), S' + S'')$. Part (1) of Theorem 3.6, combined with Parts (2), (3) and (4) of Theorem 1.18, then implies that for the arbitrarily given triangulation τ one has that for any non-degenerate potential S on $Q(\tau)$, the algebra $\mathcal{P}(Q(\tau), S)$ is tame.

The spheres with less than 6 punctures and the annulus with exactly two marked points are treated separately.

For Part (2), consider the quivers

$T_1 =$ 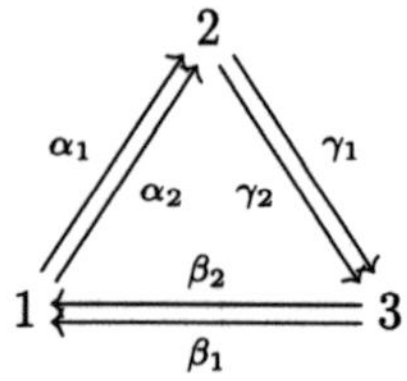

and $T_2 =$ 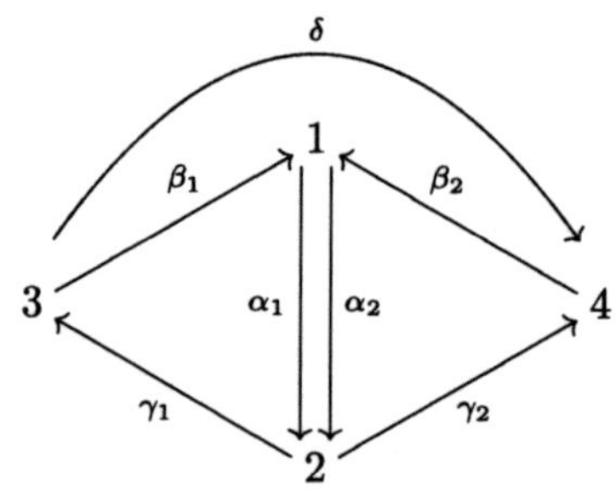

It is straightforward to see that if $(\Sigma, \mathbb{M})$ is a torus with exactly one marked point and τ is any triangulation of $(\Sigma, \mathbb{M})$, then $Q(\tau)$ is isomorphic to T_1 if $\partial\Sigma = \varnothing$, and to T_2 if $\partial\Sigma \neq \varnothing$. It is also easy to see that for any vertex i of T_ℓ ($\ell = 1, 2$), the quiver $\mu_i(T_\ell)$ is isomorphic to T_ℓ via a quiver isomorphism $\pi_i : T_\ell \to \mu_i(T_\ell)$ that acts as a uniquely determined permutation on the (common) vertex set (the quiver isomorphism is unique for $\ell = 2$, and there is a little choice involved at the arrow level for $\ell = 1$). Let

$$W_1 = \alpha_1\beta_1\gamma_2 + \alpha_1\beta_2\gamma_1 + \alpha_2\beta_1\gamma_1 \in R\langle\langle T_1\rangle\rangle \tag{3.1}$$

and

$$W_2 = \alpha_1\beta_1\gamma_1 + \alpha_1\beta_2\gamma_2 + \alpha_2\beta_2\delta\gamma_1 \in R\langle\langle T_2\rangle\rangle. \tag{3.2}$$

Direct computation shows that for any vertex i of T_ℓ, the image of W_ℓ under the isomorphism π_i is precisely the potential of the QP-mutation $\mu_i(T_\ell, W_\ell)$. This readily implies the non-degeneracy of (T_ℓ, W_ℓ). Using techniques of Galois coverings, it is shown in **[23]** that the Jacobian algebra $\mathcal{P}(T_\ell, W_\ell)$ is wild for $\ell = 1, 2$. □

The next result says that the quivers of the form $Q(\tau)$ are pretty much the only quivers for which we can find non-degenerate potentials with tame Jacobian algebras. That is, if we take an arbitrary quiver Q, not necessarily arising from a triangulation of a surface, such that $\mathcal{P}(Q, S)$ is a tame algebra for some non-degenerate potential $S \in R\langle\langle Q\rangle\rangle$, then we can be almost certain that Q arises from a triangulation of a surface with marked points. Here is the precise statement:

THEOREM 3.7. **[23]** *Let Q be any 2-acyclic quiver. If there exists a non-degenerate potential $S \in R\langle\langle Q\rangle\rangle$ such that the Jacobian algebra $\mathcal{P}(Q, S)$ is tame, then either Q is the quiver associated to a triangulation of a surface with marked points, or Q belongs to the quiver mutation class of one of the following nine exceptional quivers: E_6, E_7, E_8, $\widetilde{E}_6$, $\widetilde{E}_7$, $\widetilde{E}_8$, $E_6^{(1,1)}$, $E_7^{(1,1)}$, $E_8^{(1,1)}$.*

ON THE PROOF. To prove Theorem 3.7 we make use of Theorem 3.5 Part (1), Drozd's famous tame/wild dichotomy theorem **[14]**, and Felikson-Shapiro-Tumarkin's crucial classification of the quiver whose quiver mutation class is finite **[15]**. □

REMARK 3.8. The quivers E_6, E_7, E_8, $\widetilde{E}_6$, $\widetilde{E}_7$, $\widetilde{E}_8$, $E_6^{(1,1)}$, $E_7^{(1,1)}$ and $E_8^{(1,1)}$ are not very complicated and can be found in **[15]** or **[23]**.

4. Uniqueness of non-degenerate potentials

For quivers of the form $Q(\tau)$ with τ a (tagged) triangulation of $(\Sigma, \mathbb{M})$, **[23]** classifies all non-degenerate potentials on $Q(\tau)$ in practically all cases. It turns out that almost all quivers of the form $Q(\tau)$ admit exactly one non-degenerate potential. We give the precise statements in Theorems 4.1 and 4.2 below.

THEOREM 4.1. **[23]** *Let $(\Sigma, \mathbb{M})$ be a surface with non-empty boundary and any number of punctures, and let τ be any tagged triangulation of $(\Sigma, \mathbb{M})$.*

1. *If $(\Sigma, \mathbb{M})$ is not the unpunctured torus with exactly one marked point, then, up to right-equivalence, $S(\tau)$ is the only non-degenerate potential on $Q(\tau)$.*
2. *If $(\Sigma, \mathbb{M})$ is the unpunctured torus with exactly one marked point, then $Q(\tau)$ admits exactly two non-degenerate potentials up to right-equivalence, namely $S(\tau)$ and another potential W such that $\mathcal{P}(Q(\tau), W)$ is wild (W is the potential referred to in Theorem 3.6).*

ON THE PROOF. Given an arbitrary 2-acyclic quiver Q, [**23**, Theorem 8.19] gives a sufficiency criterion for a potential on Q to be the only non-degenerate potential on Q. For $(\Sigma, \mathbb{M})$ with non-empty boundary and different from the unpunctured torus with exactly one marked point, in the proof of [**23**, Theorem 8.20] a triangulation σ is constructed whose associated potential $S(\sigma)$ satisfies the hypothesis of the alluded criterion. Part (1) then follows from Theorem 2.6 and the fact that right-equivalent QPs have right-equivalent QP-mutations (the latter is a result of Derksen-Weyman-Zelevinsky).

In the case of the unpunctured torus with exactly one marked point, we already know that $(Q(\tau), S(\tau))$ and $(Q(\tau), W)$ are non-degenerate, where $W = W_2$ is the potential given in 3.2. That they are not right-equivalent can be proved either directly, or by means of the following argument: We know that $\mathcal{P}(Q(\tau), S(\tau))$ is tame and $\mathcal{P}(Q(\tau), W)$ is wild, and a result of Derksen-Weyman-Zelevinsky tells us that right-equivalent QPs have isomorphic Jacobian algebras; thus $(Q(\tau), S(\tau))$ and $(Q(\tau), W)$ cannot be right-equivalent by Drozd's tame/wild dichotomy theorem.

Given an arbitrary 2-acyclic quiver Q, [**23**, Lemma 8.18] gives a sufficiency criterion for a potential $S \in R\langle\langle Q\rangle\rangle$ and a positive integer m to satisfy the property that for every potential $S' \in \mathfrak{m}^{m+1}$ the QP $(Q, S+S')$ be right-equivalent to (Q, S). Direct computation shows that $S(\tau)$ satisfies the criterion with $m = 3$, and $W = W_2$ satisfies the criterion with $m = 4$. Finally, using basic linear algebra, it is shown that every non-degenerate potential on $Q(\tau)$ is right-equivalent to $S(\tau) + S'$ or to $W + S'$ for some potential $S' \in \mathfrak{m}^5$. Part (2) follows. □

THEOREM 4.2. *Let $(\Sigma, \mathbb{M})$ be a surface with empty boundary, and let τ be any tagged triangulation of $(\Sigma, \mathbb{M})$.*

(1) [**23**] *If the genus of Σ is positive and the number $|\mathbb{M}|$ of punctures is at least three, then any non-degenerate potential on $Q(\tau)$ is right-equivalent to a non-zero scalar multiple of $S(\tau)$.*
(2) [**35**] *If $(\Sigma, \mathbb{M})$ is a positive-genus surface with exactly one puncture, then the degree-3 component of $S(\tau)$ (that is, the part of $S(\tau)$ that arises from the triangles of τ) is a non-degenerate potential which is not right-equivalent to any scalar multiple of $S(\tau)$.*
(3) [**23**] *If $(\Sigma, \mathbb{M})$ is a sphere with at least five punctures, then any non-degenerate potential on $Q(\tau)$ is right-equivalent to a non-zero scalar multiple of $S(\tau)$.*

ON THE PROOF. Part (1) of Theorem 4.2 follows from a combination of several facts (the proofs of some of which are rather technical):

(a) Any surface as in (1) admits a triangulation σ satisfying that no arc in σ is a loop, that $Q(\sigma)$ does not have double arrows[2], and that each puncture is an endpoint of at least four arcs of σ.
(b) For any triangulation σ as in (a), any potential S' not involving any cycle cyclically-equivalent to a term appearing $S(\sigma)$, and any collection $\mathbf{x} = (x_p)_{p \in \mathbb{P}}$ of non-zero scalars indexed by the punctures of $(\Sigma, \mathbb{M})$, the QP $(Q(\sigma), S(\sigma, \mathbf{x}) + S')$ is right-equivalent to $(Q(\sigma), S(\sigma, \mathbf{x}))$, where $S(\sigma, \mathbf{x})$ is the potential obtained from $S(\sigma)$ by multiplying each cycle surrounding a puncture p by x_p. (The proof of this fact is rather technical and somewhat delicate: the right-equivalence between $(Q(\sigma), S(\sigma, \mathbf{x}) + S')$ and

[2]We believe that the absence of double arrows can be deduced from the absence of loops

$(Q(\sigma), S(\sigma, \mathbf{x}))$ given in **[23]** is defined as the composition of three algebra automorphisms of $R\langle\langle Q(\sigma)\rangle\rangle$ that are not given explicitly, but rather obtained as limits of certain sequences of automorphisms. The convergence of these sequences is a delicate issue.)

(c) For any triangulation σ as in (a), every non-degenerate potential on $Q(\sigma)$ is right-equivalent to $S(\sigma, \mathbf{x}) + S'$ for some S' and $\mathbf{x} = (x_p)_{p \in \mathbb{P}}$ as in (b).

(d) For any triangulation as in (a) and any collection $\mathbf{x} = (x_p)_{p \in \mathbb{P}}$ as in (b), the QP $(Q(\sigma), S(\sigma, \mathbf{x}))$ is right-equivalent to $(Q(\sigma), \lambda S(\sigma))$ for some non-zero scalar $\lambda \in \mathbb{C}$. (For this we use the fact that $\mathbb{C}$ is algebraically closed: there is an n for which the ground field must have n^{th} roots of all its elements.)

(e) For arbitrary potentials W and W' on a given quiver, and any non-zero scalar λ, if (Q, W) is right-equivalent to $(Q, \lambda W')$, then the QP-mutation $\mu_i(Q, W)$ is right-equivalent to $(\overline{Q}, \lambda\overline{W'})$, where $(\overline{Q}, \overline{W'}) = \mu_i(Q, W')$.

For the proof of Part (2) of Theorem 4.2, Ladkani notes that, for once-punctured surfaces with empty boundary and positive genus, the proof of the first statement of Theorem 2.5, given in **[33]** for the QPs of the form $(Q(\tau), S(\tau))$ with τ ideal triangulation, can be easily adapted to show the following: If τ and σ are ideal triangulations of a once-punctured surface with empty boundary and positive genus, and τ and σ are related by the flip of an arc i, then $\mu_i(Q(\tau), S(\tau)^{(3)})$ is right-equivalent to $(Q(\sigma), S(\sigma)^{(3)})$, where $S(\tau)^{(3)}$ (resp. $S(\sigma)^{(3)}$) denotes the degree-3 component of $S(\tau)$ (resp. $S(\sigma)$). This fact has the non-degeneracy of $(Q(\tau), S(\tau)^{(3)})$ as a straightforward consequence. That $(Q(\tau), S(\tau)^{(3)})$ is not right-equivalent to $(Q(\tau), \lambda S(\tau))$ for any $\lambda \in \mathbb{C}^*$ follows from the fact that the former QP has infinite-dimensional Jacobian algebra, whereas, due to Ladkani's result Theorem 3.2, the latter QP has finite-dimensional Jacobian algebra.

The proof of part (3) is identical to the proof of part (1) if $(\Sigma, \mathbb{M})$ is a sphere with at least six punctures. The sphere with five punctures is treated separately, but with an argument which is similar to the proof part (1). □

For positive-genus surfaces with empty boundary and exactly two punctures we have the following:

CONJECTURE 4.3. **[23]** *If τ is a tagged triangulation of a positive-genus surface $(\Sigma, \mathbb{M})$ with empty boundary and exactly two punctures, then any non-degenerate potential on $Q(\tau)$ is right-equivalent to a non-zero scalar multiple of $S(\tau)$.*

The proof we have sketched of part (1) of Theorem 4.2 cannot be applied to prove this conjecture, since the surfaces in the conjecture do not have triangulations without loops.

For the once-punctured torus, we have the following result by Geuenich:

THEOREM 4.4. **[27]** *Let τ be a triangulation of the once-punctured torus. Then:*

(1) *An arbitrary potential $S \in Q(\tau)$ is non-degenerate if and only if, up to a change of arrows, the degree-3 component of S is equal to either $a_1b_1c_1 + a_2b_2c_2$ or $a_1b_1c_2 + a_1b_2c_1 + a_2b_1c_1$.*

(2) *There exists an infinite sequence $(S_n)_{n\geq 0}$ of non-degenerate potentials on $Q(\tau)$, with the property that $\lim_{n\to\infty} \dim_{\mathbb{C}}(\mathcal{P}(Q(\tau), S_n)) = \infty$. Hence this sequence has a subsequence $(S_{n_m})_{m\geq 0}$ such that for $m_1 \neq m_2$, the QP*

$(Q(\tau), S_{n_{m_1}})$ is not right-equivalent to $(Q(\tau), \lambda S_{n_{m_2}})$ for any non-zero scalar λ.

We expect that Part (2) of Theorem 4.4 can be generalized to any once-punctured surface without boundary.

5. Some applications

Quivers with potentials can be thought of as a tool to obtain categories. Besides the module categories of Jacobian algebras, there are other categories associated to QPs that are of interest not only to representation-theorists, but to authors from other areas as well. We give a very rough description of a couple of these categories, the reader is referred to [**3**] and [**4**, Section 3] for precise definitions, statements and citations.

Let (Q, S) be a non-degenerate QP, and let $\overline{Q}$ be the graded quiver whose vertex set is Q_0 and whose arrows are:

In degree 0: All arrows of Q;
In degree -1: an arrow $a^* : i \to j$ for each arrow $a : j \to i$ of Q;
In degree -2: an arrow $t_i : i \to i$ for each vertex $i \in Q_0$.

Every path on $\overline{Q}$ has non-positive degree (defined as the sum of the degrees of its constituent arrows), and for $\ell \in \mathbb{Z}$ an element $u \in R\langle\langle\overline{Q}\rangle\rangle$ is said to have degree ℓ if all its constituent paths have degree ℓ. Letting $\widehat{\Gamma}^{(\ell)}$ denote the set of all degree-ℓ elements of $R\langle\langle\overline{Q}\rangle\rangle$, we see that $\widehat{\Gamma}^{(\ell)}$ consists of all possibly infinite $\mathbb{C}$-linear combinations of paths of degree ℓ, that

$$\widehat{\Gamma}(Q, S) := \bigoplus_{\ell \in \mathbb{Z}} \widehat{\Gamma}^{(\ell)} \tag{5.1}$$

is a subalgebra of $R\langle\langle\overline{Q}\rangle\rangle$, and that $\widehat{\Gamma}^{(0)} = R\langle\langle Q\rangle\rangle$. There is a differential d on $\widehat{\Gamma}(Q, S)$ that can be defined as the unique degree-1 continuous $\mathbb{C}$-linear map that satisfies the rules

$$d(a) = 0 \quad \text{and} \quad d(a^*) = \partial_a(S) \quad \text{for all } a \in Q_1,$$

$$d(t_i) = e_i \left(\sum_{a \in Q_1} [a, a^*] \right) e_i \quad \text{for all } i \in Q_0,$$

together with the Leibniz rule $d(uv) = (du)v + (-1)^{\ell} u dv$ for all homogeneous $u \in \widehat{\Gamma}^{(\ell)}$ and v. The algebra $\widehat{\Gamma}(Q, S)$, together with the grading (5.1) and the differential d above, is called the *complete Ginzburg DG algebra*[3] of (Q, S).

Via the *derived category* of the category of DG modules over $\widehat{\Gamma}(Q, S)$, one arrives at a *3-Calabi-Yau triangulated* category $\mathcal{D}\widehat{\Gamma}(Q, S)$. When the Jacobian algebra $\mathcal{P}(Q, S)$ is finite-dimensional, Amiot defines the *generalized cluster category* $\mathcal{C}(Q, S)$ as certain quotient of a subcategory of $\mathcal{D}\widehat{\Gamma}(Q, S)$. The generalized cluster category turns out to be Hom-*finite 2-Calabi-Yau triangulated.* Amiot and Keller-Yang show that QP-mutations induce equivalences of categories:

THEOREM 5.1. *Let (Q, S) be a non-degenerate QP and $i \in Q_0$. Then*

(1) *(Q, S) induces a canonical* t-structure *on $\mathcal{D}\widehat{\Gamma}(Q, S)$ whose heart $\mathcal{A}_{(Q,S)}$ is equivalent to the module category of $\mathcal{P}(Q, S)$;*

[3] *DG* for *differential graded*

(2) [**28**] *there are two canonical equivalences of triangulated categories* $\Phi_i^+, \Phi_i^- : \mathcal{D}\widehat{\Gamma}(\mu_i(Q,S)) \to \mathcal{D}\widehat{\Gamma}(Q,S)$, *such that the images of the* t-*structure induced by* $\mu_i(Q,S)$ *on* $\mathcal{D}\widehat{\Gamma}(\mu_i(Q,S))$ *are the* right *and* left tilts *of the* t-*structure induced by* (Q,S) *on* $\mathcal{D}\widehat{\Gamma}(Q,S)$;
(3) [**3,4**] *there is a* cluster-tilting *object* $T_{(Q,S)}$ *of* $\mathcal{C}(Q,S)$ *canonically attached to* (Q,S);
(4) [**3,4,28**] $\mathcal{C}(\mu_i(Q,S))$ *is equivalent to* $\mathcal{C}(Q,S)$ *by means of an equivalence of triangulated categories that sends* $T_{\mu_i(Q,S)}$ *to the cluster-tilting object of* $\mathcal{C}(Q,S)$ *obtained from* $T_{(Q,S)}$ *by* IY-mutation[4] *with respect to* i.

The hearts of $\mathcal{D}\widehat{\Gamma}(Q,S)$ that can be obtained from $\mathcal{A}_{(Q,S)}$ by finite sequences of tilts are often called *canonical hearts*, whereas the cluster-tilting objects of $\mathcal{C}_{(Q,S)}$ that can be obtained from $T_{(Q,S)}$ by finite-sequences of IY-mutations are called *reachable* cluster-tilting objects.

REMARK 5.2. Since mutations of QPs are only defined up to right-equivalence, Parts (2) and (4) of Theorem 5.1 make an implicit use of the following fact: up to an equivalence of triangulated categories that takes t-structures to t-structures and tilts to tilts (resp. cluster-tilting objects to cluster-tilting objects and IY-mutations to IY-mutations), the category $\mathcal{D}\widehat{\Gamma}(Q,S)$ (resp. $\mathcal{C}(Q,S)$) does not change if we replace (Q,S) with a QP which is right-equivalent to it. Actually, something stronger is true: $\mathcal{D}\widehat{\Gamma}(Q,S)$ (resp. $\mathcal{C}(Q,S)$) does not change if we replace (Q,S) with a QP which is right-equivalent to $(Q,\lambda S)$ for some non-zero scalar λ.

By Theorems 3.1 and 3.2, every Jacobian algebra of the form $\mathcal{P}(Q(\tau),S(\tau))$ is finite-dimensional. Thus, a combination of Theorems 2.6 and 5.1 yields:

THEOREM 5.3. *Let* $(\Sigma,\mathbb{M})$ *be a surface with marked points. If* $(\Sigma,\mathbb{M})$ *is not a sphere with less than 5 punctures, then there exist:*
(1) *a 3-Calabi-Yau triangulated category* $\mathcal{D}(\Sigma,\mathbb{M})$, *with canonical hearts and tilts of canonical hearts combinatorially interpreted as tagged triangulations and flips of tagged triangulations, respectively.*
(2) *A* Hom-*finite 2-Calabi-Yau triangulated category* $\mathcal{C}(\Sigma,\mathbb{M})$, *with reachable cluster-tilting objects and IY-mutations of reachable cluster-tilting objects combinatorially interpreted as tagged triangulations and flips of tagged triangulations, respectively*[5]; *cf.* [**4**, Section 3.4], [**9**, Theorem 4.10].

Indeed, one defines $\mathcal{D}(\Sigma,\mathbb{M}) = \mathcal{D}\widehat{\Gamma}(Q(\tau),S(\tau))$ and $\mathcal{C}(\Sigma,\mathbb{M}) = \mathcal{C}(Q(\tau),S(\tau))$ for any tagged triangulation τ of $(\Sigma,\mathbb{M})$. Up to equivalences of triangulated categories, $\mathcal{D}(\Sigma,\mathbb{M}) = \mathcal{D}\widehat{\Gamma}(Q(\tau),S(\tau))$ and $\mathcal{C}(\Sigma,\mathbb{M}) = \mathcal{C}(Q(\tau),S(\tau))$ are independent of τ by Theorems 2.6 and 5.1 (see also Remark 5.2).

Recent work [**6**] of Bridgeland-Smith shows that spaces of *Bridgeland stability conditions* on the categories $\mathcal{D}(\Sigma,\mathbb{M})$ can be realized as spaces of *quadratic differentials* on the Riemann surface Σ. When Σ has empty boundary, Smith [**42**] has furthermore shown that $\mathcal{D}(\Sigma,\mathbb{M})$ can be interpreted as the *Fukaya category* of a symplectic 6-manifold underlying certain Calabi-Yau-3 varieties that fiber over the surface Σ.

---

[4] *IY* after *Iyama-Yoshino*

[5] When $(\Sigma,\mathbb{M})$ is a once-punctured surface with empty boundary, this statement needs a slight refinement.

In physics, the QPs $(Q(\tau), S(\tau))$, as well as their QP-mutation compatibility with flips, have been used by Alim-Cecotti-Cordova-Espahbodi-Rastogi-Vafa [**1**], [**2**], and Cecotti [**8**], in their study of $N = 2$ *quantum field theories* and associated *BPS quivers and spectra.*

Let us give a (extremely rough) sketch of the passage from quadratic differentials to stability conditions when $\partial\Sigma = \varnothing$, our main reference being [**6**]. We start with how quadratic differentials gives rise to a well-defined triangulated category. A *generic* quadratic differential ϕ which is holomorphic on $\Sigma \setminus \mathbb{M}$ and has poles of order 2 at each $p \in \mathbb{M}$, induces a *horizontal foliation* on Σ. A typical curve in this foliation joins either two poles or a pole and a zero of ϕ. On the set of curves of the foliation that join poles of ϕ, one can define an equivalence relation by setting two curves to be equivalent if they are parallel. By picking a system of representatives for this equivalence relation we obtain a triangulation τ_ϕ of $(\Sigma, \mathbb{M})$. This triangulation τ_ϕ gives rise to the category $\mathcal{D}(\Sigma, \mathbb{M})$ via its associated QP $(Q(\tau_\phi), S(\tau_\phi))$.

When we said "generic" in the previous paragraph, we meant that the behavior of ϕ described takes place in an open subset of the space of meromorphic quadratic differentials on Σ that have poles of order two precisely at the points of $\mathbb{M}$. The space of generic quadratic differentials has several connected components, called *chambers.* If we allow ϕ to vary inside a given chamber, the associated triangulation τ_ϕ does not change. However if we allow ϕ to move from one chamber to another, the triangulation τ_ϕ changes. The change suffered by τ_ϕ is either a flip or a *pop.* If it is a flip, by Theorems 2.6 and 5.1 we see that the category $\mathcal{D}(\Sigma, \mathbb{M})$ does not change (but undergoes a tilt of canonical t-structure). If it is a pop, the situation becomes somewhat subtle, but again, the category $\mathcal{D}(\Sigma, \mathbb{M})$ does not change.

So, when we pick a generic quadratic differential ϕ, we are picking a t-structure on $\mathcal{D}(\Sigma, \mathbb{M})$. The simples of the heart of this t-structure are in one-to-one correspondence with the arcs in τ_ϕ. Each of these arcs is transverse to a curve that joins two zeros of ϕ. We integrate $\sqrt{\phi}$ along each of these curves, thus obtaining an assignment of a complex number to each arc in τ_ϕ. This assignment extends uniquely to a group homomorphism from the Grothendieck group $K_0(\mathcal{D}(\Sigma, \mathbb{M}))$ to $\mathbb{C}$. This group homomorphism is actually a stability condition on $\mathcal{D}(\Sigma, \mathbb{M})$.

We warn the reader that things are very far from being as simple as we have just described (for example, ϕ above is not any quadratic differential, but has to be a *complete and saddle-free GMN differential*). In reality, numerous non-trivial considerations are needed and the picture is a lot more complex (and beautiful) than we have made it seem here, see [**6**].

We end the paper with a few remarks regarding the uniqueness of the categories $\mathcal{C}(\Sigma, \mathbb{M})$ and $\mathcal{D}(\Sigma, \mathbb{M})$.

REMARK 5.4. The fact that the category $\mathcal{C}(\Sigma, \mathbb{M}) = \mathcal{C}(Q(\tau), S(\tau))$ is independent of τ does not mean that there cannot exist other generalized cluster categories, defined through other potentials on the same quivers $Q(\tau)$, whose reachable cluster-tilting objects can be parameterized by tagged triangulations, with IY-mutations interpreted as flips of tagged triangulations. In other words, the sole fact that $\mathcal{C}(\Sigma, \mathbb{M})$ is well-defined does not imply that it is the unique generalized cluster category one can associate to a surface. The same comment goes for $\mathcal{D}(\Sigma, \mathbb{M})$.

EXAMPLE 5.5. Suppose $(\Sigma, \mathbb{M})$ is a torus with exactly one boundary component and exactly one marked point. For each triangulation τ of $(\Sigma, \mathbb{M})$, let $W(\tau)$ be the potential given by (3.2). Then $\mathcal{P}(Q(\tau), W(\tau))$ is finite-dimensional, and for any two triangulations τ and σ related by the flip of an arc i, the QPs $\mu_i(Q(\tau), W(\tau))$ and $(Q(\sigma), W(\sigma))$ are right-equivalent. Hence, the categories $\mathcal{D}_{\text{wild}} = \mathcal{D}\widehat{\Gamma}(Q(\tau), W(\tau))$ and $\mathcal{C}_{\text{wild}} = \mathcal{C}(Q(\tau), W(\tau))$ are independent of τ, and satisfy the assertions made in Theorem 5.3. However, since $\mathcal{P}(Q(\tau), W(\tau))$ is wild and $\mathcal{P}(Q(\tau), S(\tau))$ is tame, the categories $\mathcal{C}_{\text{wild}}$ and $\mathcal{C}(\Sigma, \mathbb{M})$ are most likely not equivalent.

REMARK 5.6. Let $(\Sigma, \mathbb{M})$ be a once-punctured surface with empty boundary and positive genus. The potentials $S(\tau)^{(3)}$ from Part (2) of Theorem 4.2 also satisfy that triangulations related by a flip have QPs related by QP-mutation. Thus they also give rise to categories $\mathcal{D}\widehat{\Gamma}(Q(\tau), S(\tau)^{(3)})$ that are actually independent from τ. However, since the Jacobian algebras $\mathcal{P}(Q(\tau), S(\tau)^{(3)})$ are infinite-dimensional, in order to obtain generalized cluster categories one has to apply Plamondon's construction **[39]** rather than that of Amiot. In the case of the once-punctured torus, these comments also apply for the potential given by (3.1).

Acknowledgements

The first version of the present survey article was mainly based on a talk contributed by the author to the *Taller de Vinculación Matemáticos Mexicanos Jóvenes en el Mundo*, held at the *Centro de Investigación en Matemáticas* (CIMAT), Guanajuato, México, in August 2012. I am grateful to the organizers of this meeting, Noé Bárcenas-Torres, Fernando Galaz-García and Mónica Moreno-Rocha, for the opportunity of presenting my work.

The subsequent versions of this survey incorporate some recent results that did not appear in the first version. Quite a few of these results were obtained in joint work with Christof Geiss and Jan Schröer, to whom I am grateful for many interesting discussions. I owe thanks to Jan Schröer for the enthusiastic working atmosphere while I was a member of his research group as a postdoc at the Mathematisches Institut of Universität Bonn.

References

[1] M. Alim, S. Cecotti, C. Córdova, S. Espahbodi, A. Rastogi, and C. Vafa, *BPS quivers and spectra of complete $\mathcal{N} = 2$ quantum field theories*, Comm. Math. Phys. **323** (2013), no. 3, 1185–1227, DOI 10.1007/s00220-013-1789-8. MR3106506

[2] M. Alim, S. Cecotti, C. Córdova, S. Espahbodi, A. Rastogi, and C. Vafa, *$\mathcal{N} = 2$ quantum field theories and their BPS quivers*, Adv. Theor. Math. Phys. **18** (2014), no. 1, 27–127. MR3268234

[3] C. Amiot, *Cluster categories for algebras of global dimension 2 and quivers with potential* (English, with English and French summaries), Ann. Inst. Fourier (Grenoble) **59** (2009), no. 6, 2525–2590. MR2640929 (2011c:16026)

[4] C. Amiot, *On generalized cluster categories*, Representations of algebras and related topics, EMS Ser. Congr. Rep., Eur. Math. Soc., Zürich, 2011, pp. 1–53, DOI 10.4171/101-1/1. MR2931894

[5] I. Assem, T. Brüstle, G. Charbonneau-Jodoin, and P.-G. Plamondon, *Gentle algebras arising from surface triangulations*, Algebra Number Theory **4** (2010), no. 2, 201–229, DOI 10.2140/ant.2010.4.201. MR2592019 (2011h:16045)

[6] T. Bridgeland, I. Smith. *Quadratic differentials as stability conditions.* arXiv:1302.7030

[7] P. Caldero, F. Chapoton, and R. Schiffler, *Quivers with relations arising from clusters (A_n case)*, Trans. Amer. Math. Soc. **358** (2006), no. 3, 1347–1364, DOI 10.1090/S0002-9947-05-03753-0. MR2187656 (2007a:16025)

[8] S. Cecotti, *Categorical tinkertoys for $\mathcal{N} = 2$ gauge theories*, Internat. J. Modern Phys. A **28** (2013), no. 5-6, 1330006, 124, DOI 10.1142/S0217751X13300068. MR3036676

[9] G. Cerulli Irelli and D. Labardini-Fragoso, *Quivers with potentials associated to triangulated surfaces, Part III: tagged triangulations and cluster monomials*, Compos. Math. **148** (2012), no. 6, 1833–1866, DOI 10.1112/S0010437X12000528. MR2999307

[10] W. Crawley-Boevey, *Tameness of biserial algebras*, Arch. Math. (Basel) **65** (1995), no. 5, 399–407, DOI 10.1007/BF01198070. MR1354686 (96i:16021)

[11] H. Derksen, J. Weyman, A. Zelevinsky. *Quivers with potentials and their representations I: Mutations.* Selecta Math. 14 (2008), no. 1, 59–119. arXiv:0704.0649

[12] H. Derksen, J. Weyman, and A. Zelevinsky, *Quivers with potentials and their representations II: applications to cluster algebras*, J. Amer. Math. Soc. **23** (2010), no. 3, 749–790, DOI 10.1090/S0894-0347-10-00662-4. MR2629987 (2012c:16044)

[13] P. Dowbor and A. Skowroński, *On the representation type of locally bounded categories*, Tsukuba J. Math. **10** (1986), no. 1, 63–72. MR846416 (88a:16057)

[14] Ju. A. Drozd, *Tame and wild matrix problems*, Representation theory, II (Proc. Second Internat. Conf., Carleton Univ., Ottawa, Ont., 1979), Lecture Notes in Math., vol. 832, Springer, Berlin-New York, 1980, pp. 242–258. MR607157 (83b:16024)

[15] A. Felikson, M. Shapiro, and P. Tumarkin, *Skew-symmetric cluster algebras of finite mutation type*, J. Eur. Math. Soc. (JEMS) **14** (2012), no. 4, 1135–1180, DOI 10.4171/JEMS/329. MR2928847

[16] V. V. Fock and A. B. Goncharov, *Dual Teichmüller and lamination spaces*, Handbook of Teichmüller theory. Vol. I, IRMA Lect. Math. Theor. Phys., vol. 11, Eur. Math. Soc., Zürich, 2007, pp. 647–684, DOI 10.4171/029-1/16. MR2349682 (2008k:32033)

[17] S. Fomin, M. Shapiro, and D. Thurston, *Cluster algebras and triangulated surfaces. I. Cluster complexes*, Acta Math. **201** (2008), no. 1, 83–146, DOI 10.1007/s11511-008-0030-7. MR2448067 (2010b:57032)

[18] S. Fomin, D. Thurston. *Cluster algebras and triangulated surfaces. Part II: Lambda lengths.* arXiv:1210.5569

[19] S. Fomin and A. Zelevinsky, *Cluster algebras. I. Foundations*, J. Amer. Math. Soc. **15** (2002), no. 2, 497–529 (electronic), DOI 10.1090/S0894-0347-01-00385-X. MR1887642 (2003f:16050)

[20] S. Fomin and A. Zelevinsky, *Y-systems and generalized associahedra*, Ann. of Math. (2) **158** (2003), no. 3, 977–1018, DOI 10.4007/annals.2003.158.977. MR2031858 (2004m:17010)

[21] M. Freedman, J. Hass, and P. Scott, *Closed geodesics on surfaces*, Bull. London Math. Soc. **14** (1982), no. 5, 385–391, DOI 10.1112/blms/14.5.385. MR671777 (84b:53042)

[22] C. Geiss and R. González-Silva, *Tubular Jacobian Algebras*, Algebr. Represent. Theory **18** (2015), no. 1, 161–181, DOI 10.1007/s10468-014-9486-7. MR3313594

[23] C. Geiss, D. Labardini-Fragoso, J. Schröer. *The representation type of Jacobian algebras.* arXiv:1308.0478

[24] Ch. Geiss, B. Leclerc, and J. Schröer, *Cluster algebras in algebraic Lie theory*, Transform. Groups **18** (2013), no. 1, 149–178, DOI 10.1007/s00031-013-9215-z. MR3022762

[25] M. Gekhtman, M. Shapiro, and A. Vainshtein, *Cluster algebras and Poisson geometry* (English, with English and Russian summaries), Mosc. Math. J. **3** (2003), no. 3, 899–934, 1199. {Dedicated to Vladimir Igorevich Arnold on the occasion of his 65th birthday}. MR2078567 (2005i:53104)

[26] M. Gekhtman, M. Shapiro, and A. Vainshtein, *Cluster algebras and Weil-Petersson forms*, Duke Math. J. **127** (2005), no. 2, 291–311, DOI 10.1215/S0012-7094-04-12723-X. MR2130414 (2006d:53103)

[27] J. Geuenich. *Quivers with potentials for quivers with 3 vertices.* Master Thesis, University of Bonn (2013).

[28] B. Keller and D. Yang, *Derived equivalences from mutations of quivers with potential*, Adv. Math. **226** (2011), no. 3, 2118–2168, DOI 10.1016/j.aim.2010.09.019. MR2739775 (2012d:16043)

[29] Y. Kodama and L. K. Williams, *KP solitons, total positivity, and cluster algebras*, Proc. Natl. Acad. Sci. USA **108** (2011), no. 22, 8984–8989, DOI 10.1073/pnas.1102627108. MR2813307 (2012f:37142)

[30] Y. Kodama and L. Williams, *KP solitons and total positivity for the Grassmannian*, Invent. Math. **198** (2014), no. 3, 637–699, DOI 10.1007/s00222-014-0506-3. MR3279534

[31] Y. Kodama and L. Williams, *Combinatorics of KP solitons from the real Grassmannian*, Algebras, quivers and representations, Abel Symp., vol. 8, Springer, Heidelberg, 2013, pp. 155–193, DOI 10.1007/978-3-642-39485-0_8. MR3183885

[32] M. Kontsevich, Y. Soibelman. *Stability structures, motivic Donaldson-Thomas invariants and cluster transformations.* arXiv:0811.2435

[33] D. Labardini-Fragoso, *Quivers with potentials associated to triangulated surfaces*, Proc. Lond. Math. Soc. (3) **98** (2009), no. 3, 797–839, DOI 10.1112/plms/pdn051. MR2500873 (2010b:16033)

[34] D. Labardini-Fragoso, *Quivers with potentials associated to triangulated surfaces*, Proc. Lond. Math. Soc. (3) **98** (2009), no. 3, 797–839, DOI 10.1112/plms/pdn051. MR2500873 (2010b:16033)

[35] S. Ladkani. *Which mutation classes of quivers have constant number of arrows?* arXiv:1104.0436

[36] S. Ladkani. *On Jacobian algebras from closed surfaces.* arXiv:1207.3778

[37] L. Mosher, *Tiling the projective foliation space of a punctured surface*, Trans. Amer. Math. Soc. **306** (1988), no. 1, 1–70, DOI 10.2307/2000830. MR927683 (89f:57014)

[38] K. Nagao, Y. Terashima, M. Yamazaki. *Hyperbolic 3-manifolds and Cluster Algebras.* arXiv:1112.3106

[39] P.-G. Plamondon, *Cluster algebras via cluster categories with infinite-dimensional morphism spaces*, Compos. Math. **147** (2011), no. 6, 1921–1954, DOI 10.1112/S0010437X11005483. MR2862067

[40] I. Reiten. *Tilting theory and cluster algebras.* arXiv:1012.6014

[41] R. Schiffler, *A geometric model for cluster categories of type D_n*, J. Algebraic Combin. **27** (2008), no. 1, 1–21, DOI 10.1007/s10801-007-0071-6. MR2366159 (2008k:16025)

[42] I. Smith. *Quiver algebras as Fukaya categories.* arXiv:1309.0452

[43] S. Trepode, Y. Valdivieso-Díaz. *On finite dimensional Jacobian Algebras.* arXiv:1207.1917

INSTITUTO DE MATEMÁTICAS, UNIVERSIDAD NACIONAL AUTÓNOMA DE MÉXICO
E-mail address: labardini@matem.unam.mx

Riemann-Roch without denominators for oriented cohomology theories

José Malagón-López

ABSTRACT. Riemann-Roch without denominators for any oriented cohomology theory for smooth schemes over a field of characteristic zero is proved, generalizing the corresponding theorem for the case of Chow groups.

Introduction

The Riemann-Roch theorem without denominators was first stated and proved in characteristic zero by Grothendieck in [**1**], and later proved in any characteristic by Jouanolou in [**7**]. The theorem state that if $f : Z \to X$ is a closed embedding of smooth schemes and $E \to Z$ is a vector bundle, then the image of f_*E under the Chern polynomial is computed in terms of a certain universal polynomial. The main result of this note is the extension of the Riemann-Roch without denominators in characteristic zero when we consider any oriented cohomology theory instead of the Chow ring.

We follow the approach taken in [**2**] after observing that the proof for the Chow ring theory carries-out to the case of algebraic cobordism (Theorem 4.1). This is possible since the main computation can be obtained into the context of formal group laws (Lemma 3.2). Then, using the universality of Ω^*, it is shown that the result holds for any oriented cohomology theory.

In the first section we recall the concept of oriented cohomology theories and algebraic cobordism and touch on the properties of such theories that we need. In particular, we check that the statement and proof of the localized Chern class formula for the Chow ring in [**3**] extends for algebraic cobordism. In the next two sections we recover the identites required about the Chern polynomial and prove our main computational tool. In section § 4 we state and prove our main result. We finish with some examples.

Conventions. Throughout k will denote a field of characteristic zero. Let $\mathbf{Sch}_k$ denote the category whose objects are separated, quasi-projective schemes of finite type over k, and let $\mathbf{Sm}_k$ be the full-subcategory of smooth quasi-projective k-schemes. All the schemes to be considered are assumed to be irreducible.

2010 *Mathematics Subject Classification.* Primary 14C40, 14F43.

Key words and phrases. Oriented cohomology theories, integral Riemann-Roch theorem.

1. Oriented Cohomology Theories

An oriented cohomology theory is a contravariant functor from the category of smooth quasi-projective k-schemes to the category of graded rings satisfying certain axioms: existence of push-forwards maps for projective morphisms, base change (i.e. compatability of the variances in Cartesian squares), homotopy invariance and projective bundle formula. A morphism of oriented cohomology theories is a natural transformation of contravariant functors that commutes with the push-forwards maps. In characteristic zero, Levine and Morel constructed the universal oriented cohomology called algebraic cobordism and denoted by Ω^*; see [**8**] for details.

Any oriented cohomology theory A^* is endowed with a theory of Chern classes c_i^A and a formal group law $F_A(u,v) \in A^*(k)[[u,v]]$ which describe the behavior of the first Chern class with respect to the tensor product of line bundles [**8**, Lemma 1.1.3]. Namely,

$$c_1^A(L_1 \otimes L_2) = F_A\left(c_1^A(L_1), c_1^A(L_2)\right) \tag{1.1}$$

for any line bundles L_1, L_1 over a k-scheme X. In the case of algebraic cobordism, the induced morphism from the Lazard ring $\mathbb{L}$, the universal formal group law, to the ring $\Omega^*(k)$ is an isomorphim [**8**, Theorem 1.2.7].

Localized Chern Class. Given a regular embedding $i : Z \to X$ of codimension d and a morphism $f : Y \to X$ in $\mathbf{Sch}_k$, the refined Gysin morphism [**8**, §6.6.2] will be denoted as

$$i^! : \Omega^*(Y) \to \Omega^*(Y \times_X Z).$$

PROPOSITION 1.1. *Let Y be irreducible smooth k-scheme, and $p : Q \to Y$ be a vector bundle of rank d with zero section s_Q. Let $s : Y \to Q$ be a section with zero scheme $i : Z(s) \to Y$. Consider the fiber transverse square*

$$\begin{array}{ccc} Z(s) & \longrightarrow & Y \\ {\scriptstyle i}\downarrow & & \downarrow{\scriptstyle s} \\ Y & \xrightarrow[s_Q]{} & Q. \end{array}$$

Then $i_*\left(s_Q^!(\alpha)\right) = c_d(Q) \cdot \alpha$ *for any* $\alpha \in \Omega^*(Y)$.

PROOF. Applying base change for refined Gysing morphisms [**8**, Proposition 6.6.3(1)] to the fiber square above we get $i_*\left(s_Q^!(\alpha)\right) = s_Q^* s_*(\alpha)$. Since s_Q^* and s^* are both inverse to p^* we have $s_Q^* s_*(\alpha) = s^* s_*(\alpha)$. Applying base change for refined Gysing morphisms and excess intersection formula [**8**, Theorem 6.6.9] to the fiber square

$$\begin{array}{ccc} Y & = & Y \\ \Vert & & \downarrow{\scriptstyle s} \\ Y & \xrightarrow[s]{} & Q \end{array}$$

we conclude $s^* s_*(\alpha) = s^!(\alpha) = c_d(Q) \cdot \alpha$. □

2. Chern Roots

Fix an oriented cohomology theory A^* with formal group law F. Let $\imath$ denote the formal inverse of F, which is the power series in $A^*(k)[[u]]$ such that $F(u, \imath(u)) = 0$. We will use the notation $u +_F v$ for $F(u,v)$. If n is any integer, then $[n]_F \cdot u$ will denote $u +_F \cdots +_F u$, with n terms. Let X be a fixed k-scheme, so that all the vector bundles considered in this section are over X. We will consider the factorization of the Chern polynomial of a vector bundle E in term of its Chern roots: $c_t(E) = \prod_{i=1}^r (1 + \alpha_i t)$.

The following identities [**8**, Lemma 6.2] are a consequence of the Whitney formula and the splitting principle (cf. [**3**, Remark 3.2.3]).

Dual Bundles. Given a line bundle $L \to X$, let $L^\vee$ denote the dual bundle of L. Since $c_1(L^\vee) = \imath(c_1(L))$ and the dual of a direct sum of line bundles is the direct sum of of their duals, we have

$$c_t(E^\vee) = \prod_{i=1}^r (1 + \imath(\alpha_i)\, t)\,. \tag{2.1}$$

Tensor Products. Let $\alpha_1, \dots, \alpha_r$ and $\beta_1, \dots, \beta_s$ be the Chern roots of E and F respectively. From (1.1) it follows that the Chern roots of $E \otimes F$ are $F(\alpha_i, \beta_j)$, i.e.

$$c_t(E \otimes F) = \prod_{i,j} (1 + (\alpha_i +_F \beta_j)\, t)\,. \tag{2.2}$$

Exterior Product. If $\alpha_1, \dots, \alpha_r$ are the Chern roots of E then

$$c_t(\wedge^p E) = \prod_{i_1 < \cdots < i_p} \left(1 + \left(\alpha_{i_1} +_F \cdots +_F \alpha_{i_p}\right) t\right). \tag{2.3}$$

Indeed, assume that for any vector bundle $E \to X$ of rank r the equation (2.3) holds. Let $L \to X$ be a line bundle and set $\alpha_{r+1} = c_1(L)$. By [**6**, §4.13] we have a short exact sequence of vector bundles

$$0 \longrightarrow \wedge^{p-1}(E) \otimes L \longrightarrow \wedge^p (E \oplus L) \longrightarrow \wedge^p E \longrightarrow 0.$$

The multiplicity of the Chern polynomial implies

$$c_t(\wedge^p (E \oplus L)) = c_t\left(\wedge^{p-1}(E) \otimes L\right) \cdot c_t(\wedge^p E)\,.$$

For $p \leq r$, by Equation (2.2) and induction we have

$$\begin{aligned} c_t(\wedge^p (E \oplus L)) &= c_t\left(\wedge^{p-1}(E) \otimes L\right) \cdot c_t(\wedge^p E) \\ &= \prod_{1 \leq i_1 < \cdot < i_{p-1} \leq r} \left(1 + \left(\alpha_{i_1} +_F \cdots +_F \alpha_{i_{p-1}} +_F \alpha_{r+1}\right) t\right) \\ &\qquad \prod_{1 \leq i_1 < \cdot < i_p \leq r} \left(1 + \left(\alpha_{i_1} +_F \cdots +_F \alpha_{i_p}\right) t\right) \\ &= \prod_{1 \leq i_1 < \cdot < i_p \leq r+1} \left(1 + \left(\alpha_{i_1} +_F \cdots +_F \alpha_{i_p}\right) t\right). \end{aligned}$$

For $p = r+1$, since $\wedge^N E = 0$ for all $N > \operatorname{rank}(E)$, Equation (2.2) and induction implies

$$c_t\left(\wedge^{r+1}(E \oplus L)\right) = c_t\left(\wedge^r(E) \otimes L\right) = 1 + (\alpha_1 +_F \cdots +_F \alpha_{r+1})\, t.$$

3. A Universal Polynomial

Throughtout this section we will assume the following: *G and D are vector bundles over a k-scheme X of rank r and d respectively, and such that*

$$\wedge^\bullet D^\vee \otimes G: \quad 0 \to \wedge^d D^\vee \otimes G \to \cdots \to \wedge^2 D^\vee \otimes G \to D^\vee \otimes G \to G$$

is a resolution of a vector bundle E over X.

By the Whitney formula and the multiplicity of the Chern polynomial we have

$$\mathrm{c}_t(E) = \mathrm{c}_t\left(\sum_{p=0}^{d}(-1)^p \left(\wedge^p D^\vee \otimes G\right)\right) = \prod_{p=0}^{d} \mathrm{c}_t\left(\wedge^p D^\vee \otimes G\right)^{(-1)^p}.$$

Set $\mathrm{c}_t\left(\wedge^\bullet D^\vee \otimes G\right) = \mathrm{c}_t\left(\sum_{p=0}^{d}(-1)^p \left(\wedge^p D^\vee \otimes G\right)\right)$. Let $\alpha_1, \ldots, \alpha_r$ and $\eta_1, \ldots, \eta_d$ be the Chern roots of G and D respectively. Using the identities from the previous section we obtain

$$\mathrm{c}_t\left(\wedge^\bullet D^\vee \otimes G\right) = \prod_{p=0}^{d} \prod_{i_1 < \cdots < i_p} \prod_{j=1}^{r} \left(1 + [\alpha_j +_F \imath(\eta_{i_1}) +_F \cdots +_F \imath(\eta_{i_p})]t\right)^{(-1)^p}. \tag{3.1}$$

REMARK 3.1. For fixed j, $1 \leq j \leq r$, and for $0 \leq p \leq d$, the power series $\mathrm{c}_t\left(\wedge^\bullet D^\vee \otimes G\right)$ has $\binom{d}{p}$ factors of the form

$$\prod_j(i_1, \ldots, i_p) := \prod_{j=1}^{r} \left(1 + [\alpha_j +_F \imath(\eta_{i_1}) +_F \cdots +_F \imath(\eta_{i_p})]\, t\right)^{(-1)^p}.$$

Moreover, for a given index i_0 there are $\binom{d-1}{p}$ factors $\prod_j(i_1, \ldots, i_p)$ with $i_0 \neq i_l$, $1 \leq l \leq p$; and $\binom{d-1}{p-1}$ factors $\prod_j(i_1, \ldots, i_p)$ with $i_0 = i_l$, $1 \leq l \leq p$.

LEMMA 3.2. *Let G and D be vector bundles as above. Then there is a unique power series $P(x_1, \ldots, x_d; u_1, \ldots, u_r)$ with coefficients in $A^*(k)$ such that*

$$\mathrm{c}_t\left(\wedge^\bullet D^\vee \otimes G\right) = 1 + \mathrm{c}_d(D) \cdot \textstyle\sum_{l \geq 1} P_l\left(D; G\right) t^l,$$

where $P_l(D; G)$ denotes homogeneous part of degree l of the power series P evaluated on the Chern classes of D and G.

PROOF. From Equation (3.1) we have that $\mathrm{c}_t\left(\wedge^\bullet D^\vee \otimes G\right) - 1$ is a power series over t with constant term zero, and whose higher degree terms are homogeneous polynomials on the α_j's and $\imath(\eta_i)$'s with coefficients in $A^*(k)$. Since $\mathrm{c}_d(D) = \eta_1 \cdots \eta_d$, it suffices to show that each η_i is a factor of $\mathrm{c}_t\left(\wedge^\bullet D^\vee \otimes G\right) - 1$.

Fix $1 \leq i_0 \leq d$ and consider the Remark 3.1. Since η_{i_0} is a factor of $\imath(\eta_{i_0})$, if $\eta_{i_0} = 0$ then the $\binom{d-1}{p-1}$ factors $\prod_j(i_1, \ldots, i_p)$ for which $i_0 = i_l$ will cancel out with the $\binom{d-1}{p-1}$ factors $\prod_j(i_1, \ldots, i_{p-1})$ for which $i_0 \neq i_l$. Thus, $\mathrm{c}_t\left(\wedge^\bullet D^\vee \otimes G\right) - 1 = 0$ if $\eta_{i_0} = 0$. □

Our next step is to obtain an explicit description of $P_d(D, G)$.

REMARK 3.3. Since $\mathrm{c}_d(D) = \eta_1 \cdots \eta_d$, Lemma 3.2 implies the following.

(1) $P_l(D, G) = 0$ for $1 \leq l \leq d - 1$.
(2) None of the Chern roots of G is a factor of $P_d(D, G)$.

Since any formal group law $F(u,v)$ is of the form $u+v+\sum_{i,j\geq 1} a_{ij}u^iv^j$, from Remark 3.3 it follows that in the computation of $P_d(D,G)$ it suffices to consider

$$\prod_{p=0}^{d} \prod_{1\leq i_1<\cdots<i_p\leq d} \prod_{j=1}^{r} \left(1+[\alpha+\imath(\eta_{i_1})+\cdots+\imath(\eta_{i_p})]\,t\right)^{(-1)^p}.$$

Finally, it is known (see for instance [**3**, Example 15.3.1]) that the coefficient of t^d of the expression above is $-r\,(d-1)!\,\imath(\eta_1)\cdots\imath(\eta_d)$. We conclude the following.

PROPOSITION 3.4. *Let G and D be vector bundles over a k-scheme X of rank r and d respectively. Assume that $\wedge^\bullet D^\vee\otimes G$ is a resolution of a vector bundle E over X. Let $\eta_1,\ldots,\eta_d$ be the Chern roots of D. Then*

$$\begin{aligned}\mathrm{c}_t\left(\wedge^\bullet D^\vee\otimes G\right) &= 1 - r\,(d-1)!\,\imath(\eta_1)\cdots\imath(\eta_d)\,t^d + \textit{higher terms}\\ &= 1 + (-1)^{d-1}\,r\,(d-1)!\,\mu(\eta_1)\cdots\mu(\eta_d)\,\mathrm{c}_d(D)\,t^d + \textit{higher terms},\end{aligned}$$

where $\mu(u)$ is the power series such that $\imath(u)=-u\mu(u)$.

4. Main Theorem

We will state and prove first the main result of this note for algebraic cobordism, and then, using the universality of algebraic cobordism, we will prove it for any oriented cohomology theory.

THEOREM 4.1 (Riemann-Roch without Denominators). *Let c^Ω denote the total Chern class operator for algebraic cobordism. Let $f: Z\to X$ be a closed embedding of codimension d of irreducible smooth schemes with normal bundle N_f. Let E be a vector over Z of rank r. Then*

$$\mathrm{c}^\Omega\left(f_*E\right) = 1 + f_*\left(P(N_f;E)\right).$$

The idea of the proof is to reduce it to the case of the following proposition by means of deformation to the normal bundle.

PROPOSITION 4.2. *Let D be a vector bundle of rank d over a smooth scheme X. Consider the projective bundle space $\mathbb{P}=\mathbb{P}(D\oplus\mathcal{O}_X)$. Let $p:\mathbb{P}\to X$ denote the projection map and $s_\mathbb{P}: X\to\mathbb{P}$ its zero section. Let E be a vector bundle of rank r over X. Then*

$$\mathrm{c}^\Omega\left(s_{\mathbb{P}*}E\right) = 1 + s_{\mathbb{P}*}\left(P(D;E)\right).$$

PROOF. Since D is vector bundle, we have that $s_\mathbb{P}$ is regular with normal bundle D [**4**, Corollary IV.3.3]. Consider the tautological exact sequence over $\mathbb{P}$

$$0\longrightarrow\mathcal{Q}\longrightarrow p^*(D\oplus\mathcal{O}_X)\longrightarrow\mathcal{O}_\mathbb{P}(1)\longrightarrow 0.$$

Let $s:\mathcal{O}_\mathbb{P}\to\mathcal{Q}^\vee$ be the canonical section which is induced by the dual of the composition $\mathcal{Q}\to p^*(D\oplus\mathcal{O}_X)\to p^*\mathcal{O}_X=\mathcal{O}_\mathbb{P}$. We have that the zero scheme of s is $s_\mathbb{P}(X)$ and $s_\mathbb{P}^*\mathcal{Q}^\vee\cong D$; see [**4**, §IV] for details. By regularity of $s_\mathbb{P}$, the dual of the section s extends to a resolution

$$0\longrightarrow\wedge^d\mathcal{Q}\longrightarrow\cdots\longrightarrow\wedge^2\mathcal{Q}\longrightarrow\mathcal{Q}\longrightarrow\mathcal{O}_\mathbb{P}\longrightarrow\mathcal{O}_{s_\mathbb{P}(X)}\longrightarrow 0.$$

Since $s_{\mathbb{P}*}\mathcal{O}_X\otimes p^*E\cong s_{\mathbb{P}*}E$, tensoring by p^*E we obtain a resolution $\wedge^\bullet\mathcal{Q}\otimes p^*E$ of $s_{\mathbb{P}*}E$. Applying Lemma 3.2 we get

$$\mathrm{c}^\Omega\left(s_{\mathbb{P}*}E\right) = \mathrm{c}^\Omega\left(\wedge^\bullet\mathcal{Q}\otimes p^*E\right) = 1 + \mathrm{c}^\Omega_d\left(\mathcal{Q}^\vee\right)\,P\left(\mathcal{Q}^\vee, p^*E\right).$$

Now, Proposition 1.1 implies $s_{\mathbb{P}*}(1_X) = c_d^{\Omega}(\mathcal{Q}^\vee)$, and since $s_{\mathbb{P}}^*\mathcal{Q}^\vee \cong D$, we conclude

$$1 + c_d^{\Omega}(\mathcal{Q}^\vee)\, P(\mathcal{Q}^\vee, p^*E) = 1 + s_{\mathbb{P}*}(1_X)\, P(\mathcal{Q}^\vee, p^*E) = 1 + s_{\mathbb{P}*}(P(D, E)).$$

□

PROOF OF THEOREM 4.1. Consider the diagram associated to the deformation to the normal bundle

$$\begin{array}{ccccc} Z & \xrightarrow{g_0'} & \tilde{Z} & \xleftarrow{g_1'} & Z \\ {\scriptstyle f}\big\downarrow & (I) & {\scriptstyle \tilde{f}}\big\downarrow & (II) & \big\downarrow{\scriptstyle s_{\mathbb{P}}} \\ X & \xrightarrow[g_0]{} & \tilde{X} & \xleftarrow[g_1]{} & \mathbb{P}, \end{array}$$

where $\mathbb{P} = \mathbb{P}(N_f \oplus \mathcal{O}_Z)$, $\tilde{X} = \mathrm{Bl}_{Z\times\{0\}}(X \times \mathbb{P}^1) - \mathrm{Bl}_{Z\times\{0\}}(X \times \{0\})$, $\tilde{Z} = Z \times \mathbb{A}^1$, $g_k : Z \to Z \times \mathbb{A}^1$ is the embedding given by $z \mapsto (z, k)$, for $k = 0, 1$, and $s_{\mathbb{P}}$ is the zero section of the projection map $\mathbb{P} \to Z$; see [**3**, § 5.1] for details.

Consider $\tilde{E} = q^*E$, where $q : \tilde{Z} \to Z$ is the canonical projection. Let $N_{\tilde{f}}$ be the normal bundle associated to $\tilde{f}$. Since the Cartesian square (I) is transverse we have that $g_0'^*(N_{\tilde{f}}) = N_f$. Now, functoriality of Chern classes and base change for algebraic K-theory imply

$$g_0^*\left(c^{\Omega}\left(\tilde{f}_*q^*E\right)\right) = c^{\Omega}\left(g_0^*\tilde{f}_*q^*E\right) = c^{\Omega}\left(f_*g_0'^*q^*E\right) = c^{\Omega}\left(f_*E\right).$$

Similarly, functoriality of Chern classes and base change for Gysin morphisms in algebraic cobordism [**8**, Proposition 6.5.4] imply

$$g_0^*\left(1 + \tilde{f}_*(P(N_{\tilde{f}}; q^*E))\right) = 1 + f_*g_0'^*(P(N_{\tilde{f}}; q^*E)) = 1 + f_*(P(N_f; D)).$$

Thus, it suffices to show the theorem for $\tilde{f}$ and q^*E. Proceeding as above, from the transverse Cartesian square (II) we have

$$g_1^*\left(c^{\Omega}\left(\tilde{f}_*q^*E\right)\right) = c^{\Omega}\left(s_{\mathbb{P}*}E\right),$$

and

$$g_1^*\left(1 + \tilde{f}_*\left(P\left(N_{\tilde{f}}; q^*E\right)\right)\right) = 1 + s_{\mathbb{P}*}\left(P(N_f; E)\right).$$

Here we are using that N_f is isomorphic to the normal bundle of $s_{\mathbb{P}}$, which is a consequence of being $s_{\mathbb{P}}$ a regular embedding.

Let $\tilde{j} : U \to \tilde{X}$ denote the open complement of $\tilde{f}$. By Proposition 4.2 the theorem holds for $s_{\mathbb{P}}$ and E, and since $\ker \tilde{j} \cap \ker g_1^* = 0$ [**9**, Lemma 1.4.2] we have that the theorem will hold for $\tilde{f}$ and q^*E if

$$c^{\Omega}\left(\tilde{f}_*q^*E\right) - \left(1 + \tilde{f}_*\left(P\left(N_{\tilde{f}}; q^*E\right)\right)\right) \in \ker\left(\tilde{j}^*\right). \tag{4.1}$$

But (4.1) follows from the localization sequence associated to the morphisms $\tilde{f}$ and $\tilde{j}$. □

COROLLARY 4.3. *Let A^* be an oriented cohomology theory. Let $f : Z \to X$ be a closed embedding of codimension d of irreducible smooth schemes with normal bundle N_f. Let E be a vector over X of rank r. Then* $c\,(f_*E) = 1 + f_*\,(P(N, E))$.

PROOF. A direct consequence of Theorem 4.1 since the canonical morphism $\Omega^* \to A^*$ sends Chern classes to Chern classes. □

EXAMPLE 4.4 (Chow Ring). The formal inverse of the additive formal group law is $\imath(u) = -u$, so the power series $P(D, G)$ coincides with the power series of [**3**, Lemma 15.3]. Thus, Corollary 4.3 for CH^* is the classical Riemman-Roch without denominators.

References

[1] *Théorie des intersections et théorème de Riemann-Roch*, Lecture Notes in Mathematics, Vol. 225, Springer-Verlag, (1971). Séminaire de Géométrie Algébrique du Bois-Marie 1966–1967 (SGA 6), Dirigé par P. Berthelot, A. Grothendieck et L. Illusie.

[2] P. Baum, W. Fulton and R. MacPherson, *Riemann-Roch for singular varieties* , Inst. Hautes Études Sci. Publ. Math., **45** (1975), pp. 101-145.

[3] W. Fulton, *Intersection Theory*, Springer-Verlag, 2nd edition, 1998.

[4] W. Fulton and S. Lang, *Riemann-Roch algebra*, Grundlehren der Mathematischen Wissenschaften [Fundamental Principles of Mathematical Sciences], vol. 277, Springer-Verlag, New York, 1985. MR801033 (88h:14011)

[5] A. Grothendieck, *La théorie des classes de Chern* (French), Bull. Soc. Math. France **86** (1958), 137–154. MR0116023 (22 #6818)

[6] F. Hirzebruch, *Topological methods in algebraic geometry*, Classics in Mathematics, Springer-Verlag, Berlin, 1995. Translated from the German and Appendix One by R. L. E. Schwarzenberger; With a preface to the third English edition by the author and Schwarzenberger; Appendix Two by A. Borel; Reprint of the 1978 edition. MR1335917 (96c:57002)

[7] J. P. Jouanolou, *Riemann-Roch sans dénominateurs*, Invent. Math. **11** (1970), 15–26. MR0332789 (48 #11115)

[8] M. Levine and F. Morel, *Algebraic cobordism*, Springer Monographs in Mathematics, Springer, Berlin, 2007. MR2286826 (2008a:14029)

[8] J. Malagón-López, K. Zainoiulline and C. Zhong, *Invariants, exponents and formal group laws*, arXiv:1207.1880, 2012.

[9] I. Panin, *Riemann-Roch theorems for oriented cohomology*, Axiomatic, enriched and motivic homotopy theory, NATO Sci. Ser. II Math. Phys. Chem., vol. 131, Kluwer Acad. Publ., Dordrecht, 2004, pp. 261–333, DOI 10.1007/978-94-007-0948-5_8. MR2061857 (2005g:14025)

DEPARTMENT OF MATHEMATICAL AND COMPUTATIONAL SCIENCES, UNIVERSITY OF TORONTO AT MISSISSAUGA, CANADA

E-mail address: jose.malagon.lopez@mail.utoronto.ca

A survey on the Lyubeznik numbers

Luis Núñez-Betancourt, Emily E. Witt, and Wenliang Zhang

ABSTRACT. The Lyubeznik numbers are invariants of a local ring containing a field that capture ring-theoretic properties, but also have numerous connections to geometry and topology. We discuss basic properties of these integer-valued invariants, as well as describe some significant results and recent developments (including certain generalizations) in the area.

CONTENTS

1. Introduction

Since the introduction of the Lyubeznik numbers **[Lyu93]**, the study of these invariants of local rings containing a field has grown in several compelling directions. One aim of this paper is to present definitions used in the study of the Lyubeznik numbers, along with examples and applications of these invariants, to those new to them. For those familiar with them, we also present recent results on, and generalizations of, the Lyubeznik numbers, as well as open problems in the area.

Given a module M over a ring S, if $I^\bullet$ is a minimal injective resolution of M, then each I^i is isomorphic to a direct sum of indecomposable injective modules $E_S(S/\mathfrak{p})$, $\mathfrak{p}$ is a prime ideal of S; *i.e.*, injective hulls over S of $S/\mathfrak{p}$. The number

2010 *Mathematics Subject Classification.* Primary 13D45, 13N10, 13H99.

Key words and phrases. Local cohomology, Lyubeznik numbers, Stanley-Reisner rings, D-modules, simplicial complexes.

The third author was partially supported by the NSF grant DMS #1247354/#1068946.

of copies of $E_S(S/\mathfrak{p})$ in I^i is the *i-th Bass number of M with respect to* $\mathfrak{p}$, denoted $\mu_i(\mathfrak{p}, M)$ and equals $\dim_{S_\mathfrak{p}/\mathfrak{p}S_\mathfrak{p}} \operatorname{Ext}^i_S(S_\mathfrak{p}/\mathfrak{p}S_\mathfrak{p}, M_\mathfrak{p})$.

Huneke and Sharp proved that if S is a regular ring of characteristic $p > 0$ and I is an ideal of S, then the Bass numbers of the local cohomology modules of the form $H^j_I(S)$, $j \in \mathbb{N}$, are finite, raising the analogous question in the characteristic zero case [**HS93**]. Utilizing D-module theory, Lyubeznik proved the same statement for regular local rings of characteristic zero containing a field [**Lyu93**].

Relying on the finiteness of the Bass numbers of local cohomology modules, Lyubeznik introduced a family of integer-valued invariants associated to a local ring containing a field, now called *Lyubeznik numbers*. They are defined as follows: Suppose that $(R, \mathfrak{m}, K)$ is a local ring admitting a surjection from an n-dimensional regular local ring $(S, \mathfrak{n}, K)$ containing a field, and let I denote the kernel of the surjection. Given $i, j \in \mathbb{N}$, the *Lyubeznik number of R with respect to $i, j \in \mathbb{N}$* is defined as $\dim_K \operatorname{Ext}^i_S\left(K, H^{n-j}_I(S)\right)$, and is denoted $\lambda_{i,j}(R)$. Notably, this invariant depends only on R, i, and j; in particular, it is independent of the choice of S and of the surjection. Moreover, if $(R, \mathfrak{m})$ is any local ring containing a field, then defining $\lambda_{i,j}(R)$ as $\lambda_{i,j}(\widehat{R})$, where $\widehat{R}$ denotes the completion of R at $\mathfrak{m}$, extends the definition.

Lyubeznik numbers are indicators of certain ring-theoretic properties. For example, let R be a local ring containing a field, and let $d = \dim(R)$. Then $\lambda_{d,d}(R)$ equals the number of connected components of the Hochster-Huneke graph of the completion of the strict Henselization of R [**Lyu06a, Zha07**]. Consequently,

(1) $\lambda_{d,d}(R) \neq 0$ for $d = \dim(R)$ [**Lyu93**];
(2) $\lambda_{d,d}(R) = 1$ if R is Cohen-Macaulay [**Kaw02**], or even if R satisfies Serre's condition S_2 (*cf.* Theorem 4.6); and
(3) $\lambda_{d,d}(R) = 1$ if R is analytically normal [**Lyu93**].

Strikingly, the Lyubeznik numbers, which are defined in a purely algebraic context, have extensive connections with geometry and topology. For example, if R is the local ring of an isolated singularity of a complex space of pure dimension at least two, then every Lyubeznik number of R equals the $\mathbb{C}$-vector space dimension of a certain singular cohomology group [**GLS98**]. Blickle and Bondu give similar connections between Lyubeznik numbers and étale cohomology in the positive characteristic setting [**BB05**], which is generalized by Blickle in [**Bli07**].

The Lyubeznik numbers of a Stanley-Reisner ring capture properties of the associated simplicial and combinatorial structure [**ÀM00, ÀM04a, ÀMGLZA03, Mus00, ÀMV14, Yan01a**]. For instance, the Lyubeznik numbers are a measure of the acyclicity of certain linear strands of the Alexander dual of the simplicial complex associated to the ring. In addition, there are several algorithms for computing the Lyubeznik numbers of these rings [**ÀM00, ÀMV14**].

For rings of characteristic zero, there are algorithms for computing Lyubeznik numbers that employ the D-module structure of local cohomology modules [**Wal99, ÀML06**]. In [**ÀM00, ÀML06**], these numbers are computed as multiplicities of the characteristic cycles associated to certain local cohomology modules.

Due to their wide range of applications, several variants and generalizations of the Lyubeznik numbers have been created. Analogous invariants of a simplicial normal Gorenstein semigroup ring modulo a squarefree monomial ideal are proven well defined in [**Yan01a**]. There is a version for projective varieties over a field of prime characteristic, and it is not known whether the definition can be extended to the equal characteristic zero case [**Zha11**]. In characteristic zero, there is a family of invariants closely related to the Lyubeznik numbers defined in terms of certain characteristic cycle multiplicities from D-module theory [**ÀM04b**]. For all local rings containing a field, there is a generalization of the Lyubeznik numbers defined using D-modules, with several ideals and a coefficient field as parameters; these invariants can be used to measure singularities in positive characteristic [**Bli04, NBW14**]. Recently, an alternate version of the Lyubeznik numbers has been defined for, in particular, local rings of mixed characteristic [**NBW13**].

2. Preliminaries

2.1. Local cohomology. Before defining the Lyubeznik numbers, we need to introduce *local cohomology modules.* We include some results on this topic that will be used as tools later in this survey. Please see, among many beautiful references, [**BS98, Lyu02, ILL+07**] for more details.

Let S be a Noetherian ring, and fix $f_1, \dots, f_\ell \in S$. Consider the Čech-like complex, $\check{\mathrm{C}}^\bullet(\underline{f}; S)$:

$$0 \to S \to \bigoplus_{1 \le i \le \ell} S_{f_i} \to \bigoplus_{1 \le i < j \le \ell} S_{f_i f_j} \to \cdots \to S_{f_1 \cdots f_\ell} \to 0,$$

where $\check{\mathrm{C}}^i(\underline{f}; S) = \bigoplus\limits_{1 \le j_1 < \ldots < j_i \le \ell} S_{f_{j_1} \cdots f_{j_i}}$, and on each summand, the homomorphism $\check{\mathrm{C}}^i(\underline{f}; S) \to \check{\mathrm{C}}^{i+1}(\underline{f}; S)$ is a localization map with an appropriate sign. For instance, if $\ell = 2$, the complex is

$$0 \to S \to S_{f_1} \oplus S_{f_2} \to S_{f_1 f_2} \to 0,$$

where $S \to S_{f_1} \oplus S_{f_2}$ maps $s \mapsto (\frac{s}{1}, \frac{s}{1})$ and $S_{f_1} \oplus S_{f_2} \to S_{f_1 f_2}$ maps $\left(\frac{s}{f_1^\alpha}, \frac{r}{f_2^\beta}\right) \mapsto \frac{s}{f_1^\alpha} - \frac{r}{f_2^\beta}$. We point out that

$$\check{\mathrm{C}}^\bullet(\underline{f}; S) = \varinjlim_t \mathcal{K}(\underline{f}^t; S),$$

where $\mathcal{K}(\underline{f}^t; S)$ is the *Koszul complex* associated to $f_1^t \dots, f_\ell^t$, and the map $\mathcal{K}(\underline{f}^t; S) \to \mathcal{K}(\underline{f}^{t+1}; S)$ is induced by $S/(f_1^t, \dots f_\ell^t) \xrightarrow{\cdot f_1 \cdots f_\ell} S/(f_1^{t+1}, \dots f_\ell^{t+1})$.

DEFINITION 2.1 (Local cohomology). Let $I = (f_1, \dots, f_\ell)$ be an ideal of a Noetherian ring S, let M be an S-module, and fix $i \in \mathbb{N}$. We define the *i-th local cohomology of M with support in I*, denoted $H^i_I(M)$, as the i-th cohomology of the complex $\check{\mathrm{C}}^\bullet(\underline{f}; S) \otimes_S M$; *i.e.*,

$$H^i_I(M) := H^i(\check{\mathrm{C}}^\bullet(\underline{f}; S) \otimes_S M) = \frac{\operatorname{Ker}\left(\check{\mathrm{C}}^i(\underline{f}; S) \otimes_S M \to \check{\mathrm{C}}^{i+1}(\underline{f}; S) \otimes_S M\right)}{\operatorname{Im}\left(\check{\mathrm{C}}^{i-1}(\underline{f}; S) \otimes_S M \to \check{\mathrm{C}}^i(\underline{f}; S) \otimes_S M\right)}$$

REMARK 2.2. The local cohomology module $H^i_I(M)$ does not depend on the choice of generators for I, $f_1, \dots, f_\ell$. Moreover, it depends only on the radical of I; *i.e.*, $H^i_I(M) = H^i_{\sqrt{I}}(M)$. For this reason, if $X = \operatorname{Spec} S$ is the affine scheme associated to S and $Z = \mathbb{V}(I) = \{P \in X \mid I \subseteq P\}$ is the Zariski closed subset defined by I, then $H^i_I(M)$ can be written as $H^i_Z(M)$.

Note that there are several ways to define local cohomology. In fact, the definition we chose is not the most natural, but will be advantageous in our discussion due to the interactions between the cited complex $\check{C}^\bullet(\underline{f}; S)$ and D-modules (see Section 2.2). The local cohomology module $H^i_I(M)$ can also be defined as the i-th right derived functor of $\Gamma_I(-)$, where $\Gamma_I(M) = \{v \in M \mid I^j v = 0 \text{ for some } j \in \mathbb{N}\}$. It can also be defined as the direct limit $\varinjlim_t \operatorname{Ext}^i_S(S/I^t, M)$, whose maps are induced by the natural surjections $S/I^{t+1} \twoheadrightarrow S/I^t$.

Next, we collect some basic facts on local cohomology modules that will be used later.

REMARK 2.3 (Graded Local Duality). Let K be a field, and consider $R = K[x_0, \ldots, x_n]$ under the standard grading; *i.e.*, $\deg(x_i) = 1$ for $0 \le i \le n$. If f is a homogeneous polynomial in R and M is a graded R-module, then M_f is naturally a graded R-module. Hence, it follows from Definition 2.1 that each local cohomology module $H^j_I(M)$ is naturally graded whenever M is graded and I is homogeneous. If $\mathfrak{m} = (x_0, \ldots, x_n)$ is the homogeneous maximal ideal of R, then for each finitely generated graded R-module M, there is a functorial isomorphism

$$H^t_{\mathfrak{m}}(M)_\ell \cong \operatorname{Hom}_K\left(\operatorname{Ext}^{n+1-t}_R(M, R(-n-1))_{-\ell}, K\right)$$

for all integers t and ℓ (*cf.* [**BS98**, 13.4.6]).

For any graded R-module M, the graded Matlis dual $D(M)$ of M is defined to be the graded R-module defined by $D(M)_\ell = \operatorname{Hom}_K(M_{-\ell}, K)$. Graded Local Duality states that there is a degree-preserving isomorphism

$$H^t_{\mathfrak{m}}(M) \cong D\left(\operatorname{Ext}^{n+1-t}_R(M, R(-n-1))\right)$$

for all integers t and all finitely generated graded R-modules M. See [**BS98**, 13.3 and 13.4] for details.

REMARK 2.4 (Connection between local cohomology and sheaf cohomology). Take R as in Remark 2.3. Let M be a finitely generated graded R-module, and let $\widetilde{M}$ be the sheaf on $\mathbb{P}^n$ associated to M. Then there are a functorial isomorphisms (*cf.* [**Eis95**, A4.1], [**ILL$^+$07**, Lecture 13])

$$H^t_{\mathfrak{m}}(M) \cong \bigoplus_{\ell \in \mathbb{Z}} H^{t-1}(\mathbb{P}^n, \widetilde{M}(\ell)) \text{ when } t \ge 2,$$

and an exact sequence (functorial in M) of degree-preserving maps

$$0 \to H^0_{\mathfrak{m}}(M) \to M \to \bigoplus_{\ell \in \mathbb{Z}} H^0(\mathbb{P}^n, \widetilde{M}(\ell)) \to H^1_{\mathfrak{m}}(M) \to 0.$$

REMARK 2.5 (Mayer-Vietoris sequence for local cohomology). Given ideals I and J of a Noetherian ring R, and an R-module M, there exists a long exact sequence

$$0 \to H^0_{I+J}(M) \to H^0_I(M) \oplus H^0_J(M) \to H^0_{I\cap J}(M) \to$$
$$\to H^1_{I+J}(M) \to H^1_I(M) \oplus H^1_J(M) \to H^1_{I\cap J}(M) \to \cdots,$$

functorial in M.

REMARK 2.6. The local cohomology modules $H^i_I(M)$ are usually not finitely generated, even when M is. For instance, if $(S, \mathfrak{m}, K)$ is an n-dimensional regular local ring, then $H^n_\mathfrak{m}(S) \cong E_S(K)$, the injective hull of K over S, which is not finitely generated unless S is a field.

2.2. D-modules. Let R be a Noetherian ring, and let $S = R[\![x_1, \ldots, x_n]\!]$. Let $D(S, R)$ denote the ring of R-linear differential operators of S, the subring of $\operatorname{End}_R(S)$ given by

$$D(S,R) = S\left\langle \frac{1}{t!}\frac{\partial^t}{\partial x_1^t}, \ldots, \frac{1}{t!}\frac{\partial^t}{\partial x_n^t} \right\rangle_{t \in \mathbb{N}},$$

where $\frac{1}{t!}\frac{\partial^t}{\partial x_i^t}$ is the $R[\![x_1, \ldots, x_{i-1}, x_{i+1}, \ldots, x_n]\!]$-linear endomorphism of S induced by

$$\frac{1}{t!}\frac{\partial^t}{\partial x_i^t} \cdot x_j^v = \begin{cases} 0 & \text{if } t > v, \text{and} \\ \binom{v}{t}\, x^{v-t} & \text{otherwise.} \end{cases}$$

For example, when $t = 1$, $\frac{\partial^t}{\partial x_i^t} f = \frac{\partial}{\partial x_i} f$ is the formal derivative of f with respect to x_i.

Since $D(S, R) \subseteq \operatorname{End}_R(S)$, $\theta \in D(S, R)$ acts on $s \in S$ by $\theta \cdot s = \theta(s)$. In addition, $D(S, R)$ also acts naturally on S_f for every $f \in S$, and with this action, the localization map $S \to S_f$ is a homomorphism of $D(S, R)$-modules (see [**Lyu93, Lyu00a**] for details). If $t = 1$, this action is given by the quotient rule for derivatives. As a consequence, the local cohomology modules are $D(S, R)$-modules. Moreover, for all $i_1, \ldots, i_\ell \in \mathbb{N}$ and all ideals $I_1, \ldots, I_\ell$ of S, the iterated local cohomology modules $H^{i_\ell}_{I_\ell} \cdots H^{i_2}_{I_2} H^{i_1}_{I_1}(S)$ are also $D(S, R)$-modules [**Lyu93**].

When $R = K$ is a field of characteristic zero, it was proven in [**Bjö79**] that each localization S_f has finite length as a $D(S, K)$-module. Consequently, each local cohomology module of the form $H^{i_\ell}_{I_\ell} \cdots H^{i_2}_{I_2} H^{i_1}_{I_1}(S)$, where $i_1, \ldots, i_\ell \in \mathbb{N}$ and $I_1, \ldots, I_\ell$ are ideals of S, also has finite length as a $D(S, K)$-module. In [**Lyu93**], Lyubeznik used this fact to show that the Bass numbers $\dim_K \operatorname{Ext}^i_S(K, H^j_I(S))$ are finite for all $i, j \in \mathbb{N}$ and all ideals $I \subseteq S$. Subsequently, they generalized these results to fields of positive characteristic in [**Lyu97, Lyu00a**].

2.3. A key functor. Here, we introduce a functor motivated by work of Kashiwara and used by Lyubeznik to prove that the Lyubeznik numbers are well defined [**Lyu93**]. Further properties of this functor were developed by the first two authors to define the generalized Lyubeznik numbers [**NBW14**]. Let K be a field, let $R = K[\![x_1, \ldots, x_n]\!]$, and let $S = R[\![x_{n+1}]\!]$. Let G denote the functor from the category of R-modules to that of S-modules given by, for an R-module M,

$$G(M) = M \otimes_R S_{x_{n+1}}/S.$$

This functor satisfies the following properties; we refer to [**NBW14**, Section 3] for details.

PROPERTIES 2.7. With G the functor defined above, the following hold.

(1) G is an equivalence of categories from the category of $D(R, K)$-modules to that of $D(S, K)$-modules supported at $\mathbb{V}(x_{n+1}S)$, the Zariski closed set given by x_{n+1}, with inverse functor $N \mapsto \operatorname{Ann}_N(x_{n+1}S)$;

(2) M is a finitely generated R-module if and only if $G(M)$ is a finitely generated $D(S, K)$-module;

(3) M is an injective R-module if and only if $G(M)$ is an injective S-module;
(4) $\operatorname{Ass}_S G(M) = \{(P, x_{n+1})S \mid P \in \operatorname{Ass}_R M\}$;
(5) $G\left(H_I^j(M)\right) = H^{j+1}_{(I,x_{n+1})S}(M \otimes_R S)$ for every ideal $I \subseteq R$ and every $j \in \mathbb{N}$;
(6) $G\left(H_{I_s}^{j_s} \cdots H_{I_2}^{j_2} H_{I_1}^{j_1}(M)\right) \cong H^{j_s}_{(I_s,x_{n+1})S} \cdots H^{j_2}_{(I_2,x_{n+1})S} H^{j_1+1}_{(I_1,x_{n+1})S}(M \otimes_R S)$ for all ideals I_i of R and all $j_i \in \mathbb{N}$, $1 \le i \le s$; and
(7) $\operatorname{Ext}^i_S(M, G(N)) = \operatorname{Ext}^i_R(M, N)$ for all R-modules M and N, and all $i \in \mathbb{N}$.

Moreover, this functor preserves certain properties over rings of differential operators. Namely,

(1′) M is a finitely generated $D(R, K)$-module if and only if $G(M)$ is a finitely generated $D(S, K)$-module, and
(2′) $\operatorname{length}_{D(R,K)} M = \operatorname{length}_{D(S,K)} G(M)$.

3. Definition and first properties

Referring to local cohomology modules, which were reviewed in Section 2.1, we may now define the Lyubeznik numbers [**Lyu93**].

THEOREM/DEFINITION 3.1 (Lyubeznik numbers). Let $(R, \mathfrak{m}, K)$ be a local ring admitting a surjection from an n-dimensional regular local ring (S, η, K) containing a field, $\pi : S \twoheadrightarrow R$. Let $I = \ker(\pi)$. The *Lyubeznik number of R with respect to $i, j \in \mathbb{N}$ is defined as*

$$\lambda_{i,j}(R) := \dim_K \operatorname{Ext}^i_S\left(K, H_I^{n-j}(S)\right),$$

and depends only on R, i, and j; *i.e.*, this number is independent of the choice of S and of π. If $(R, \mathfrak{m}, K)$ is *any* local ring containing a field, then $\lambda_{i,j}(R) := \lambda_{i,j}(\widehat{R})$, where $\widehat{R}$ is the completion of R with respect to $\mathfrak{m}$.

SKETCH OF PROOF. We have that $\dim_K \operatorname{Ext}^i_S(K, H_I^{n-j}(S))$ is finite [**Lyu93**, **HS93**], so it remains to prove that these numbers are well defined. Let $\pi' : S' \twoheadrightarrow R$ be another surjection, where S' is a regular local ring of dimension n' that contains a field. Set $I' = \operatorname{Ker}(\pi')$, and let $\mathfrak{m}'$ be the maximal ideal of S'. Since the Bass numbers with respect to the maximal ideal are not affected by completion, we may assume that R, S, and S' are complete, so that we can take $S = K[\![x_1, \ldots, x_n]\!]$ and $S' = K[\![y_1, \ldots, y_{n'}]\!]$. Let $S'' = K[\![z_1, \ldots, z_{n+n'}]\!]$, and let $\pi'' : S'' \twoheadrightarrow R$ be the surjective map defined by $\pi''(z_j) = \pi(x_j)$ for $1 \le j \le n$ and $\pi''(z_j) = \pi'(y_{j-n})$ for $n < j \le n + n'$. Let $I'' = \operatorname{Ker}(\pi'')$. By using Properties 2.7 (5) and (7), it can be shown that:

$$\dim_K \operatorname{Ext}^i_{S''}(K, H^{n+n'-j}_{I''}(S'')) = \dim_K \operatorname{Ext}^i_S(K, H_I^{n-j}(S)) \text{ and}$$
$$\dim_K \operatorname{Ext}^i_{S''}(K, H^{n+n'-j}_{I''}(S'')) = \dim_K \operatorname{Ext}^i_{S'}(K, H^{n'-j}_{I'}(S')),$$

and we are done. □

Some basic properties of the Lyubeznik numbers are the following [**Lyu93**].

PROPERTIES 3.2. If $(R, \mathfrak{m}, K)$ is a d-dimensional local ring containing a field, then the following hold.

(1) $\lambda_{i,j}(R) = 0$ if either $j > d$ or $i > j$,
(2) $\lambda_{d,d}(R) \neq 0$, and

(3) $\lambda_{d,d}(R) = 1$ if R is analytically normal.

Due to Property 3.2 (1), we can record all nonzero Lyubeznik numbers in the following matrix.

DEFINITION 3.3 (Lyubeznik table). The *Lyubeznik table of* R is defined as the following $(d+1) \times (d+1)$ matrix:

$$\Lambda(R) := (\lambda_{i,j}(R))_{0 \le i,j \le d} =$$
$$\begin{pmatrix} \lambda_{0,0}(R) & \lambda_{0,1}(R) & \cdots & \lambda_{0,d-2}(R) & \lambda_{0,d-1}(R) & \lambda_{0,d}(R) \\ 0 & \lambda_{1,1}(R) & \cdots & \lambda_{1,d-2}(R) & \lambda_{1,d-1}(R) & \lambda_{1,d}(R) \\ 0 & 0 & \ddots & \vdots & \vdots & \vdots \\ \vdots & \vdots & \ddots & \lambda_{d-2,d-2}(R) & \lambda_{d-2,d-1}(R) & \lambda_{d-2,d}(R) \\ 0 & 0 & \cdots & 0 & \lambda_{d-1,d-1}(R) & \lambda_{d-1,d}(R) \\ 0 & 0 & \cdots & 0 & 0 & \lambda_{d,d}(R) \end{pmatrix}.$$

Note that all entries below the diagonal are zero by Property 3.2 (1).

There are algorithms to compute Lyubeznik numbers in certain cases in characteristic zero, which rely on the D-module structure of local cohomology modules [**Wal99, ÀML06**].

4. The highest Lyubeznik number

Some basic properties of the Lyubeznik numbers are that, for a d-dimensional local ring R that contains a field, $\lambda_{d,d}(R) \neq 0$, and $\lambda_{i,j}(R) = 0$ if either $i > d$ or $j > d$. (see Properties 3.2). Hence, the following definition is natural.

DEFINITION 4.1 (Highest Lyubeznik number). Given a d-dimensional local ring R that contains a field, $\lambda_{d,d}(R)$ is called the *highest Lyubeznik number of* R.

Consider the following simple example.

EXAMPLE 4.2. Let R be a d-dimensional complete local ring containing a field. Suppose that R can be written as S/I where S is a regular local ring that contains field and I is an ideal of S such that

$$H^j_I(S) \neq 0 \text{ if and only if } j = \operatorname{ht}(I). \tag{4.1}$$

One can see that the spectral sequence $E_2^{p,q} = H^p_{\mathfrak{m}}(H^q_I(S)) \underset{p}{\Rightarrow} H^{p+q}_{\mathfrak{m}}(S) = E^{p,q}_\infty$ degenerates at the E_2-page and, hence, we have that

$$\lambda_{i,j}(R) = \begin{cases} 1 & \text{if } i = j = d, \text{and} \\ 0 & \text{otherwise.} \end{cases} \tag{4.2}$$

Ideals I that satisfy (4.1) are called *cohomologically complete intersection ideals* [**HS08**]. Examples of cohomologically complete intersection ideals include ideals generated by regular sequences, and Cohen-Macaulay ideals in positive characteristic.

We say that a ring has *a trivial Lyubeznik table* if the entries of its Lyubeznik table follow the formula given in (4.2). There are rings that are not cohomologically complete intersections and still have a trivial Lyubeznik table; *e.g.*, sequentially Cohen-Macaulay rings in prime characteristic [**ÀM14**].

Example 4.2 indicates that the highest Lyubeznik number, $\lambda_{d,d}(R)$, stands out in that it is never zero, but all other Lyubeznik numbers may vanish. Our goal of this section is to give a characterization of $\lambda_{d,d}(R)$. To this end, we will consider the Hochster-Huneke graph of a local ring [**HH94a**, Definition 3.4].

DEFINITION 4.3 (Hochster-Huneke graph). Let B be a local ring. The *Hochster-Huneke graph* Γ_B *of* B is defined as follows: its vertices are the top-dimensional minimal prime ideals of B, and two distinct vertices P and Q are joined by an edge if and only if $\mathrm{ht}_B(P+Q)=1$.

Before we can state the main result in this section, we want to point out that, since $\lambda_{i,j}(R)$ does not change under any faithfully flat extension, one *cannot* use topological information about $\mathrm{Spec}(R)$ to characterize $\lambda_{d,d}(R)$. It turns out that one must consider a faithfully flat extension B of R that is complete and has a separably closed residue field (*a fortiori* strictly Henselian); more specifically, one may choose $B=\widehat{R^{sh}}$, the completion of the strict Henselization of R. (As pointed out in [**Lyu06a**], the graph Γ_B with $B=\widehat{R^{sh}}$ can be realized by a substantially smaller ring. Indeed, if k is a coefficient field of $\widehat{R}$, then there exists a finite separable extension K of k such that the graphs Γ_B and $\Gamma_{\widehat{R}\otimes_k K}$ are isomorphic. Since this finite separable extension K of k is not explicitly constructed, we choose to work with $B=\widehat{R^{sh}}$.)

To get a sense of the difference between the graph of R and that of $B=\widehat{R^{sh}}$, we consider the following example.

EXAMPLE 4.4. Let $S=\mathbb{Q}[X,Y,Z,W]_{(X,Y,Z,W)}$, define

$$I=(Z^2-3X^2,W^2-3Y^2,ZW-3XY,XW-3YZ),$$

and let $R=S/I$. Let x,y,z, and w denote the images of X,Y,Z, and W in R, respectively. One can verify that $R\cong\mathbb{Q}\left[x,y,\sqrt{3}x,\sqrt{3}y\right]_{(x,y,\sqrt{3}x,\sqrt{3}y)}$; hence, R is a domain. Therefore, the Hochster-Huneke graph of R consists solely of one vertex, so that Γ_R has only one connected component.

On the other hand, $B=\widehat{R^{sh}}=\frac{\mathbb{Q}^{sep}[\![X,Y,Z,W]\!]}{(Z-\sqrt{3}X,W-\sqrt{3}Y)(Z+\sqrt{3}X,W+\sqrt{3}Y)}$. Clearly, there are 2 vertices in the graph of B which correspond to the minimal prime ideals $(Z-\sqrt{3}X,W-\sqrt{3}Y)$ and $(Z+\sqrt{3}X,W+\sqrt{3}Y)$, respectively, and the height of the sum of these ideals is two in B. Therefore, Γ_B consists of two vertices and no edges, and Γ_B has two connected components.

It is not hard to check that $\lambda_{2,2}(R)=\lambda_{2,2}(B)=2$.

Now we can state the main result of this section. The following theorem was first proved by Lyubeznik in characteristic $p>0$ in [**Lyu06a**], and later a characteristic-free proof using different methods was given in [**Zha07**].

THEOREM 4.5. *Let R be a d-dimensional local ring that contains a field. Then $\lambda_{d,d}(R)$ is equal to the number of connected components of the Hochster-Huneke graph of $\widehat{R^{sh}}$.*

SKETCH OF PROOF. Let $B=\widehat{R^{sh}}$. Since $\lambda_{d,d}(R)$ does not change under faithfully flat extensions, we may replace R by B, and assume that R is a complete local ring that contains its separably closed coefficient field. Let Γ denote the Hochster-Huneke graph of R.

Let $\Gamma_1, \dots, \Gamma_t$ denote the connected components of Γ. For $1 \le j \le t$, let I_j be the intersection of the minimal primes of R that are vertices of Γ_j. Using the Mayer-Vietoris sequence of local cohomology (see Remark 2.5), one can prove that

$$\lambda_{d,d}(R) = \sum_{j=1}^{t} \lambda_{d,d}(R/I_j).$$

Hence, we are reduced to proving that $\lambda_{d,d}(R) = 1$ when R is equidimensional (since each R/I_j is) and Γ is connected.

We now proceed by induction on $d := \dim(R)$. When $d = 2$, the theorem has already been established by Walther in [**Wal01**] using the "Second Vanishing Theorem" of local cohomology [**HL90**, Theorem 1.1].

Suppose now that $d \ge 3$, and assume that the theorem holds for all rings of dimension less than d. Since R is complete, we can write $R = S/I$, where $(S, \mathfrak{m})$ is an n-dimensional complete regular local ring that contains a separably closed field, and I is an ideal of S. We will choose a special element $s \in S$ as follows: Note that there are only finitely many minimal elements of $\operatorname{Supp}_S(H_I^{n-d+1}(S))$ ([**Lyu93**, Corollary 3.6]). If $\operatorname{Ass}_S\left(H_I^{n-d+1}(S)\right) \neq \{\mathfrak{m}\}$, then by prime avoidance, we can fix an element $s \in \mathfrak{m}$ that is not in any minimal prime of I, nor in any minimal element of $\operatorname{Supp}_S\left(H_I^{n-d+1}(S)\right)$. On the other hand, if $\operatorname{Ass}_S\left(H_I^{n-d+1}(S)\right) = \{\mathfrak{m}\}$, then fix $s \in \mathfrak{m}$ that is not in any minimal prime of I.

Note that by our choice of s, if $\bar{s}$ denotes the image of s in R, the ring $R/\bar{s}R$ is equidimensional and $\dim(R/\bar{s}R) = d-1$. Moreover, one can prove that $\bar{s}$ satisfies the following two properties:

(1) $\lambda_{d,d}(R) = \lambda_{d-1,d-1}\left(R/\sqrt{\bar{s}R}\right)$, and

(2) The Hochster-Huneke graph of $R/\sqrt{\bar{s}R}$ is connected.

Once these are established, we have $\lambda_{d,d}(R) = \lambda_{d-1,d-1}\left(R/\sqrt{\bar{s}R}\right) = 1$ by the inductive hypotheses, completing the sketch. □

As an application of Theorem 4.5, we will give a much simpler proof of the following theorem, originally proved by Kawasaki in [**Kaw02**] using a spectral sequence argument.

THEOREM 4.6. *Let S be a regular local ring that contains a field, let I be an ideal of S, and let $d = \dim(S/I)$. If S/I satisfies Serre's S_2-condition, then $\lambda_{d,d}(S/I) = 1$.*

PROOF. Let $R = S/I$, and let $B = \widehat{R^{sh}}$. According to Theorem 4.5, it suffices to show that the Hochster-Huneke graph Γ_B is connected. As S is regular, so is also Cohen-Macaulay, it is universally catenary (*cf.* [**Mat89**, Theorem 17.9]). Thus, B is catenary. Since R satisfies the S_2 condition, so does B. Then [**Har62**, Remark 2.4.1] implies that B is equidimensional. Therefore, Γ_B must be connected by [**HH94a**, Theorem 3.6]. □

It turns out that the converse of Theorem 4.6 does not hold, as the following example from [**Kaw02**] illustrates.

EXAMPLE 4.7. Let K be a field, and let $S = K[x_1, x_2, x_3, x_4, x_5, x_6]_{(x_1, \dots, x_6)}$; moreover, let

$$I = (x_1, x_2, x_3) \cap (x_2, x_3, x_4) \cap (x_3, x_4, x_5) \cap (x_4, x_5, x_6) \cap (x_5, x_6, x_1). \tag{4.3}$$

It is clear that $\dim(S/I) = 3$, and one can check that $\lambda_{3,3}(S/I) = 1$. If we set $P = (x_1, x_2, x_3, x_5, x_6)$, then the depth of $(S/I)_P$ is 1, but the height of P in S/I is 2. Hence, S/I does not satisfy the S_2 condition.

We end this section by pointing out a related recent result of Schenzel. Suppose that S is a complete regular local ring of equal characteristic and that I is an ideal of S of height c, and let $d = \dim(S/I)$. Schenzel proved that $B := \mathrm{Hom}_S(H^c_I(S), H^c_I(S))$ is a free S-module of rank $\lambda_{d,d}(S/I)$ [**Sch11**].

5. Lyubeznik numbers for projective schemes

Throughout this section, we fix the following notation.

NOTATION 5.1. Fix a projective scheme X over a field K. Let $S = K[x_0, \cdots, x_n]$, and consider the standard grading on S; *i.e.*, $\deg(x_i) = 1$ for $0 \le i \le n$. Let $\mathfrak{m}$ denote its homogeneous maximal ideal, and let I be a homogeneous ideal of S. Under an embedding $\iota : X \hookrightarrow \mathbb{P}^n_K$, we may write $X = \mathrm{Proj}(S/I)$. Let $R = (S/I)_{\mathfrak{m}}$, the local ring at the vertex of the affine cone over X, which clearly contains a field.

In our setting, we may consider the Lyubeznik numbers of R. Theorem 4.5 has the following interesting consequence.

THEOREM 5.2. *In our setting, if K^{sep} denotes the separable closure of K, let $X_1, \ldots, X_t$ denote the d-dimensional irreducible components of $X^{sep} := X \times_K K^{sep}$. Let Γ_X be the graph on vertices $X_1, \ldots, X_t$, where X_i and X_j are joined by an edge if and only if $\dim(X_i \cap X_j) = d - 1$. Then $\lambda_{d+1,d+1}(R)$ is the number of connected components of Γ_X. Consequently, $\lambda_{d+1,d+1}(R)$ is an invariant of X, independent of the choice of embedding ι.*

This theorem prompts the following question.

QUESTION 5.3. In our setting, is it true that $\lambda_{i,j}(R)$ depends only on X, i, and j; *i.e.*, it is independent of the choice of embedding $\iota : X \hookrightarrow \mathbb{P}^n_K$?

When K has characteristic $p > 0$, the answer to Question 5.3 turns out to be affirmative[1]. The main goal of this section is to sketch the proof of the following theorem [**Zha11**, Theorem 1.1].

THEOREM 5.4. *In our setting, assume that K has characteristic $p > 0$. Then each $\lambda_{i,j}(R)$ depends only on X, i, and j; in particular, it does not depend on the choice of embedding $\iota : X \hookrightarrow \mathbb{P}^n_K$.*

Noting Remark 2.4, one can see that if the index i equals zero or one, the connection between local cohomology and sheaf cohomology is more complicated than in the other cases; consequently, the proof of Theorem 5.4 is more technical.

For this reason, we will sketch the proof of Theorem 5.4 in the case that $i \ge 2$, and will, accordingly, make this assumption for the rest of the section. Since field extensions do not affect the Lyubeznik numbers, we will also assume that K is algebraically closed. For each S-module M, set

$$\mathcal{E}^{i,j}(M) := \mathrm{Ext}^{n+1-i}_S\left(\mathrm{Ext}^{n+1-j}_S(M, S), S\right).$$

[1] Recently, it has been proved in [**Swi14**] that Question 5.3 has a positive answer for a nonsingular projective variety X in characteristic zero.

Let $\omega_X^\bullet$ denote the dualizing complex on X induced by the structure morphism $X \to K$ **[Har66]**, and for each sheaf of $\mathcal{O}_X$-modules $\mathcal{G}$, set

$$E^{i,j}(X,\mathcal{G}) := \operatorname{Ext}_X^{1-i}\left(\mathcal{E}xt^{1-j}(\mathcal{G},\omega_X^\bullet),\omega_X^\bullet\right).$$

Consider the following characteristic-free result.

PROPOSITION 5.5. *The degree zero piece of $\mathcal{E}^{i,j}(S/I)$, $\mathcal{E}^{i,j}(S/I)_0$, depends only on $X, i,$ and j; in particular, it is independent of the choice of embedding $\iota : X \hookrightarrow \mathbb{P}_k^n$.*

PROOF. If $\omega_X^\bullet$ denotes the dualizing complex on X induced by the structure morphism $X \to K$, we have the following isomorphisms.

$$\begin{aligned}
\mathcal{E}^{i,j}(S/I)_0 &= \operatorname{Ext}_S^{n+1-i}\left(\operatorname{Ext}_S^{n+1-j}(S/I,S(-n-1)),S(-n-1)\right)_0 \\
&\cong D\left(H_{\mathfrak{m}}^i\left(\operatorname{Ext}_S^{n+1-j}\left(S/I,S(-n-1)\right)\right)_0\right) && \text{by Remark 2.3} \\
&\cong D\left(H^{i-1}(\mathbb{P}_k^n,\mathcal{E}xt^{n+1-j}(\mathcal{O}_X,\omega_{\mathbb{P}_k^n}))\right) && \text{by Remark 2.4} \\
&\cong \operatorname{Ext}_{\mathbb{P}_k^n}^{n+1-i}\left(\mathcal{E}xt^{n+1-j}(\mathcal{O}_X,\omega_{\mathbb{P}_k^n}),\omega_{\mathbb{P}_k^n}\right) && \text{by Serre Duality} \\
&\cong E^{i,j}(X,\mathcal{O}_X),
\end{aligned}$$

where the last isomorphism is a consequence of Grothendieck Duality for finite morphisms. □

Now we make the transition to characteristic $p > 0$ setting. The fundamental tool in the study of rings S of characteristic $p > 0$ is the *Frobenius endomorphism*

$$F : S \to S, \text{ given by } F(s) = s^p \tag{5.1}$$

and its iterates $F^e : S \to S$ for each $e \geq 1$, given by $F^e(s) = s^{p^e}$. Note that some discussion on singularities defined using these maps appear in Section 7.

Very often one must distinguish the source ring from the target ring in the e-th iterate of the Frobenius endomorphism. We will do so by denoting the target ring S by $S^{(e)}$; *i.e.*, $S^{(e)}$ is the same as S as an abelian group, but its S-module structure is given via F^e: for each $s \in S$ and $s' \in S^{(e)}$, $s \cdot s' = s^{p^e}s'$. Using this notation, we can introduce the *e-th iterated Frobenius functor* (also called the *Peskine-Szpiro functor*), a functor from the category of S-modules to itself. It is denoted $\mathcal{F}^e$, and given an S-module M, $\mathcal{F}^e(M) = S^{(e)} \otimes_S M$. Since S is a polynomial ring in our setting, it is straightforward to check that $\mathcal{F}^e$ is an exact functor, as the image of F^e is S^{p^e}, and clearly S is a free module over S^{p^e}. One can also check the following facts about $\mathcal{F}^e$.

PROPOSITION 5.6 (Basic facts on $\mathcal{F}^e$). *The Frobenius functor $\mathcal{F}^e$ satisfies the following properties for all integers t and all $e \geq 1$, where each isomorphism is degree-preserving.*

(1) $\mathcal{F}^e(S) \cong S$.
(2) $\mathcal{F}^e(S/I) \cong S/I^{[p^e]}$.
(3) $\mathcal{F}^e\left(\operatorname{Ext}_S^t(M,N)\right) \cong \operatorname{Ext}_S^t\left(\mathcal{F}^e(M),\mathcal{F}^e(N)\right)$.

It follows from Proposition 5.6 that, for $i, j \in \mathbb{N}$, we have the composition of natural maps

$$\mathcal{E}^{i,j}(S/I) \to \mathcal{F}^e(\mathcal{E}^{i,j}(S/I)) = S^{(e)} \otimes_S \mathcal{E}^{i,j}(S/I) \xrightarrow{\sim} \mathcal{E}^{i,j}(S/I^{[p^e]}) \to \mathcal{E}^{i,j}(S/I),$$

where the first map is given by $z \mapsto 1 \otimes z$ for each $z \in \mathcal{E}^{i,j}(S/I)$, and the last map is induced by the natural surjection $S/I^{[p^e]} \twoheadrightarrow S/I$. We can restrict the composition to $\mathcal{E}^{i,j}(S/I)_0$, and we call the restriction α. Note that α is not K-linear, but α satisfies $\alpha(cz) = c^p\alpha(z)$ for every $c \in K$ and $z \in \mathcal{E}^{i,j}(S/I)_0$. Such a map is called a *p-linear structure.* Given this p-linear structure α on $\mathcal{E}^{i,j}(S/I)_0$, we consider its stable part,

$$\left(\mathcal{E}^{i,j}(S/I)_0\right)_s := \bigcap_{e \geq 1} \alpha^e\left(\mathcal{E}^{i,j}(S/I)_0\right).$$

SKETCH OF THE PROOF OF THEOREM 5.4 WHEN $i \geq 2$. One can prove that

$$\lambda_{i,j}(R) = \dim_K \left(\mathcal{E}^{i,j}(S/I)_0\right)_s.$$

Since $\mathcal{E}^{i,j}(S/I)_0$ is independent of the embedding by Proposition 5.5, it remains to show that the p-linear structure, α, on $\mathcal{E}^{i,j}(S/I)_0$ is also independent of the embedding. To this end, one can consider the the natural p-linear structure β on $E^{i,j}(X, \mathcal{O}_X)$ induced by the Frobenius endomorphism $F_X : X \to X$, and prove that the diagram

$$\begin{array}{ccc} \mathcal{E}^{i,j}(S/I)_0 & \xrightarrow{\sim} & E^{i,j}(X, \mathcal{O}_X) \\ \downarrow{\scriptstyle \alpha} & & \downarrow{\scriptstyle \beta} \\ \mathcal{E}^{i,j}(S/I)_0 & \xrightarrow{\sim} & E^{i,j}(X, \mathcal{O}_X) \end{array}$$

commutes. Thus,

$$\dim_K \left(\mathcal{E}^{i,j}(S/I)_0\right)_s = \dim_K E^{i,j}(X, \mathcal{O}_X)_s,$$

where $E^{i,j}(X, \mathcal{O}_X)_s$ is the stable part of $E^{i,j}(X, \mathcal{O}_X)$ under β, which depends only on X, i, and j, completing the proof. □

6. Further geometric and topological properties

Suppose that V is a scheme of finite type over $\mathbb{C}$ with an isolated singularity at $p \in V$; moreover, let $R = \mathcal{O}_{V,p}$. Lyubeznik remarked in **[Lyu93]** that by combining results of Ogus on connections of local cohomology modules with de Rham cohomology in **[Ogu73]** with results of Hartshorne relating de Rham cohomology and singular cohomology **[Har75]**, one can show that if $j \leq \dim(R) - 1$, then $\lambda_{0,j}(R) = \dim_{\mathbb{C}} H^j_p(V, \mathbb{C})$, where $H^j_p(V, \mathbb{C})$ denotes the j-th singular cohomology group of V with complex coefficients and support in p.

In **[GLS98]**, García López and Sabbah extend this result, computing all the Lyubeznik numbers in this case in terms of vector space dimensions of singular cohomology groups. This work employs the Riemann-Hilbert correspondence and duality for holonomic D-modules.

THEOREM 6.1. *Let V be a scheme of finite type over $\mathbb{C}$ with an isolated singularity at the point x_0. Let $R = \mathcal{O}_{V,x_0}$ and let $d = \dim(R)$. If $d = 1$, then $\lambda_{1,1}(R) = 1$ and all other $\lambda_{i,j}(R) = 0$. On the other hand, if $d \geq 2$, then if $H^j_{x_0}(V, \mathbb{C})$ denotes the j-th singular cohomology group of V with coefficients in $\mathbb{C}$ and support in x_0, the following hold.*

(1) $\lambda_{0,j}(R) = \dim_{\mathbb{C}} H^j_{x_0}(V, \mathbb{C})$ *if* $1 \leq j \leq d-1$,
(2) $\lambda_{i,d}(R) = \dim_{\mathbb{C}} H^{i+d}_{x_0}(V, \mathbb{C})$ *if* $2 \leq i \leq d$, *and*
(3) *All other* $\lambda_{i,j}(R) = 0$.

In fact, part (2) of Theorem 6.1 is deduced from part (1) via Poincaré duality, as illustrated by the Lyubeznik table:

$$\Lambda(R) = \begin{pmatrix} 0 & \lambda_{0,1}(R) & \lambda_{0,2}(R) & \cdots & \lambda_{0,d-1}(R) & 0 \\ 0 & 0 & 0 & \cdots & 0 & 0 \\ 0 & 0 & 0 & \cdots & 0 & \lambda_{0,d-1}(R) \\ 0 & 0 & 0 & \cdots & 0 & \lambda_{0,d-2}(R) \\ \vdots & \vdots & \vdots & \ddots & \vdots & \vdots \\ 0 & 0 & 0 & \cdots & 0 & \lambda_{0,2}(R) \\ 0 & 0 & 0 & \cdots & 0 & \lambda_{0,1}(R)+1 \end{pmatrix}.$$

Blickle and Bondu have found interpretations of Lyubeznik numbers in terms of étale cohomology [**BB05**], which is closely related to Theorem 6.1. Results in [**GLS98**] and [**BB05**] have been generalized in [**Bli07**]. The result is as follows.

THEOREM 6.2. *Let K be a separably closed field, and given a closed d-dimensional K-subvariety Y of a smooth n-dimensional variety X, let $x \in Y$, and let $A = \mathcal{O}_{Y,x}$. Suppose that the modules $H^{n-i}_{[Y]}(\mathcal{O}_X)$ are supported in the point x for all $i \neq d$. Then the following hold.*

(1) *For $2 \leq a \leq d$, one has that*
 (a) $\lambda_{a,d}(A) = \lambda_{0,d-a+1}(A)$ *for $a \neq d$,*
 (b) $\lambda_{d,d}(A) = \lambda_{0,1}(A) + 1$, *and*
 (c) *All other $\lambda_{a,i}(A) = 0$.*

(2) *If $\delta_{a,d}$ denotes the Kroneker delta function, then*

$$\lambda_{a,d}(R) - \delta_{a,d} = \begin{cases} \dim_{\mathbb{F}_p} H^{d-a+1}_{\{x\}}(Y_{\acute{e}t}, \mathbb{F}_p) & \text{if } K \text{ has characteristic } p, \text{ and} \\ \dim_{\mathbb{C}} H^{d-a+1}_{\{x\}}(Y_{an}, \mathbb{C}) & \text{if } K = \mathbb{C}. \end{cases}$$

7. Generalized Lyubeznik numbers

The *generalized Lyubeznik numbers* are a family of invariants associated to a local ring containing a field, which contains the Lyubeznik numbers. These invariants can capture more subtle information about the ring than the (original) Lyubeznik numbers can. Moreover, they can encode information about F-singularities in characteristic $p > 0$, and have connections to D-module characteristic cycle multiplicities in characteristic zero. Throughout our discussion, we use [**NBW14**] as our reference. The generalized Lyubeznik numbers are defined as follows.

THEOREM/DEFINITION 7.1 (Generalized Lyubeznik numbers). Let $(R, \mathfrak{m}, K)$ be a local ring containing a field and let $\widehat{R}$ be its completion at $\mathfrak{m}$. Given a coefficient field L of $\widehat{R}$, there exists a surjection $\pi : S \to \widehat{R}$, where $S = K[\![x_1, \ldots, x_n]\!]$ for some $n \geq 1$, and such that $\pi(K) = L$. For $1 \leq j \leq \ell$, fix $i_j \in \mathbb{N}$ and ideals $I_j \subseteq R$, and let $J_j = \pi^{-1}\left(I_j\widehat{R}\right) \subseteq S$. The *generalized Lyubeznik number of R with respect to L, the I_j, and the i_j* is defined as

$$\lambda^{i_\ell,\ldots,i_1}_{I_\ell,\ldots,I_1}(R; L) := \operatorname{length}_{D(S,K)} H^{i_\ell}_{J_\ell} \cdots H^{i_2}_{J_2} H^{n-i_1}_{J_1}(S).$$

This number is finite and depends only on R, L, the I_j, and the i_j (*i.e.*, it is independent of the choice of S and of π).

When $\widehat{R}$ has a unique coefficient field (*e.g.*, when K is a perfect field of characteristic $p > 0$), or when the choice of L is clear in the context, this invariant is denoted $\lambda^{i_\ell,\ldots,i_1}_{I_\ell,\ldots,I_1}(R)$.

In the definition, we may take $I_1 \subseteq \ldots \subseteq I_\ell$ without losing information: For any R-module M satisfying $H_I^0(M) = M$ for some ideal I of R, then for every ideal J of S, $H_J^i(M) \cong H_{I+J}^i(M)'$ for all $i \in \mathbb{N}$. Moreover, all $\lambda_{I_\ell,\ldots,I_1}^{i_\ell,\ldots,i_1}(R,L) = \lambda_{I_\ell,\ldots,I_2,0}^{i_\ell,\ldots,i_1}(R/I_1,L)$.

We note that Àlvarez Montaner has also given a generalization of the Lyubeznik numbers in the characteristic zero setting in [**ÀM04b**]; his invariants are defined in terms of D-module characteristic cycles.

In general, to avoid the dependence on the choice of coefficient field of $\widehat{R}$ in the definition of generalized Lyubeznik numbers, one would need to answer the following question, asked by Lyubeznik.

QUESTION 7.2 ([**Lyu02**]). Let S be a complete regular local ring of equal characteristic. For $1 \leq j \leq \ell$, fix $i_j \in \mathbb{N}$ and ideals $J_j \subseteq S$. Given any two coefficient fields K and L of S, is

$$\operatorname{length}_{D(S,K)} H_{J_\ell}^{i_\ell} \cdots H_{J_2}^{i_2} H_{J_1}^{n-i_1}(S) = \operatorname{length}_{D(S,L)} H_{J_\ell}^{i_\ell} \cdots H_{J_2}^{i_2} H_{J_1}^{n-i_1}(S)?$$

To the best of our knowledge, this question is open even in the case that $s = 1$.

As their nomenclature indicates, the generalized Lyubeznik numbers are, in fact, generalizations of the Lyubeznik numbers, as the following proposition makes explicit.

PROPOSITION 7.3. *If $(R, \mathfrak{m}, K)$ is a local ring containing a field, then $\lambda_{i,j}(R) = \lambda_{\mathfrak{m},0}^{i,j}(R; L)$ for any coefficient field L of $\widehat{R}$.*

SKETCH OF PROOF. Since the Bass numbers with respect to the maximal ideal are not affected by completion, R is, without loss of generality, complete.

Consider a surjection $\pi : S := K[\![x_1, \ldots, x_n]\!] \twoheadrightarrow R$, where $\pi(K) = L$. If $\mathfrak{n}$ denotes the maximal ideal of S, then $H_\mathfrak{n}^i H_I^{n-j}(S)$ is injective by [**Lyu93**, Corollary 3.6]. Thus, by [**Lyu93**, Lemma 1.4],

$$\dim_K \operatorname{Hom}(K, H_\mathfrak{n}^i H_I^{n-j}(S)) = \dim_K \operatorname{Ext}_S^i(K, H_I^{n-j}(S)),$$

which, in turn, equals $\lambda_{i,j}(R)$. As $H_\mathfrak{n}^i H_I^{n-j}(S)$ is supported at $\mathfrak{m}$, it is isomorphic to $E_S(K)^{\oplus \ell} \cong E_S(L)^{\oplus \ell}$ for some $\ell \in \mathbb{N}$. Since $E_S(L)$ is a simple $D(S,L)$-module, we have that $\dim_K \operatorname{Hom}(K, H_\mathfrak{n}^i H_I^{n-j}(S)) = \ell = \operatorname{length}_{D(S,L)} H_\mathfrak{n}^i H_I^{n-j}(S) = \lambda_{\mathfrak{m},0}^{i,j}(R; L)$. □

The proof that the generalized Lyubeznik numbers are well defined critically uses the key functor G defined in Subsection 2.3; we now sketch this proof.

SKETCH OF PROOF OF THEOREM/DEFINITION 7.1. First, note that the $D(S,K)$-module length of the iterated local cohomology modules is finite by [**Bjö79, Lyu93, Lyu97**]. Again, without loss of generality, we may assume that R is complete. Along with the surjection $\pi : S = K[\![x_1, \ldots, x_n]\!] \twoheadrightarrow R$, fix $S' = K[\![y_1, \ldots, y_{n'}]\!]$ and take a surjection $\pi' : S' \twoheadrightarrow R$ satisfying $\pi'(K) = L$. Set $I' = \operatorname{Ker}(\pi')$, and let $\mathfrak{m}'$ be the maximal ideal of S'. Let $J_j{}' = (\pi')^{-1}(I_j)$ for $1 \leq j \leq s$.

Let $S'' = K[\![x_1, \dots, x_n, y_1, \dots, y_{n'}]\!]$ and let π'' denote the surjection $\pi'' : S'' \twoheadrightarrow R$ defined by

$$x_i \mapsto \pi(x_i) \quad \text{for } 1 \le i \le n, \text{ and}$$
$$y_j \mapsto \pi'(y_j) \quad \text{for } 1 \le j \le n'.$$

Let $J_j'' = (\pi'')^{-1}(I_j)$ for each $1 \le j \le s$.

Since π is surjective, there exist $\sigma_j \in S$ such that $\pi(\sigma_j) = \pi''(y_j)$ for each $1 \le j \le n'$. Let $\varphi : S'' \to S$ denote the map defined by $\varphi(x_i) = x_i$ for $1 \le i \le n$ and $\varphi(y_j) = \sigma_j$ for $1 \le j \le n'$, so that φ splits the inclusion $S \hookrightarrow S''$. Since for $1 \le j \le n'$, $y_1 - \sigma_1 \in \mathrm{Ker}(\varphi)$, and the map $S''/(y_1 - \sigma_1, \dots, y_{n'} - \sigma_{n'}) \to S$ induced by φ is an isomorphism, $\mathrm{Ker}(\varphi) = (y_1 - \sigma_1, \dots, y_{n'} - \sigma_{n'})$. Thus, $J_j'' = J_j + (y_1 - \sigma_1, \dots, y_{n'} - \sigma_{n'})$. Moreover, the elements $x_1, \dots, x_n, y_1 - \sigma_1, \dots, y_{n'} - \sigma_{n'}$ form a regular system of parameters for S'', by Properties 2.7 (6) and (2′) of the functor G defined in Subsection 2.3, we obtain that

$$\mathrm{length}_{D(S,K)}\, H^{i_\ell}_{J_\ell} \cdots H^{i_2}_{J_2} H^{n-i_1}_{J_1}(S) = \mathrm{length}_{D(S'',K)}\, H^{i_\ell}_{J''_\ell} \cdots H^{i_2}_{J''_2} H^{n'+n-i_1}_{J''_1}(S''),$$

which, by an analogous argument, equals $\mathrm{length}_{D(S',K)}\, H^{i_\ell}_{J'_\ell} \cdots H^{i_2}_{J'_2} H^{n'-i_1}_{J'_1}(S')$ as well. □

Some vanishing properties of the generalized Lyubeznik numbers are as follows; *cf.* Properties 3.2.

PROPERTIES 7.4. Given any local ring R containing a field, ideals $I_1 \subseteq \ldots \subseteq I_\ell$ of R, $i_j \in \mathbb{N}$ for $1 \le j \le \ell$, and a coefficient field L of $\widehat{R}$, the following hold.

(1) $\lambda^{i_1}_{I_1}(R; L) \neq 0$ if $i_1 = \dim(R/I_1)$.
(2) $\lambda^{i_2, i_1}_{I_2, I_1}(R; L) = 0$ if $i_2 > i_1$, and is nonzero if $i_1 = \dim(R/I_1)$ and $i_2 = \dim(R/I_1) - \dim(R/I_2)$.
(3) $\lambda^{i_\ell, \dots, i_1}_{I_\ell, \dots, I_1}(R; L) = 0$ if either $i_1 > \dim(R/I_1)$, or if $i_j > \dim(R/I_{j-1})$ and $2 \le j \le \ell$.

The generalized Lyubeznik numbers behave in the following way under finite field extensions: if $K \subseteq K'$ is a finite field extension, $R = K[\![x_1, \dots, x_n]\!]$, and $R' = K'[\![x_1, \dots, x_n]\!]$, then for all ideals I_j of R and all $i_j \in \mathbb{N}$, $1 \le j \le \ell$, $\lambda^{i_\ell, \dots, i_1}_{I_\ell, \dots, I_1}(R) = \lambda^{i_\ell, \dots, i_1}_{I_\ell R', \dots, I_1 R'}(R')$.

Although the Lyubeznik numbers cannot distinguish between complete intersection rings (see Properties 3.2), or between one-dimensional rings, the generalized Lyubeznik numbers can, as the following properties show.

PROPERTIES 7.5. The following hold.

(1) Fix a complete local ring $(R, \mathfrak{m}, K)$ of dimension one and a coefficient field L of R, and let $P_1, \dots P_t$ denote the minimal primes of R. Then $\lambda^1_0(R; L) = \lambda^1_0(R/P_1; L) + \ldots + \lambda^1_0(R/P_t; L) + t - 1$.
(2) If K is a field and $S = K[\![x_1, \dots, x_n]\!]$, let $f = f_1^{\alpha_1} \cdots f_t^{\alpha_t}$, where $f_1, \dots, f_t \in S$ are irreducible and each $\alpha_i \in \mathbb{N}$. Then $\lambda^{n-1}_0(S/f) \ge \lambda^{n-1}_0(S/f_1) + \ldots + \lambda^{n-1}_0(S/f_t) + t - 1$.

Proposition 7.5 indicates that the generalized Lyubeznik numbers of the form $\lambda^j_0(R; L)$ are of specific interest. One way they can be used is to define the following invariant of a local ring containing a field.

DEFINITION 7.6 (Lyubeznik characteristic). Let R be a d-dimensional local ring containing a field, and fix a coefficient field L of $\widehat{R}$. The *Lyubeznik characteristic of R with respect to L* is defined as

$$\chi_\lambda(R;L) := \sum_{i=0}^{d}(-1)^i \lambda_0^i(R;L).$$

As a consequence of the Mayer-Vietoris sequence for local cohomology (see Remark 2.5), given ideals I and J of a local ring R containing a field, $\chi_\lambda(R/I;L)+\chi_\lambda(R/J;L) = \chi_\lambda(R/(I+J);L) + \chi_\lambda(R/I\cap J;L)$ for any coefficient field L of $\widehat{R}$. We now discuss some connections of certain generalized Lyubeznik numbers with singularities in positive characteristic.

Note that the Lyubeznik characteristic of a Stanley-Reinser ring is computed in Theorem 8.14. Suppose that R is a ring of characteristic $p > 0$. In this case, we have the Frobenius endomorphism $F : R \to R$ defined by $F(r) = r^p$, as discussed in Section 5. A variety of types of singularities can be defined via Frobenius (*e.g.*, *F-injective*, *F-pure*, and *F-regular* singularities).

If R is reduced, let $R^{1/p^e} = \{r^{1/p^e} \mid r \in R\}$, the ring obtained by adjoining the p^e-th roots of elements of R. We say that R is *F-finite* if $R^{1/p}$ is a finitely generated R-module. Throughout this discussion, we assume that R is reduced and F-finite.

We now recall some basic definitions from the theory of tight closure developed by Hochster and Huneke [**HH90, HH89**]; we also refer to [**Hun96**] for a reference. If I is an ideal of R, the *tight closure of I*, I^*, is the ideal of R consisting of all elements $z \in R$ for which there exists some $c \in R$ that is not in any minimal prime of R, such that

$$cz^q \in I^{[q]} \text{ for all } q = p^e \gg 0.$$

An ideal I is *tightly closed* if $I^* = I$. A ring is called *F-rational* if every parameter ideal is tightly closed, and is called *weakly F-regular* if every ideal of the ring is tightly closed. A ring is called *F-regular* if every localization of the ring is weakly F-regular. It is not known whether every weakly F-regular ring is F-regular. We point out that tight closure does not commute with localization in general [**BM10**].

The *test ideal* of R is defined by

$$\tau(R) = \bigcap_{J\subseteq R \text{ ideal}} (J : J^*).$$

The test ideal of $\tau(R)$ is an important object in understanding how "far" a ring is from being weakly F-regular. In particular, $\tau(R) = R$ if and only if R is weakly F-regular.

If the R-module inclusion $R \hookrightarrow R^{1/p}$ splits, R is called *F-pure*. The F-purity of a ring simplifies computations for cohomology groups and implies vanishing properties of these groups [**HR76, Lyu06b**]. If for every $c \in R$ not contained in any minimal prime of R, there exists $e \geq 1$ such that the R-module map $R \to R^{1/p^e}$ defined by $1 \mapsto c^{1/p^e}$ splits, then R is called *strongly F-regular*. A local ring $(R, \mathfrak{m})$ is *F-injective* if the action of Frobenius on the local cohomology module $H^i_{\mathfrak{m}}(R)$ is injective for every $i \in \mathbb{N}$. The relations among these properties are the following:

$$\left\{\begin{matrix}F\text{-injective}\\ \text{rings}\end{matrix}\right\} \supseteq \left\{\begin{matrix}F\text{-pure}\\ \text{rings}\end{matrix}\right\} \supseteq \left\{\begin{matrix}\text{weakly}\\ F\text{-regular}\\ \text{rings}\end{matrix}\right\} \supseteq$$

$$\left\{\begin{matrix}F\text{-regular}\\ \text{rings}\end{matrix}\right\} \supseteq \left\{\begin{matrix}\text{strongly}\\ F\text{-regular}\\ \text{rings}\end{matrix}\right\} \supseteq \left\{\begin{matrix}\text{regular}\\ \text{rings}\end{matrix}\right\}.$$

The generalized Lyubeznik numbers of the form $\lambda_0^j(R)$ can provide information on singularities in characteristic $p > 0$. In particular, results of Blickle indicate that they encode information on F-regularity and F-rationality. We present these results, as done in [**NBW14**].

THEOREM 7.7. [**Bli04**] *Suppose that $(R, \mathfrak{m}, K)$ is a d-dimensional complete local domain of characteristic $p > 0$. If R is F-injective and $\lambda_0^d(R; L)$, for L a coefficient field of R, then R is F-rational. Moreover, if K is perfect and R is F-rational, then $\lambda_0^d(R) = 1$.*

EXAMPLE 7.8. Suppose that K is a field, and let $R = K[X]$ be the polynomial ring over K in the entries of an $r \times r$ square matrix X of indeterminates, $r > 1$. Let $\mathfrak{m}$ denote the homogeneous maximal ideal of R, and let $\det X$ denote the determinant of X. When K is a perfect field of characteristic $p > 0$, $R_{\mathfrak{m}}/(\det X)R_{\mathfrak{m}}$ is strongly F-regular, so also F-rational, and $\lambda_0^{r^2-1}(R_{\mathfrak{m}}/(\det X)R_{\mathfrak{m}}) = 1$ [**HH94b**]. However, when K has characteristic zero, we have that $\lambda_0^{r^2-1}(R_{\mathfrak{m}}/(\det X)R_{\mathfrak{m}}) \geq 2$ [**Wal05**].

Work in [**NBP13**] indicates that the generalized Lyubeznik numbers can measure how "far" an F-pure hypersurface ring is from being F-regular. Take an element f of an F-finite regular local ring R of characteristic $p > 0$ such that $R/(f)$ is F-pure. Let τ_1 be the pullback of the test ideal of $R/(f)$ to R, and for $i > 1$, let τ_i denote the pullback of the test ideal of R/τ_{i-1} to R. Vassilev showed that the chain $0 \subsetneq \tau_1 \subsetneq \tau_2 \subsetneq \cdots \subsetneq \tau_s = R$ is finite and each R/τ_i is F-pure [**Vas98**]. In fact, $\lambda_0^{\dim(R/(f))}(R/(f); L) \geq s$ for any coefficient field L of $\widehat{R/(f)}$. In particular, if $\lambda_0^{\dim(R/(f))}(R/(f); L) = 1$, then $R/(f)$ is F-regular, which recovers the result of Blickle in this case (*cf.* Theorem 7.7).

QUESTION 7.9. What further geometric and algebraic properties are captured by the generalized Lyubeznik numbers of a local ring containing a field? In particular, do these numbers measure singularities in characteristic zero?

8. Lyubeznik numbers of Stanley-Reisner rings

DEFINITION 8.1 (Simplicial complex, face/simplex, dimension of a simplicial complex or face, facet). A *simplicial complex* Δ on the vertex set $[n] = \{1, \ldots, n\}$ is a collection of subsets, called *faces* or *simplices*, that is closed under taking subsets. The *dimension* of a face $\sigma \in \Delta$ is defined as $\dim(\sigma) = |\sigma| - 1$. The *dimension* of Δ, $\dim(\Delta)$, is the maximum of the dimensions of its faces; by convention, $\dim(\Delta) := -1$ if $\Delta = \varnothing$. We let $F_i(\Delta)$ denote the set of faces of Δ dimension i. A *facet* of Δ is a face of Δ that is not strictly contained in any other face.

If Δ_1 and Δ_2 are simplicial complexes on the vertex set $[n]$, then $\Delta_1 \cap \Delta_2$ and $\Delta_1 \cup \Delta_2$ are also simplicial complexes. Moreover, a simplicial complex is determined by its facets.

DEFINITION 8.2 (Alexander dual of a simplicial complex, link). The *Alexander dual of a simplicial complex* Δ on $[n]$ is defined as

$$\Delta^\vee := \{[n] \setminus \sigma \mid \sigma \in \Delta\}.$$

Given a simplex $\sigma \in \Delta$, we define its *link* inside Δ by $\text{link}_\Delta = \{\tau \in \Delta \mid \tau \cup \sigma \in \Delta \text{ and } \tau \cap \sigma = \varnothing\}$

NOTATION 8.3. Let Δ be a simplicial complex on the vertex set $[n]$, and $\sigma \in \Delta$. Then x^σ denotes the monomial $\prod_{i \in \sigma} x_i \in S = K[x_1, \ldots, x_n]$.

DEFINITION 8.4 (Stanley-Reisner ideal and ring). Given a field K, the *Stanley-Reisner ideal associated to the simplicial complex* Δ of $K[x_1, \ldots, x_n]$ is the squarefree monomial ideal $I_\Delta = \langle x^\sigma \mid \sigma \notin \Delta \rangle$. The *Stanley-Reisner ring of* Δ is $K[x_1, \ldots, x_n]/I_\Delta$.

THEOREM 8.5 (see [**MS05**, Theorem 1.7]). *Given a field K, the map sending Δ to the ideal I_Δ of $S = K[x_1, \ldots, x_n]$ defines a bijection from simplicial complexes on the vertex set $[n]$ to squarefree monomial ideals of S.*

EXAMPLE 8.6. Given a field K, let $S = K[x_1, \ldots, x_6]$ and let

$$I = (x_1x_2x_3, x_1x_6, x_2x_4, x_2x_5, x_2x_6, x_3x_6).$$

Then $I = I_\Delta$, where Δ is the following simplicial complex:

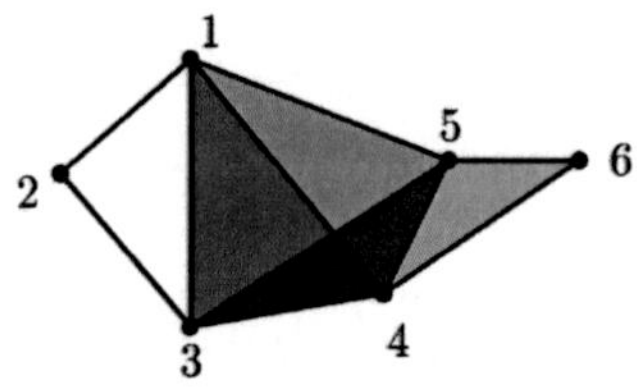

We now fix notation for the following discussion.

NOTATION 8.7. Suppose that K is a field, let $S = K[x_1, \ldots, x_n]$, and let $\mathfrak{m}$ denote its homogeneous maximal ideal $(x_1, \ldots, x_n)$. Fix the standard grading on S; *i.e.*, $\deg(x_i) = 1$ for $1 \le i \le n$. Moreover, fix a simplicial complex Δ on vertex set $[n]$, and set $I = I_\Delta$.

Suppose that I is an squarefree monomial ideal in S, and let $R = S/I$. Since I is a homogeneous ideal, $H_I^{n-j}(S)$ and $H_\mathfrak{m}^i H_I^{n-j}(S)$ are graded S-modules for all $i, j \in \mathbb{N}$ (see Remark 2.3). Results of Lyubeznik imply that $H_\mathfrak{m}^i H_I^{n-j}(S)$ is isomorphic, as S-modules, with a direct sum of $\lambda_{i,j}(R_\mathfrak{m})$ copies the injective hull $E_S(K)$ [**Lyu00c**]. Results of Y. Zhang in characteristic $p > 0$, and extended to arbitrary characteristic by Ma and the third author, imply that, in fact, $H_\mathfrak{m}^i H_I^{n-j}(S) \cong H_\mathfrak{m}^n(S)^{\oplus \lambda_{i,j}(R_\mathfrak{m})}$ as graded S-modules [**MZ13, Zha12**]. This was previously known for monomial ideals (see, for instance, [**Mus00, Yan01a**]). In particular, $\lambda_{i,j}(R_\mathfrak{m})$ is equal to the K-vector space dimension of the $(-n)$-degree piece of $H_\mathfrak{m}^i H_I^{n-j}(S)$. From now on in this section, for a graded ring R with homogenous maximal ideal $\mathfrak{m}$, $\lambda_{i,j}(R_\mathfrak{m})$.

Mustaţă used the $\mathbb{N}^n$-graded structures of S and of $H_I^j(S)$ to describe the structure of $H_I^j(S)$ in terms of Δ: for every $\alpha \in \mathbb{Z}^n$, $\left[H_I^j(S)\right]_{-\alpha} = \widetilde{H}^{j-2}(\Delta^\vee; K)$

[**Mus00**]. In addition, in [**Yan01a**], Yanagawa used this graded structure on the local cohomology modules to prove that

$$\lambda_{i,j}(R) = \dim_K \left[\operatorname{Ext}_S^{n-i}\left(\operatorname{Ext}_S^{n-j}(S/I,S)\right),S\right]_{(0,\dots,0)}.$$

We point out that there exist analogous formulas for the Bass numbers of local cohomology modules of simplicial normal Gorenstein semigroup rings with support in squarefree monomial ideals [**Yan01b**].

Let $F_\bullet$ denote the minimal graded free resolution of $I_{\Delta^\vee}$,

$$F_\bullet: \quad 0 \to F_\ell \xrightarrow{d_\ell} F_\ell \xrightarrow{d_{\ell-1}} \dots \longrightarrow F_1 \xrightarrow{d_1} F_0 \to I_{\Delta^\vee} \to 0,$$

where $d_i(F_i) \subseteq \mathfrak{m}F_{i-1}$. Since the resolution is graded, we write F_i as a direct sum of copies of S with degree shifts,

$$F_i = \bigoplus_{j\in\mathbb{Z}} S(-j)^{\beta_{i,j}},$$

where $\beta_{i,j}$ is called the (i,j)*-th graded Betti number,* an invariant of $I_{\Delta^\vee}$. We define the r*-linear strand of* $I_{\Delta^\vee}$ by

$$F_\bullet^{\langle r\rangle}: \quad 0 \to F_{n-r}^{\langle r\rangle} \xrightarrow{d_{n-r}} F_{n-r-1}^{\langle r\rangle} \xrightarrow{d_{n-r-1}} \dots \longrightarrow F_1^{\langle r\rangle} \xrightarrow{d_1} F_0^{\langle r\rangle} \to 0,$$

where

$$F_i^{\langle r\rangle} = \bigoplus_{|j|=i+r} S(-j)^{\beta_{i,j}}$$

and the maps correspond to the components in the resolution $F_\bullet$. We consider the complex $G_\bullet^{\langle r\rangle}$ given by tensoring $F_\bullet^{\langle r\rangle}$ with $S/(x_1-1,\dots,x_n-1)$ and taking the corresponding monomial matrices (see [**Mil00**]).

THEOREM 8.8 ([**ÀMV14**]). *Let K be a field and let $S = K[x_1,\dots,x_n]$. Fix Δ, a simplicial complex on vertex set $[n]$, and let $I = I_\Delta$ denote the Stanley-Reisner ideal of S associated to Δ; moreover, let $R = S/I_\Delta$ be the corresponding Stanley-Reisner ring. Let $I_{\Delta^\vee}$ be the Stanley-Reisner ideal associated to the Alexander dual, $\Delta^\vee$, of Δ, and let $G_\bullet^{\langle r\rangle}$ be the complex defined in the previous paragraph.*
Then

$$\lambda_{i,j}(R) = \dim_K H_{j-i}\left(G_\bullet^{\langle n-j\rangle}\right),$$

where $H_\bullet$ denotes homology.

REMARK 8.9. The form of Theorem 8.8 is not exactly as it appears in [**ÀMV14**]. Their result transposes the monomial matrices of $G_\bullet^{\langle r\rangle}$, and uses cohomology because the functor to the category of hypercubes that they consider is contravariant.

Since there are algorithms for computing $F_\bullet$, and so $G_\bullet^{\langle n-j\rangle}$, the previous theorem gives an algorithm to compute the Lyubeznik numbers of Stanley-Reisner rings in any characteristic. There are other algorithms and techniques for computing Lyubeznik numbers for such rings [**ÀM00, ÀM04a, Yan01a**].

The algorithm given by Theorem 8.8 was implemented by Fernández-Ramos [**ÀMFR13**] in `Macaulay2` [**GS**] (see [**ÀM13**] for details).

EXAMPLE 8.10. For S, I, and Δ from Example 8.6,

$$\Lambda(S/I) = \begin{pmatrix} 0 & 0 & 0 & 0 & 0 \\ 0 & 0 & 0 & 0 & 0 \\ 0 & 0 & 0 & 0 & 0 \\ 0 & 0 & 0 & 0 & 0 \\ 0 & 0 & 0 & 0 & 1 \end{pmatrix}.$$

EXAMPLE 8.11. Let K be a field, let $S = K[x_1, \dots, x_6]$, and let I be as defined in (4.3). Then

$$\Lambda(S/I) = \begin{pmatrix} 0 & 0 & 0 & 0 \\ 0 & 0 & 0 & 0 \\ 0 & 0 & 0 & 0 \\ 0 & 0 & 0 & 1 \end{pmatrix}.$$

EXAMPLE 8.12 ([**ÀMV14**]). Let $S = K[x_1, \dots, x_6]$ and let I denote the monomial ideal of S generated by

$$x_1x_2x_3, \ x_1x_2x_4, \ x_1x_3x_5, \ x_1x_4x_6, \ x_1x_5x_6,$$
$$x_2x_3x_6, \ x_2x_4x_5, \ x_2x_5x_6, \ x_3x_4x_5, \ \text{and} \ x_3x_4x_6.$$

The simplicial complex associated to I corresponds to a minimal triangulation of $\mathbb{P}^2_{\mathbb{R}}$, and the projective algebraic set that I defines in $\mathbb{P}^5_K$ has been called *Reisner's variety* since he introduced it in [**Rei76**, Remark 3]. If $K = \mathbb{Q}$, then

$$\Lambda(S/I) = \begin{pmatrix} 0 & 0 & 0 & 0 \\ 0 & 0 & 0 & 0 \\ 0 & 0 & 0 & 0 \\ 0 & 0 & 0 & 1 \end{pmatrix},$$

but if $K = \mathbb{F}_2$, then

$$\Lambda(S/I) = \begin{pmatrix} 0 & 0 & 1 & 0 \\ 0 & 0 & 0 & 0 \\ 0 & 0 & 0 & 1 \\ 0 & 0 & 0 & 1 \end{pmatrix}.$$

This example shows that the Lyubeznik numbers of Stanley-Reisner rings depend on the characteristic. We point out that Blickle has found examples for which these invariants exhibit "bad" behavior under reduction to positive characteristic [**Bli07**].

The generalized Lyubeznik numbers for Stanley-Reisner rings can also be computed as certain lengths in the category of straight modules [**Yan01a**], and in terms of characteristic cycle multiplicities in characteristic zero [**NBW14**]. Indeed, let K be a field and let $S = K[x_1, \dots, x_n]$, so that its completion at its homogeneous maximal ideal is $\widehat{S} = K[\![x_1, \dots, x_n]\!]$. Let $I_1, \dots, I_\ell$ be ideals of S generated by squarefree monomials, and fix $i_1, \dots, i_\ell \in \mathbb{N}$. Then

$$\begin{aligned} \lambda^{i_\ell, \dots, i_1}_{I_\ell, \dots, I_1}(\widehat{S}) &= \operatorname{length}_{\mathbf{Str}} H^{i_\ell}_{I_\ell} \cdots H^{i_2}_{I_2} H^{n-i_1}_{I_1}(\omega_S) \\ &= \sum_{\alpha \in \{0,1\}^n} \dim_k \left[H^{i_\ell}_{I_\ell} \cdots H^{i_2}_{I_2} H^{n-i_1}_{I_1}(\omega_S) \right]_{-\alpha}. \end{aligned}$$

Moreover, if K has characteristic zero, then $\lambda^{i_1, \dots, i_\ell}_{I_1, \dots, I_\ell}(\widehat{S}) = e\left(H^{i_\ell}_{I_\ell} \cdots H^{i_2}_{I_2} H^{n-i_1}_{I_1}(S)\right)$, where $e(M)$ denotes the characteristic cycle multiplicity of a $D(S, K)$-module M.

Therefore, certain generalized Lyubeznik numbers of a characteristic zero Stanley-Reisner ring can be computed using algorithms for calculating characteristic cycle multiplicities (see [**ÀM00, ÀMGLZA03, ÀM04a**]). In addition, one can also compute the Lyubeznik characteristic using these algorithms. We point out that the generalized Lyubeznik numbers may differ from the characteristic cycle multiplicities if the ideal is not monomial.

EXAMPLE 8.13. Take S and I from Example 8.12, and let $R = S/I$. If $K = \mathbb{Q}$, then $\lambda_0^4(R) = 31$ and all other $\lambda_0^j(R)$ vanish. If $K = \mathbb{F}_2$, then $\lambda_0^4(R) = 32$, $\lambda_0^3(R) = 1$, and all other $\lambda_0^j(R) = 0$. Notice that in both cases, $\chi_\lambda(R) = 31$.

Unlike the Lyubeznik numbers and the generalized Lyubeznik numbers (see Examples 8.12 and 8.13), the Lyubeznik characteristic is characteristic-independent in this setting, as the following result shows.

THEOREM 8.14 ([**NBW14**]). *Take a simplicial complex Δ on vertex set $[n]$. Let R denote the Stanley-Reisner ring of Δ, and let $\mathfrak{m}$ be its homogeneous maximal ideal. Then*

$$\chi_\lambda(R_\mathfrak{m}) = \sum_{i=-1}^{n} (-2)^{i+1} |F_i(\Delta)|.$$

EXAMPLE 8.15. Take S and Δ from Example 8.6, let $R = S/I_\Delta$, and let $\mathfrak{m}$ denote the homogeneous maximal ideal of R. Then by Theorem 8.14,

$$\begin{aligned}\chi_\lambda(R_\mathfrak{m}) &= |F_{-1}(\Delta)| + (-2)|F_0(\Delta)| + 4|F_1(\Delta)| - 8|F_2(\Delta)| + 16|F_3(\Delta)| \\ &= 1 - 2\cdot 6 + 4 \cdot 10 - 8 \cdot 5 + 16 \cdot 1 = 5.\end{aligned}$$

In this case, $\lambda_0^4(R_\mathfrak{m}) = 6$, $\lambda_0^3(R_\mathfrak{m}) = 2$, $\lambda_0^2(R_\mathfrak{m}) = 1$, and all other $\lambda_0^j(R_\mathfrak{m}) = 0$. On the other hand, we verify that

$$\chi_\lambda(R_\mathfrak{m}) = \lambda_0^4(R_\mathfrak{m}) - \lambda_0^3(R_\mathfrak{m}) + \lambda^2(R_\mathfrak{m}) = 6 - 2 + 1 = 5.$$

9. Lyubeznik numbers in mixed characteristic

Although the Lyubeznik numbers are defined only for local rings containing a field (*i.e.*, in equal characteristic), in [**NBW13**], our reference for this section, an alternate definition of Lyubeznik numbers is given for all local rings with characteristic $p > 0$ residue field. These invariants are called the *Lyubeznik numbers in mixed characteristic*, as they are, in particular, defined for local rings of mixed characteristic.

The definition of the Lyubeznik numbers in mixed characteristic, like the Lyubeznik numbers, relies on the Cohen Structure Theorems. For any field K of characteristic $p > 0$, there is a unique (up to isomorphism) complete (unramified) Noetherian discrete valuation ring of mixed characteristic of the form (V, pV, K). Any complete local ring $(R, \mathfrak{m}, K)$ of mixed characteristic $p > 0$ is the homomorphic image of some $V[\![x_1, \ldots, x_n]\!]$, where (V, pV, K) is the complete Noetherian DVR corresponding to K. As any complete local ring $(R, \mathfrak{m}, K)$ of equal characteristic $p > 0$ admits a surjection from some $K[\![x_1, \ldots, x_n]\!]$, the natural surjection $V \twoheadrightarrow K$ induces the surjective composition $V[\![x_1, \ldots, x_n]\!] \twoheadrightarrow K[\![x_1, \ldots, x_n]\!] \twoheadrightarrow R$.

The fact that the Lyubeznik numbers in mixed characteristic are well defined relies not only on the existence of a surjection from a regular local ring, but on details of its construction [**Coh46**]. These invariants are defined as follows.

THEOREM/DEFINITION 9.1. Let $(R, \mathfrak{m}, K)$ be a local ring such that K has characteristic $p > 0$. By the Cohen Structure Theorems, there exists a surjection $\pi : S \twoheadrightarrow \widehat{R}$, where $(S, \mathfrak{n}, K)$ is an n-dimensional unramified regular local ring of mixed characteristic. Let $I = \operatorname{Ker}(\pi)$, and take $i, j \in \mathbb{N}$. The *Lyubeznik number of R in mixed characteristic with respect to i and j* is defined as

$$\widetilde{\lambda}_{i,j}(R) := \dim_K \operatorname{Ext}^i_S \left(K, H^{n-j}_I(S)\right).$$

This number depends only on R, i, and j; *i.e.*, it is independent of the choice of S and of π. If $(R, \mathfrak{m}, K)$ is *any* local ring such that K has characteristic $p > 0$, then $\lambda_{i,j}(R) := \lambda_{i,j}(\widehat{R})$, where $\widehat{R}$ is the completion of R with respect to $\mathfrak{m}$.

SKETCH OF PROOF. We know that each $\dim_K \operatorname{Ext}^i_S \left(K, H^{n-j}_I(S)\right)$ is finite by [**Lyu00b, NB13**]. Without loss of generality, we may assume that R is complete. We will prove that the invariants are well defined in several steps.

Step 1. Take a coefficient ring W of R, and take a complete Noetherian DVR V with residue field K. Take surjections $\pi : S := V[\![x_1, \ldots, x_{n-1}]\!] \twoheadrightarrow R$ and $\pi' : S' := V[\![y_1, \ldots, y_{n'-1}]\!] \twoheadrightarrow R$ such that $\pi(V) = W = \pi'(V)$, and $\pi(v) = \pi'(v)$ for all $v \in V$. Let $I = \operatorname{Ker}(\pi)$ and let $I' = \operatorname{Ker}(\pi')$. We then have a surjection π'' : $S'' := V[\![x_1, \ldots, x_{n-1}, y_1, \ldots, y_{n'-1}]\!] \twoheadrightarrow R$ given by $x_i \mapsto \pi(x_i)$ and $y_j \mapsto \pi'(y_j)$. If $I'' = \ker(\pi'')$, then, similar to the proof of Theorem/Definition 3.1, Properties 2.7 (5) and (7) will imply that

$$\begin{aligned} \dim_K \operatorname{Ext}_S(K, H^{n-j}_I(S)) &= \dim_K \operatorname{Ext}_{S''}(K, H^{n+n'-j}_{I''}(S'')) \\ &= \dim_K \operatorname{Ext}_{S'}(K, H^{n'-j}_{I'}(S')). \end{aligned} \tag{9.1}$$

Step 2. Now consider the general case. Take V and V' complete Noetherian DVRs with residue field K, and surjections $\pi : S := V[\![x_1, \ldots, x_{n-1}]\!] \twoheadrightarrow R$ and $\pi' : S' := V'[\![x_1, \ldots, x_{n'-1}]\!] \twoheadrightarrow R$ such that $\pi(V) = W = \pi'(V')$. Let $I = \ker(\pi)$ and $I' = \ker(\pi')$. Now, let $\mu = \dim_K(\mathfrak{m}/\mathfrak{m}^2)$, and let $T = V[\![x_1, \ldots, x_{\mu-1}]\!]$ and $T' = V'[\![y_1, \ldots, y_{\mu-1}]\!]$. It can be shown that there exist surjections $\eta : T \twoheadrightarrow R$ and $\eta' : T' \twoheadrightarrow R$ and an isomorphism $\varphi : T \to T'$ such that $\eta' \circ \varphi = \eta$. Let $J = \ker(\eta)$ and $J' = \ker(\eta')$. Then by (9.1),

$$\begin{aligned} \dim_K \operatorname{Ext}^i_S(K, H^{n-j}_I(S)) &= \dim_K \operatorname{Ext}^i_T(K, H^{\mu-j}_J(T)), \text{ and} \\ \dim_K \operatorname{Ext}^i_{S'}(K, H^{n'-j}_{I'}(S')) &= \dim_K \operatorname{Ext}^i_{T'}(K, H^{\mu-j}_{J'}(T')). \end{aligned}$$

As φ is an isomorphism, $\dim_K \operatorname{Ext}^i_T(K, H^{\mu-j}_J(T)) = \dim_K \operatorname{Ext}^i_{T'}(K, H^{\mu-j}_{J'}(T'))$, and we are done. □

In general, the Lyubeznik numbers in mixed characteristic are not easy to compute. However, if (V, pV, K) is a complete DVR of unramified mixed characteristic $p > 0$, $R = V[\![x_1, \ldots, x_n]\!]$, and $r_1, \ldots, r_\ell \in R$ is a regular sequence, then $\widetilde{\lambda}_{i,j}(R/(r_1, \ldots, r_\ell)) = 1$ when $i = j = n + 1 - \ell$, and vanishes otherwise (*cf.* Properties 3.2).

Some vanishing properties of the Lyubeznik numbers in mixed characteristic are as follows (*cf.* Properties 3.2).

PROPERTIES 9.2. Let $(R, \mathfrak{m}, K)$ be an d-dimensional local ring such that K has characteristic $p > 0$. Then

(1) $\widetilde{\lambda}_{i,j}(R) = 0$ if either $i > d$, $j > d$, or $i > j + i$, and
(2) $\widetilde{\lambda}_{d,d}(R) \neq 0$.

Given a surjection $\pi : S \twoheadrightarrow \widehat{R}$ as in the definition of the Lyubeznik numbers in mixed characteristic, with $I = \ker(\pi)$ and $n = \dim(S)$, we have that $H_I^{n-j}(S) = 0$ for $j > \dim(S/I) = d$, and $\operatorname{inj.dim} H_I^{n-j}(S) \leq \dim H_I^{n-j}(S)+1 \leq j+1$ by [**Zho98**], proving Property 9.2 (1) for $j > d$ or $i > j + i$. However, vanishing for $j > d$ is more subtle, requiring a more detailed analysis. These results make possible the following definition (*cf.* Definitions 3.3 and 4.1).

DEFINITION 9.3 (Highest Lyubeznik number in mixed characteristic). Let $(R, \mathfrak{m}, K)$ be an d-dimensional local ring such that K has characteristic $p > 0$. Then $\widetilde{\lambda}_{d,d}(R)$ is called the *highest Lyubeznik number of R in mixed characteristic.*

QUESTION 9.4. What topological properties do the Lyubeznik numbers in mixed characteristic capture? In particular, what information is encoded in the highest Lyubeznik number in mixed characteristic?

QUESTION 9.5 ([**NBW13**]). Let R be a Cohen-Macaulay local ring of dimension d. Kawasaki showed that $\lambda_{d,d}(R) = 1$ if R contains a field [**Kaw02**]. If R has mixed characteristic, is $\widetilde{\lambda}_{d,d}(R) = 1$?

DEFINITION 9.6 (Lyubeznik table in mixed characteristic). Let $(R, \mathfrak{m}, K)$ be an d-dimensional local ring such that K has characteristic $p > 0$. The *Lyubeznik table of R in mixed characteristic* is the $(d+1) \times (d+1)$ matrix

$$\widetilde{\Lambda}(R) := \left(\widetilde{\lambda}_{i,j}(R)\right)_{0\leq i,j\leq d} =$$
$$\begin{pmatrix}
\widetilde{\lambda}_{0,0}(R) & \widetilde{\lambda}_{0,1}(R) & \cdots & \widetilde{\lambda}_{0,d-2}(R) & \widetilde{\lambda}_{0,d-1}(R) & \widetilde{\lambda}_{0,d}(R) \\
\widetilde{\lambda}_{1,0}(R) & \widetilde{\lambda}_{1,1}(R) & \cdots & \widetilde{\lambda}_{1,d-2}(R) & \widetilde{\lambda}_{1,d-1}(R) & \widetilde{\lambda}_{1,d}(R) \\
0 & \widetilde{\lambda}_{2,1}(R) & \cdots & \widetilde{\lambda}_{2,d-2}(R) & \widetilde{\lambda}_{2,d-1}(R) & \widetilde{\lambda}_{2,d}(R) \\
0 & 0 & \ddots & \vdots & \vdots & \vdots \\
\vdots & \vdots & \ddots & \widetilde{\lambda}_{d-1,d-2}(R) & \widetilde{\lambda}_{d-1,d-1}(R) & \widetilde{\lambda}_{d-1,d}(R) \\
0 & 0 & \cdots & 0 & \widetilde{\lambda}_{d,d-1}(R) & \widetilde{\lambda}_{d,d}(R)
\end{pmatrix}.$$

We note that all entries below the subdiagonal in each Lyubeznik table in mixed characteristic vanish; it is not known whether all entries below the diagonal must vanish.

QUESTION 9.7. For every local ring R with characteristic $p > 0$ residue field K, is $\widetilde{\lambda}_{i,j}(R) = 0$ for $i > j$?

When $(R, \mathfrak{m}, K)$ is a local ring of equal characteristic $p > 0$, both the Lyubeznik numbers in mixed characteristic and the (original) Lyubeznik numbers are defined. When R is Cohen Macaulay, or if $\dim(R) \leq 2$, these invariants coincide: $\lambda_{i,j}(R) = \widetilde{\lambda}_{i,j}(R)$ for all $i, j \in \mathbb{N}$. However, the Lyubeznik numbers in mixed characteristic do not, in general, agree with the Lyubeznik numbers for rings of equal characteristic $p > 0$, as the following example indicates.

EXAMPLE 9.8. Let S be the ring and I the ideal from Example 8.12, and let $K = \mathbb{F}_2$. If $R = \widehat{S/I}$, then

$$\Lambda(R) = \begin{pmatrix} 0 & 0 & 1 & 0 \\ 0 & 0 & 0 & 0 \\ 0 & 0 & 0 & 1 \\ 0 & 0 & 0 & 1 \end{pmatrix}, \quad \text{but } \widetilde{\Lambda}(R) = \begin{pmatrix} 0 & 0 & 0 & 0 \\ 0 & 0 & 0 & 0 \\ 0 & 0 & 0 & 0 \\ 0 & 0 & 0 & 1 \end{pmatrix}.$$

This example is calculated by utilizing the second computation of Àlvarez Montaner and Vahidi in Example 8.12, as well as a result on Bockstein homomorphisms of local cohomology modules of Singh and Walther [**SW11, ÀMV14**].

Acknowledgments

We send many thanks to Josep Àlvarez Montaner, Manuel Blickle, Daniel J. Hernández, Mel Hochster, Gennady Lyubeznik, Felipe Pérez, and Uli Walther for useful mathematical conversations and suggestions related to this work. We also thank Josep Àlvarez Montaner for computing several examples that appear in this survey. The first author also thanks Noé Bárcenas Torres, Fernando Galaz García, and Mónica Moreno Rocha for organizing the "Taller de Vinculación Matemáticos Mexicanos Jóvenes en el Mundo" in Guanajuato, México, where this project was initiated, as well as the National Council of Science and Technology of Mexico (CONACyT) for its support through Grant 210916. We thank the referee for a careful reading of this paper, and for several comments and suggestions that improved it.

References

[ÀM00] Josep Alvarez Montaner, *Characteristic cycles of local cohomology modules of monomial ideals*, J. Pure Appl. Algebra **150** (2000), no. 1, 1–25, DOI 10.1016/S0022-4049(98)00171-6. MR1762917 (2001d:13016)

[ÀM04a] Josep Àlvarez Montaner, *Characteristic cycles of local cohomology modules of monomial ideals. II*, J. Pure Appl. Algebra **192** (2004), no. 1-3, 1–20, DOI 10.1016/j.jpaa.2004.01.003. MR2067186

[ÀM04b] Josep Àlvarez Montaner, *Some numerical invariants of local rings*, Proc. Amer. Math. Soc. **132** (2004), no. 4, 981–986 (electronic), DOI 10.1090/S0002-9939-03-07177-6. MR2045412 (2005c:13021)

[ÀM13] Josep Àlvarez Montaner. Local cohomology supported on monomial ideals. In *Monomial ideals, computations and applications*, Lecture Notes in Mathematics, pages 125–198. 2013.

[ÀM14] Josep Àlvarez Montaner. Lyubeznik table of sequentially cohen-macaulay rings. *To Appear in Communications in Algebra*, 2014.

[ÀMFR13] Josep Àlvarez Montaner and Oscar Fernández-Ramos. Local cohomology using macaulay2. In *Monomial ideals, computations and applications*, Lecture Notes in Mathematics. 2013.

[ÀMGLZA03] Josep Àlvarez Montaner, Ricardo García López, and Santiago Zarzuela Armengou, *Local cohomology, arrangements of subspaces and monomial ideals*, Adv. Math. **174** (2003), no. 1, 35–56, DOI 10.1016/S0001-8708(02)00050-6. MR1959890 (2004a:13012)

[ÀML06] Josep Àlvarez Montaner and Anton Leykin, *Computing the support of local cohomology modules*, J. Symbolic Comput. **41** (2006), no. 12, 1328–1344, DOI 10.1016/j.jsc.2006.09.001. MR2271328 (2008a:13023)

[ÀMV14] Josep Àlvarez Montaner and Alireza Vahidi, *Lyubeznik numbers of monomial ideals*, Trans. Amer. Math. Soc. **366** (2014), no. 4, 1829–1855, DOI 10.1090/S0002-9947-2013-05862-X. MR3152714

[BB05] Manuel Blickle and Raphael Bondu, *Local cohomology multiplicities in terms of étale cohomology* (English, with English and French summaries), Ann. Inst. Fourier (Grenoble) **55** (2005), no. 7, 2239–2256. MR2207383 (2007d:14009)

[Bjö79] J.-E. Björk, *Rings of differential operators*, North-Holland Mathematical Library, vol. 21, North-Holland Publishing Co., Amsterdam-New York, 1979. MR549189 (82g:32013)

[Bli04] Manuel Blickle, *The intersection homology D-module in finite characteristic*, Math. Ann. **328** (2004), no. 3, 425–450, DOI 10.1007/s00208-003-0492-z. MR2036330 (2005a:14005)

[Bli07] Manuel Blickle, *Lyubeznik's invariants for cohomologically isolated singularities*, J. Algebra **308** (2007), no. 1, 118–123, DOI 10.1016/j.jalgebra.2006.06.029. MR2290913 (2008b:13024)

[BM10] Holger Brenner and Paul Monsky, *Tight closure does not commute with localization*, Ann. of Math. (2) **171** (2010), no. 1, 571–588, DOI 10.4007/annals.2010.171.571. MR2630050 (2011d:13005)

[BS98] M. P. Brodmann and R. Y. Sharp, *Local cohomology: an algebraic introduction with geometric applications*, Cambridge Studies in Advanced Mathematics, vol. 60, Cambridge University Press, Cambridge, 1998. MR1613627 (99h:13020)

[Coh46] I. S. Cohen, *On the structure and ideal theory of complete local rings*, Trans. Amer. Math. Soc. **59** (1946), 54–106. MR0016094 (7,509h)

[Eis95] David Eisenbud, *Commutative algebra*, Graduate Texts in Mathematics, vol. 150, Springer-Verlag, New York, 1995. With a view toward algebraic geometry. MR1322960 (97a:13001)

[GLS98] R. García López and C. Sabbah, *Topological computation of local cohomology multiplicities*, Collect. Math. **49** (1998), no. 2-3, 317–324. Dedicated to the memory of Fernando Serrano. MR1677136 (2000a:13029)

[GS] Daniel R. Grayson and Michael E. Stillman. Macaulay2, a software system for research in algebraic geometry.

[Har62] Robin Hartshorne, *Complete intersections and connectedness*, Amer. J. Math. **84** (1962), 497–508. MR0142547 (26 #116)

[Har66] Robin Hartshorne, *Residues and duality*, Lecture notes of a seminar on the work of A. Grothendieck, given at Harvard 1963/64. With an appendix by P. Deligne. Lecture Notes in Mathematics, No. 20, Springer-Verlag, Berlin-New York, 1966. MR0222093 (36 #5145)

[Har75] Robin Hartshorne, *On the De Rham cohomology of algebraic varieties*, Inst. Hautes Études Sci. Publ. Math. **45** (1975), 5–99. MR0432647 (55 #5633)

[HH89] Melvin Hochster and Craig Huneke, *Tight closure and strong F-regularity*, Mém. Soc. Math. France (N.S.) **38** (1989), 119–133. Colloque en l'honneur de Pierre Samuel (Orsay, 1987). MR1044348 (91i:13025)

[HH90] Melvin Hochster and Craig Huneke, *Tight closure, invariant theory, and the Briançon-Skoda theorem*, J. Amer. Math. Soc. **3** (1990), no. 1, 31–116, DOI 10.2307/1990984. MR1017784 (91g:13010)

[HH94a] Melvin Hochster and Craig Huneke, *Indecomposable canonical modules and connectedness*, Commutative algebra: syzygies, multiplicities, and birational algebra (South Hadley, MA, 1992), Contemp. Math., vol. 159, Amer. Math. Soc., Providence, RI, 1994, pp. 197–208, DOI 10.1090/conm/159/01509. MR1266184 (95e:13014)

[HH94b] Melvin Hochster and Craig Huneke, *Tight closure of parameter ideals and splitting in module-finite extensions*, J. Algebraic Geom. **3** (1994), no. 4, 599–670. MR1297848 (95k:13002)

[HL90] C. Huneke and G. Lyubeznik, *On the vanishing of local cohomology modules*, Invent. Math. **102** (1990), no. 1, 73–93, DOI 10.1007/BF01233420. MR1069240 (91i:13020)

[HR76] Melvin Hochster and Joel L. Roberts, *The purity of the Frobenius and local cohomology*, Advances in Math. **21** (1976), no. 2, 117–172. MR0417172 (54 #5230)

[HS93] Craig L. Huneke and Rodney Y. Sharp, *Bass numbers of local cohomology modules*, Trans. Amer. Math. Soc. **339** (1993), no. 2, 765–779, DOI 10.2307/2154297. MR1124167 (93m:13008)

[HS08] Michael Hellus and Peter Schenzel, *On cohomologically complete intersections*, J. Algebra **320** (2008), no. 10, 3733–3748, DOI 10.1016/j.jalgebra.2008.09.006. MR2457720 (2010a:13029)

[Hun96] Craig Huneke, *Tight closure and its applications*, CBMS Regional Conference Series in Mathematics, vol. 88, Published for the Conference Board of the Mathematical Sciences, Washington, DC; by the American Mathematical Society, Providence, RI, 1996. With an appendix by Melvin Hochster. MR1377268 (96m:13001)

[ILL$^+$07] Srikanth B. Iyengar, Graham J. Leuschke, Anton Leykin, Claudia Miller, Ezra Miller, Anurag K. Singh, and Uli Walther, *Twenty-four hours of local cohomology*, Graduate Studies in Mathematics, vol. 87, American Mathematical Society, Providence, RI, 2007. MR2355715 (2009a:13025)

[Kaw02] Ken-ichiroh Kawasaki, *On the highest Lyubeznik number*, Math. Proc. Cambridge Philos. Soc. **132** (2002), no. 3, 409–417, DOI 10.1017/S0305004101005722. MR1891679 (2003b:13026)

[Lyu93] Gennady Lyubeznik, *Finiteness properties of local cohomology modules (an application of D-modules to commutative algebra)*, Invent. Math. **113** (1993), no. 1, 41–55, DOI 10.1007/BF01244301. MR1223223 (94e:13032)

[Lyu97] Gennady Lyubeznik, *F-modules: applications to local cohomology and D-modules in characteristic $p > 0$*, J. Reine Angew. Math. **491** (1997), 65–130, DOI 10.1515/crll.1997.491.65. MR1476089 (99c:13005)

[Lyu00a] Gennady Lyubeznik, *Finiteness properties of local cohomology modules: a characteristic-free approach*, J. Pure Appl. Algebra **151** (2000), no. 1, 43–50, DOI 10.1016/S0022-4049(99)00080-8. MR1770642 (2001g:13038)

[Lyu00b] Gennady Lyubeznik, *Finiteness properties of local cohomology modules for regular local rings of mixed characteristic: the unramified case*, Comm. Algebra **28** (2000), no. 12, 5867–5882, DOI 10.1080/00927870008827193. Special issue in honor of Robin Hartshorne. MR1808608 (2002b:13028)

[Lyu00c] Gennady Lyubeznik, *Injective dimension of D-modules: a characteristic-free approach*, J. Pure Appl. Algebra **149** (2000), no. 2, 205–212, DOI 10.1016/S0022-4049(98)00175-3. MR1757731 (2001g:13029)

[Lyu02] Gennady Lyubeznik, *A partial survey of local cohomology*, Local cohomology and its applications (Guanajuato, 1999), Lecture Notes in Pure and Appl. Math., vol. 226, Dekker, New York, 2002, pp. 121–154. MR1888197 (2003b:14006)

[Lyu06a] Gennady Lyubeznik, *On some local cohomology invariants of local rings*, Math. Z. **254** (2006), no. 3, 627–640, DOI 10.1007/s00209-006-0963-5. MR2244370 (2007g:13024)

[Lyu06b] Gennady Lyubeznik, *On the vanishing of local cohomology in characteristic $p > 0$*, Compos. Math. **142** (2006), no. 1, 207–221, DOI 10.1112/S0010437X05001533. MR2197409 (2007b:13029)

[Mat89] Hideyuki Matsumura, *Commutative ring theory*, 2nd ed., Cambridge Studies in Advanced Mathematics, vol. 8, Cambridge University Press, Cambridge, 1989. Translated from the Japanese by M. Reid. MR1011461 (90i:13001)

[Mil00] Ezra Miller, *The Alexander duality functors and local duality with monomial support*, J. Algebra **231** (2000), no. 1, 180–234, DOI 10.1006/jabr.2000.8359. MR1779598 (2001k:13028)

[MS05] Ezra Miller and Bernd Sturmfels, *Combinatorial commutative algebra*, Graduate Texts in Mathematics, vol. 227, Springer-Verlag, New York, 2005. MR2110098 (2006d:13001)

[Mus00] Mircea Mustaţă, *Local cohomology at monomial ideals*, J. Symbolic Comput. **29** (2000), no. 4-5, 709–720, DOI 10.1006/jsco.1999.0302. Symbolic computation in algebra, analysis, and geometry (Berkeley, CA, 1998). MR1769662 (2001i:13020)

[MZ13] Linquan Ma and Wenliang Zhang, *Eulerian graded $\mathcal{D}$-modules*, Math. Res. Lett. **21** (2014), no. 1, 149–167, DOI 10.4310/MRL.2014.v21.n1.a13. MR3247047

[NB13] Luis Núñez-Betancourt, *Local cohomology modules of polynomial or power series rings over rings of small dimension*, Illinois J. Math. **57** (2013), no. 1, 279–294. MR3224571

[NBP13] Luis Núñez-Betancourt and Felipe Pérez. F-jumping and F-Jacobian ideals for hypersurfaces. *Preprint*, 2013. arXiv:1302.3327.

[NBW13] Luis Núñez-Betancourt and Emily E. Witt, *Lyubeznik numbers in mixed characteristic*, Math. Res. Lett. **20** (2013), no. 6, 1125–1143, DOI 10.4310/MRL.2013.v20.n6.a11. MR3228626

[NBW14] Luis Núñez-Betancourt and Emily E. Witt. Generalized Lyubeznik numbers. *to appear in Nagoya Mathematical Journal*, 2014. arXiv:1208.5500.

[Ogu73] Arthur Ogus, *Local cohomological dimension of algebraic varieties*, Ann. of Math. (2) **98** (1973), 327–365. MR0506248 (58 #22059)

[Rei76] Gerald Allen Reisner, *Cohen-Macaulay quotients of polynomial rings*, Advances in Math. **21** (1976), no. 1, 30–49. MR0407036 (53 #10819)

[Sch11] Peter Schenzel, *On the structure of the endomorphism ring of a certain local cohomology module*, J. Algebra **344** (2011), 229–245, DOI 10.1016/j.jalgebra.2011.07.014. MR2831938 (2012j:13026)

[SW11] Anurag K. Singh and Uli Walther, *Bockstein homomorphisms in local cohomology*, J. Reine Angew. Math. **655** (2011), 147–164, DOI 10.1515/CRELLE.2011.039. MR2806109 (2012d:13031)

[Swi14] Nicholas Switala. Lyubeznik numbers for nonsingular projective varieties. *Preprint*, 2014. arXiv:1403.5776.

[Vas98] Janet Cowden Vassilev, *Test ideals in quotients of F-finite regular local rings*, Trans. Amer. Math. Soc. **350** (1998), no. 10, 4041–4051, DOI 10.1090/S0002-9947-98-02128-X. MR1458336 (98m:13009)

[Wal99] Uli Walther, *Algorithmic computation of local cohomology modules and the local cohomological dimension of algebraic varieties*, J. Pure Appl. Algebra **139** (1999), no. 1-3, 303–321, DOI 10.1016/S0022-4049(99)00016-X. Effective methods in algebraic geometry (Saint-Malo, 1998). MR1700548 (2000h:13012)

[Wal01] Uli Walther, *On the Lyubeznik numbers of a local ring*, Proc. Amer. Math. Soc. **129** (2001), no. 6, 1631–1634 (electronic), DOI 10.1090/S0002-9939-00-05755-5. MR1814090 (2001m:13026)

[Wal05] Uli Walther, *Bernstein-Sato polynomial versus cohomology of the Milnor fiber for generic hyperplane arrangements*, Compos. Math. **141** (2005), no. 1, 121–145, DOI 10.1112/S0010437X04001149. MR2099772 (2005k:32030)

[Yan01a] Kohji Yanagawa, *Bass numbers of local cohomology modules with supports in monomial ideals*, Math. Proc. Cambridge Philos. Soc. **131** (2001), no. 1, 45–60, DOI 10.1017/S030500410100514X. MR1833073 (2002c:13037)

[Yan01b] Kohji Yanagawa, *Sheaves on finite posets and modules over normal semigroup rings*, J. Pure Appl. Algebra **161** (2001), no. 3, 341–366, DOI 10.1016/S0022-4049(00)00095-5. MR1836965 (2002k:13033)

[Zha07] Wenliang Zhang, *On the highest Lyubeznik number of a local ring*, Compos. Math. **143** (2007), no. 1, 82–88, DOI 10.1112/S0010437X06002387. MR2295196 (2008g:13029)

[Zha11] Wenliang Zhang, *Lyubeznik numbers of projective schemes*, Adv. Math. **228** (2011), no. 1, 575–616, DOI 10.1016/j.aim.2011.05.020. MR2822240 (2012j:13027)

[Zha12] Yi Zhang, *Graded F-modules and local cohomology*, Bull. Lond. Math. Soc. **44** (2012), no. 4, 758–762, DOI 10.1112/blms/bds008. MR2967243

[Zho98] Caijun Zhou, *Higher derivations and local cohomology modules*, J. Algebra **201** (1998), no. 2, 363–372, DOI 10.1006/jabr.1997.7245. MR1612378 (99c:13038)

Department of Mathematics, University of Virginia, Charlottesville, Virginia 22904

E-mail address: `lcn8m@virginia.edu`

Department of Mathematics, University of Kansas, Lawrence, Kansas 66045

E-mail address: `witt@ku.edu`

Department of Mathematics, Statistics, and Computer Science, University of Illinois at Chicago, Chicago Illinois 60607

E-mail address: `wlzhang@uic.edu`

Cluster value problem in infinite-dimensional spaces

Sofia Ortega Castillo

ABSTRACT. We survey cluster value problems for Banach algebras $H(B)$ of analytic functions on the open unit ball B of a Banach space X that contain X^* and 1. For the Banach spaces X we focus on, we report on cluster value theorems for a Banach algebra $H(B)$ and a point $x^{**} \in \bar{B}^{**}$. We also draw attention to a reduction of the cluster value problem for $H(B_Y)$ for any separable Banach space X that is an ℓ_1-sum of a sequence of finite-dimensional spaces. We conclude this work by describing the related $\bar{\partial}$ problem and surveying on strong pseudoconvexity as well as uniform pseudoconvexity in the context of Banach spaces.

1. Introduction

1.1. Some particular Banach algebras. This work is focused on some problems in Functional Analysis combined with Complex Analysis. To be specific, I study certain Banach algebras of bounded analytic functions on the open unit ball B of a complex Banach space X that contain X^* (the continuous linear functionals on X) and 1. Indeed, every element of X^* acts linearly and continuously on B, thus each element of X^* is a bounded analytic function on B (the definition of an analytic function is discussed in detail later on).

From here on, $H(B)$ will denote any such algebra. One example is the algebra $H^\infty(B)$ of all bounded analytic functions on B. The study of $H^\infty(B)$ is widely spread in mathematics, as in H^∞ control theory, H^∞ functional calculus, etc. Another well known example is the ball algebra, the analytic functions on the ball B of $\mathbb{C}^n$ that extend continuously to $\bar{B}$. There are two generalizations of the ball algebra to the infinite-dimensional case: $A_u(B)$, the uniformly continuous, bounded and analytic functions on B, and $A(B)$, the uniform limits on B of polynomials in the functions in X^*. These algebras satisfy the inclusions $A(B) \subset A_u(B) \subset H^\infty(B)$. In particular $A(B) = A_u(B)$ when X is finite-dimensional, while $A_u(B) \subsetneq H^\infty(B)$ for every Banach space X. Examples of infinite-dimensional spaces with $A(B) = A_u(B)$ include the $C(K)$ spaces for K compact, Hausdorff and dispersed, while examples of spaces with $A(B) \subsetneq A_u(B)$ include ℓ_1, L_1, $C(K)$ spaces for K compact, Hausdorff

---

2010 *Mathematics Subject Classification.* Primary 32C15, 46G20; Secondary 46B25, 46J15.

Key words and phrases. Banach algebra, cluster value theorem, $\bar{\partial}$ problem, strong pseudoconvexity.

Supported in part by NSF DMS 10-01321.

and not dispersed like ℓ_∞, L_∞ and ℓ_p, L_p for $1 < p < \infty$. Some of the reasons behind these facts are in [**6**, pp. 90–92], [**26**, section 3], [**13**, Proposition 2.36] and [**35**, Main Theorem].

One of the most important topics in the study of Banach algebras $H(B)$ is the study of its set of characters, the nonzero homomorphisms from $H(B)$ to $\mathbb{C}$, called the *spectrum of* $H(B)$, and denoted by $M_{H(B)}$. Since the Banach algebras $H(B)$ we have mentioned are commutative and have an identity, the spectrum $M_{H(B)}$ is a compact Hausdorff space ([**21**, Theorem 2.5]). The study of the spectrum is simplified by fibering it over $\bar{B}^{**}$ (the closed unit ball of X^{**}) via the surjective mapping $\pi : M_{H(B)} \to \bar{B}^{**}$ given by $\pi(\tau) = \tau|_{X^*}$. Indeed, π is well-defined because every $\tau \in M_{H(B)}$ is linear, continuous and of norm one; and π is surjective because $\pi(\delta_x) = x$ for every $x \in B$ (where $\delta_x : H(B) \to \mathbb{C}$ is defined by $f \mapsto f(x)$), and since $M_{H(B)}$ is compact, $\bar{B}^{**} = \bar{B}^{w^*} \subset \pi(M_{H(B)})$.

Another related topic of interest is the Gelfand Transform: Given $f \in H(B)$, the Gelfand Transform of f is the continuous map $\hat{f} : M_{H(B)} \to \mathbb{C}$ given by $\tau \mapsto \tau(f)$. The Gelfand Transform is a generalization of the Fourier Transform for $L_1(\mathbb{R})$ under convolution.

1.2. The Corona problem. One of the big open problems in the study of algebras $H(B)$ is the *Corona problem* on the ball, which asks whether the open unit ball of a Banach space X is dense (in the weak-star topology) in $M_{H(B)}$. Note that B can be seen as a subset of $M_{H(B)}$ via the mapping $\delta : B \to M_{H(B)}$ such that $x \mapsto \delta_x$. An equivalent formulation of the Corona problem (see [**24**, p. 163]) is that whenever $f_1, \cdots, f_n \in H(B)$ satisfy $|f_1| + \cdots + |f_n| \geq \epsilon > 0$ on B, there exist $g_1, \cdots, g_n \in H(B)$ such that $f_1 g_1 + \cdots + f_n g_n = 1$.

Carleson solved the Corona problem positively for the unit disk in $\mathbb{C}$ in 1962 [**9**]. In 1970 Gamelin [**17**] discussed the corona problem for other planar domains, solving it in cases that include finitely connected planar domains, while in 1985 Garnett and Jones [**23**] solved positively the Corona problem for connected open subsets of the extended complex plane whose complement is a subset of the real axis. Garnett poses corona problems for other interesting planar domains in [**22**]. In higher dimensions, around 1970 Cole constructed an open Riemann surface which is a counterexample to a Corona theorem [**20**]. Even more, in 1975 Sibony constructed a pseudoconvex domain in $\mathbb{C}^3$ which is also a counterexample to a Corona theorem, which he improved upon in 1987, obtaining a counterexample which is a pseudoconvex domain in $\mathbb{C}^2$ with sooth boundary and strictly pseudoconvex except at one point.

As Krantz summarizes in [**30**], there is no domain known in the plane $\mathbb{C}$ for which the corona problem is known to fail, and there is no domain known in $\mathbb{C}^n$, for $n \geq 2$, on which the corona problem is known to hold true. In particular, it is a challenging open problem to determine whether the Corona problem holds true for the unit ball or polydisk in $\mathbb{C}^n$ for $n \geq 2$.

1.3. Cluster value problems. This work deals mainly with another set of big open problems: cluster value problems. Cluster value problems are related to the Corona problem (a positive answer to a Corona problem gives a positive answer to the corresponding cluster value problem, or equivalently, a negative answer to a cluster value problem would yield a counterexample to a Corona theorem). Roughly speaking, cluster value problems involve the understanding of the cluster sets of a function $f \in H(B)$, that is, the limits of f over weak* convergent nets in B, when B is seen as a subset of X^{**} with the weak-star topology. They also involve comparing such limits with the spectrum of $H(B)$ evaluated at f.

Kakutani [**28**] was among the first ones to consider questions related to cluster value problems in 1955 for domains in the complex plane, followed by I. J. Schark [**37**] in 1961 for the unit disk Δ of the complex plane. I. J. Schark gave an explicit identification of the cluster values of a function f at a point x in the boundary of Δ with the fiber over x of the spectrum evaluated at f:

THEOREM 1.1. *Let $f \in H^\infty(\Delta)$ and $x \in \partial\Delta$. If $M_x = \{\tau \in M_{H^\infty(\Delta)} : \tau(id) = x\}$, then the range of $\hat{f}$ on M_x consists of those complex numbers ζ for which there is a sequence $\{\lambda_n\}$ in Δ with $\lambda_n \to x$ and $f(\lambda_n) \to \zeta$.*

In 1973 Gamelin [**18**] proved a cluster value theorem for the polydisk in $\mathbb{C}^n$. Moreover, he proved it for finite products of open sets in $\mathbb{C}$. Then McDonald proved in 1979 in [**32**] a cluster value theorem for the Euclidean unit ball in $\mathbb{C}^n$, and actually for any strongly pseudoconvex domain in $\mathbb{C}^n$ with smooth boundary. His proof relied on a solution by Kerzman from 1971 in [**29**, 342-345] to a $\bar{\partial}$ problem in a strongly pseudoconvex domain.

Being precise, we say a finite-dimensional *cluster value theorem for $B \subset \mathbb{C}^n$* holds when the following occurs: Suppose $f \in H^\infty(B)$, $x \in \partial B$ and $\alpha \in \mathbb{C}$. Let $M_x = \{\tau \in M_{H^\infty(B)} : \tau|_{A(B)} = \delta_x\}$. There exists $m \in M_x$ such that $m(f) = \alpha$ if and only if there is a sequence $\{x_k\}_k$ in B converging to x such that $f(x_k)$ tends to α.

Before discussing cluster value theorems for the unit ball of an infinite-dimensional Banach space, let us overview the basic theory of holomorphic functions on arbitrary Banach spaces.

1.4. Complex analysis on Banach spaces. Given U an open subset of a complex Banach space X, following Mujica [**33**], we say that $f : U \subset X \to \mathbb{C}$ is an analytic (or holomorphic) function on U if for every $x \in U$ there exists $r > 0$ and continuous polynomials on X, $(P^m f(x))_{m=0}^{\infty}$, where $P^m f(x)$ is m-homogeneous, such that, if $\|y - x\| < r$ then $f(y) = \sum_{m=0}^{\infty} P^m f(x)(y - x)$, and the convergence is uniform on the ball of radius r around x. The radius of convergence of f at x, $r_c f(x)$, is the supremum of the radius of balls for which the power series around x converges uniformly. Similarly, the radius of boundedness of f at x, $r_b f(x)$, is the supremum of the radius of balls centered at x and contained in U on which f is bounded.

An m-homogeneous polynomial $\hat{L}$ on X, for $m \in \mathbb{N}$, is the restriction to the diagonal of a m-linear mapping $L : X^m \to \mathbb{C}$, i.e. $\hat{L}(x) = L(x, x, \cdots, x)$ (and it is a constant function for $m = 0$).

The following two formulas are very useful (Theorems 4.3 and 7.13 in [**33**]).

PROPOSITION 1.1. *[Cauchy-Hadamard Formula] Let U be an open subset of a complex Banach space X. If $f : U \subset X \to \mathbb{C}$ is analytic then for each $x \in U$, $1/r_c f(x) = \limsup_{m\to\infty} \|P^m f(x)\|^{1/m}$.*

PROPOSITION 1.2. *Let U be an open subset of a complex Banach space X. If $f : U \subset X \to \mathbb{C}$ is analytic then for each $x \in U$, $r_b f(x) = \min\{r_c f(x), d_U(x)\}$ (where $d_U(x)$ denotes the distance from x to the boundary of U).*

Gamelin defines an analytic function in [**19**] differently (see (ii) in Theorem 1.2 below). However, Proposition 8.6 and Theorem 8.7 in [**33**] prove that these definitions are equivalent. Moreover, Theorem 13.16 in [**33**] proves that analyticity is equivalent to $\mathbb{C}$-differentiability. Let us summarize these results in the following theorem.

THEOREM 1.2. *Given an open subset U of a Banach space X and $f : U \subset X \to \mathbb{C}$, the following are equivalent:*

(i) f is analytic,

(ii) f is continuous and its restriction to every complex one-dimensional affine subspace of X is analytic, i.e. for every $x_0 \in U$ and direction $x \in X$, the function $\lambda \to f(x_0 + \lambda x)$ depends analytically on λ for $\lambda \in \{\zeta : x_0 + \zeta x \in U\}$,

(iii) f is locally bounded and its restriction to every complex one-dimensional affine subspace of X is analytic,

(iv) f is continuous and $f|_{U\cap M}$ is analytic for each finite-dimensional subspace M of X,

(v) f is Fréchet $\mathbb{C}$-differentiable, i.e. for each point $x \in U$ there exists a $\mathbb{C}$-linear mapping $L \in X^*$ such that $\lim_{y\to x} \frac{\|f(y)-f(x)-L(y-x)\|}{\|y-x\|} = 0$.

Some examples [**33**, Section 5] of analytic functions are the following:

(i) Polynomials (finite sums of n-homogeneous polynomials),

(ii) Power series $\sum_{m=0}^{\infty} P_m(x)$ with infinite radius of convergence (where each P_m is an m-homogeneous polynomial),

(iii) Power series of the form $\sum_{m=0}^{\infty}(x_m^*)^m$, where $\{x_m^*\} \subset X^*$ and $x_m^* \xrightarrow{w^*} 0$.

Also, [**33**, Sections 5,7] extends the following classical properties to the infinite-dimensional setting.

PROPOSITION 1.3. *[Identity Principle] If f is analytic on a connected open set U and identically zero on a nonvoid open set $V \subset U$, then f is identically zero on all of U.*

PROPOSITION 1.4. *[Open Mapping Principle] If f is analytic and non-constant on a connected open set U, then f is an open mapping.*

PROPOSITION 1.5. *[Maximum Principle] If f is analytic on a connected open set U and $|f|$ attains its supremum there, then f is constant.*

PROPOSITION 1.6. *[Liouville's theorem] If f is analytic and bounded on all X, then f is constant.*

PROPOSITION 1.7. *[Cauchy Integral Formula] Suppose that $f : U \to \mathbb{C}$ is holomorphic. Let $a \in U$, $t \in X$ and $r > 0$ such that $a + \zeta t \in U$ $\forall \zeta$ in the disk $\bar{\Delta}(0, r)$. Then $\forall m \in \mathbb{N}_0$: $P^m f(a)(t) = \frac{1}{2\pi i}\int_{|\zeta|=r} \frac{f(a+\zeta t)}{\zeta^{m+1}} d\zeta$.*

PROPOSITION 1.8. *[Cauchy Inequality] Suppose that $f : U \to \mathbb{C}$ is holomorphic. Let $a \in U$, $t \in X$ and $r > 0$ such that $a + \zeta t \in U$ $\forall \zeta$ in the disk $\bar{\Delta}(0, r)$. Then $\forall m \in \mathbb{N}_0$: $\|P^m f(a)(t)\| \leq r^{-m} \sup_{|\zeta|=r} \|f(a + \zeta t)\|$.*

PROPOSITION 1.9. *[The Schwarz Lemma] Suppose that $f : U \to \mathbb{C}$ is holomorphic. Let $a \in U$ and $r > 0$ such that the ball $B(a, r) \subset U$, and suppose that f is bounded on $B(a, r)$ by C. Then $\forall x \in B(a, r)$:*

$\|f(x) - f(a)\| \leq 2C\frac{\|x-a\|}{r}$.

Since $r_b f(a) = \min\{r_c f(a), d_U(a)\}$ whenever $a \in U$ and $f : U \to \mathbb{C}$ is holomorphic, there is a classical property that does not extend to the infinite dimensional setting: The following example [**33**, p. 54] exhibits that the radius of convergence of the power series of a holomorphic function around a point is not at least the distance of the point to the boundary of the domain.

PROPOSITION 1.10. *Suppose that $(x_m^*)_m \subset X^*$ is a sequence such that $\|x_m^*\| = 1$ $\forall m$ and $x_m^* \xrightarrow{w^*} 0$. Then $f = \sum_{m=0}^{\infty} (x_m^*)^m$ is holomorphic on all of X but has radius of convergence at 0 equal to 1.*

When X is infinite-dimensional, such a sequence always exists due to the Josefson-Nissenzweig Theorem [**12**, pp. 219–225].

REMARK 1.1. If X and Y are complex Banach spaces, and $U \subset X$ is open, we say that $f : U \to Y$ is holomorphic if $y^* \circ f$ is holomorphic for all $y^* \in Y^*$. One can check that Theorem 1.2 can be extended to this setting (see [**33**] for the details).

We now have enough tools to discuss the cluster value problem in arbitrary Banach spaces.

1.5. Cluster value problems for infinite-dimensional Banach spaces. For arbitrary Banach spaces X, the *cluster value theorem for $H(B)$* asserts that, for a given $x^{**} \in \bar{B}^{**}$, the sets of cluster values

$$Cl_B(f, x^{**}) := \{\lambda : f(x_\alpha) \to \lambda,\ x_\alpha \xrightarrow{w^*} x^{**}\}$$

coincides with the evaluation of the fiber $M_{x^{**}}(B) = M_{x^{**}} := \pi^{-1}(x^{**})$ at f,

$$\hat{f}(M_{x^{**}}(B)) = \{\tau(f) : \tau \in M_{x^{**}}\},$$

for all $f \in H(B)$.

Aron, Carando, Gamelin, Lasalle and Maestre observed in [**5**] that for every $x^{**} \in \bar{B}^{**}$ we have the inclusion

$$Cl_B(f, x^{**}) \subset \widehat{f}(M_{x^{**}}(B)),\ \forall f \in H(B), \tag{1.1}$$

thus to prove the cluster value theorem for $H(B)$ it is enough to show $\widehat{f}(M_{x^{**}}(B)) \subset Cl_B(f, x^{**})$, $\forall f \in H(B)$, $x^{**} \in \bar{B}^{**}$.

Aron, Carando, Gamelin, Lasalle and Maestre also showed in [**5**] that the cluster value theorem holds at every $x^{**} \in \bar{B}^{**}$ if and only if whenever $f_1, \cdots, f_{n-1} \in A(B)$ and $f_n \in H(B)$ satisfy $|f_1| + \cdots + |f_n| \geq \epsilon > 0$ on B, there exist $g_1, \cdots, g_n \in H(B)$ such that $f_1 g_1 + \cdots f_n g_n = 1$. It now remains clear that a negative answer to a cluster value problem would yield a counterexample to a Corona problem.

When B is the unit ball of an infinite-dimensional Banach space, there are no known solutions to the Corona problem. However, Aron, Carando, Gamelin, Lasalle and Maestre proved in 2012 in [**5**] a cluster value theorem at the origin for the algebra $A_u(B)$ when X has a shrinking 1-unconditional basis. Examples of such spaces X include ℓ_p for $1 < p < \infty$ and c_0, but not ℓ_1, ℓ_∞, $L_p(0,1)$ for $1 \leq p \neq 2 \leq \infty$.

THEOREM 1.3. *If X is a Banach space with a shrinking 1-unconditional basis, then the cluster value theorem holds for $A_u(B)$ at $x = 0$,*

$$Cl_B(f, 0) = \hat{f}(M_0), \quad f \in A_u(B).$$

Aron, Carando, Gamelin, Lasalle and Maestre [5] also proved a cluster value theorem at all points of the closed unit ball of X for the algebra $A_u(B)$ when X is a Hilbert space. The main idea is to translate the problem to 0 via certain automorphisms of B, then to use the cluster value theorem at 0 for Banach spaces with a shrinking 1-unconditional basis, and to apply certain peak functions for points in ∂B.

THEOREM 1.4. *If X is a Hilbert space, then the cluster value theorem holds for $A_u(B)$ at every $x \in \bar{B}$,*

$$Cl_B(f,x) = \hat{f}(M_x), \quad f \in A_u(B_X), x \in \bar{B}.$$

The same authors [5] proved a cluster value theorem at all points of the closed unit ball of X^{**} for the algebra of bounded analytic functions, denoted by $H^\infty(B)$, when X is c_0 (the space of null sequences). The proof repeatedly uses a lemma based on a solution to a $\bar{\partial}$ equation.

THEOREM 1.5. *If X is the Banach space c_0 of null sequences, then the cluster value theorem holds for $H^\infty(B)$ at every $x \in \bar{B}^{**}$,*

$$Cl_B(f,x) = \hat{f}(M_x), \quad f \in H^\infty(B), x \in \bar{B}^{**}.$$

Note that the balls of Hilbert space and c_0^{**} are infinite-dimensional analogues of the unit ball and the polydisk of Euclidean space, respectively.

It is open whether L_p satisfies the cluster value problem, for $A_u(B_{L_p})$ or $H^\infty(B_{L_p})$, at any point of B_{L_p}, $1 \leq p \neq 2 \leq \infty$. Nevertheless, Lemma 4.4 in [15] implies a cluster value theorem for each point in ∂B and the algebra $A_u(B)$, when B is the unit ball of a uniformly convex Banach space, like ℓ_p and L_p, for $1 < p < \infty$. Also, by [1, Theorem 2.6] and [5, Corollary 2.5], there is a cluster value theorem for the algebra $A_u(B_{\ell_1})$ and each point in ∂B_{ℓ_1}, because for each boundary point there is a function in $A(B_{\ell_1})$ peaking at it. We would like to investigate what happens at interior points.

What follows is a survey of our recent investigations of the cluster value problem, joint with W. B. Johnson.

2. Recent cluster value theorems

2.1. A look at finite-dimensional subspaces of a Banach space. Generalizing the ideas and techniques in [5], the following cluster value theorem is proved in [26]:

THEOREM 2.1. *Suppose that for each finite-dimensional subspace E of X^* and $\epsilon > 0$ there exists a finite rank operator S on X so that $\|(I - S^*)|_E\| < \epsilon$ and $\|I - S\| = 1$. Then the cluster value theorem holds for $A_u(B)$ at 0.*

For example, Banach spaces with a shrinking reverse monotone Finite Dimensional Decomposition (FDD) satisfy the conditions of the previous theorem. Recall that a Banach space X has a FDD if there exist finite-dimensional subspaces $(E_n)_n$ of X such that $\forall x \in X$, $\exists!$ $(x_n)_n$ with $x_n \in E_n$ and $x = \sum x_n$; the FDD is reverse monotone if the maps $Q_n(x) = \sum_{i=n}^\infty x_i$ satisfy $\|Q_n\| = 1$ $\forall n$, and shrinking if $\forall x^* \in X^*$, $\lim_{n\to\infty} \|x^*|_{\text{span}(\cup_{j=n}^\infty E_j)}\| = 0$.

The proof of Theorem 2.1 relies on the following proposition in [26] (a generalization of Lemma 3.4 in [5]):

PROPOSITION 2.1. *Let S be a finite rank operator on X such that $P = I - S$ has norm one. If $\phi \in M_0(B) = \{\tau \in A_u(B) : \tau|_{A(B)} = \delta_0\}$, then $\hat{f}(\phi) = \widehat{f \circ P}(\phi)$, for all $f \in A_u(B)$.*

The previous proposition, however, does not hold for $H^\infty(B)$, as exhibited in the following example by Aron [**26**, Example 2.3]:

EXAMPLE 2.1. *There exists a finite rank operator S on ℓ_2 so that $P = I - S$ has norm one, and there exist $\phi \in M_0(B_{\ell_2}) = \{\tau \in H^\infty(B) : \tau|_{A(B)} = \delta_0\}$ as well as $f \in H^\infty(B_{\ell_2})$ so that $\hat{f}(\phi) \neq \widehat{f \circ P}(\phi)$.*

Theorem 2.1 and Proposition 2.1 suggest a relationship between the cluster value problem in a Banach space and its finite-codimensional subspaces. Johnson and I [**26**] established the following relationship with the help of Aron and Maestre:

PROPOSITION 2.2. *If Y is a closed finite-codimensional subspace of X and $f \in A_u(B)$, then $Cl_B(f, 0) = Cl_{B_Y}(f|_Y, 0)$, where B_Y is the unit ball of Y.*

Since c_0 satisfies a cluster value theorem for $A_u(B_{c_0})$ and c_0 is 1-codimensional in c then Proposition 2.2 suggests that c satisfies a cluster value theorem, but [**26**, Example 2.7] shows that c does not satisfy the hypothesis of Theorem 2.1:

EXAMPLE 2.2. *Let $L \in B_{c^*}$ be given by*

$$L((c_n)_n) = \lim_{n\to\infty} c_n.$$

If $S : c \to c$ is a finite rank operator with $||(S^* - I_{c^*})L|| < \epsilon$, then $||S - I_c|| \geq 2 - \epsilon$.

However, c is isomorphic to c_0 and $A_u(B_{c_0}) = A(B_{c_0})$, so by [**26**, Lemma 2.9], $A_u(B_c) = A(B_c)$, so c satisfies a cluster value theorem for $A_u(B_c)$. Moreover, using results in [**7**], [**35**] and [**3**, Section 5.4], it is further obtained in [**26**] that $A_u(B) = A(B)$ for all $X = C(K)$, where K is compact, Hausdorff and dispersed, so X satisfies a cluster value theorem for $A_u(B)$.

2.2. A cluster value theorem for spaces of continuous functions. Following the line of proof of Theorem 5.1 in [**5**], and using that $C(K)^* = \ell_1(K)$ when K is compact, Hausdorff and dispersed, in [**26**] a cluster value theorem is obtained for $H^\infty(B)$ when $X = C(K)$, and K is compact, Hausdorff and dispersed.

THEOREM 2.2. *If X is the Banach space $C(K)$, for K compact, Hausdorff and dispersed, then the cluster value theorem holds for $H^\infty(B)$ at every $x \in \bar{B}^{**}$.*

We do not know the answer to the cluster value problem for other spaces $C(K)$, however, we can give a partial answer to the following modification of the cluster value problem.

2.3. The cluster value problem for $H^\infty(B)$ over $A_u(B)$. In [**26**] is considered the following cluster value problem: Given $f_0^{**} \in \overline{B}^{**}$, the cluster value problem for $H^\infty(B)$ over $A_u(B)$ at f_0^{**} asks whether for all $\psi \in H^\infty(B)$ and $\tau \in \mathcal{M}_{f_0^{**}}(B)$ ($\mathcal{M}_{f_0^{**}}$ is $\pi^{-1}(\delta_{f_0^{**}})$ for the restriction map $\pi : M_{H^\infty(B)} \to M_{A_u(B)}$), can we find a net $(f_\alpha) \subset B$ such that $\psi(f_\alpha) \to \tau(\psi)$ and f_α converges to f_0^{**} in the polynomial-star topology, i.e. the smallest topology that makes the canonical extension of a polynomial on X to X^{**} continuous (that we denote by $\tau(\psi) \in \mathsf{Cl}_B(\psi, f_0^{**})$)? It is easy to check that $\mathsf{Cl}_B(\psi, f_0^{**}) \subset \widehat{\psi}(\mathcal{M}_{f_0^{**}}(B))$, $\forall \psi \in H^\infty(B)$.

In [**26**] it is checked that the cluster value problem for $H^\infty(B)$ over $A_u(B)$ coincides with the cluster value problem for $H^\infty(B)$ when $A_u(B) = A(B)$, and it is shown that the fiber $\mathcal{M}_0(B_{C(K)})$ contains a copy of B_{ℓ_∞}, by using results in [**3**, Sections 4.3, 2.5], [**10**], [**4**], and [**19**].

To study the cluster value problem for $H^\infty(B)$ over $A_u(B)$ when $B = B_{C(K)}$ of an arbitrary $C(K)$ space, in [**26**] the following family of automorphisms $(T_{f_0})_{f_0 \in B}$ is described:

PROPOSITION 2.3. *Let $f_0 \in B = B_{C(K)}$. $T_{f_0} : B \to B$ given by*

$$T_{f_0}(f) = \frac{f - f_0}{1 - \overline{f_0} \cdot f} \quad \forall f \in B,$$

is biholomorphic.

The previous conclusion is extended in [**26**] to the open unit ball of the second dual of $C(K)$ by rewriting $\frac{f-f_0}{1-\overline{f_0}\cdot f}$ as $(f - f_0)\sum_{n=0}^\infty (\overline{f_0} f)^n$, and using results in [**14**] and [**36**, Section 1.17] that show that $C(K)^{**}$ is a commutative C^*−algebra that extends the C^* structure of $C(K)$.

PROPOSITION 2.4. *Given $f_0^{**} \in B_{C(K)^{**}}$, let $T_{f_0^{**}} : B_{C(K)^{**}} \to B_{C(K)^{**}}$ be given by*

$$T_{f_0^{**}}(f^{**}) = (f^{**} - f_0^{**}) \sum_{n=0}^{\infty} (\overline{f_0^{**}} f^{**})^n \quad ,\forall f^{**} \in B_{C(K)^{**}}.$$

Then $T_{f_0^{**}}$ is biholomorphic.

Imitating Lemmas 4.3 and 4.4 in [**5**], it is shown in [**26**] that for an arbitrary $C(K)$ space, the family of automorphisms $(T_{f_0})_{f_0 \in B}$ of B, given by $T_{f_0} : f \to \frac{f-f_0}{1-\overline{f_0} f}$ satisfies the following proposition:

PROPOSITION 2.5. *For each $f_0 \in B$, the biholomorphic function T_{f_0} induces a homeomorphism $\hat{T}_{f_0}$ on the spectrum $M_{H(B)}$, where H denotes either the algebra A_u or the algebra H^∞, that maps $\mathcal{M}_{f_0}(B)$ homeomorphically onto $\mathcal{M}_0(B)$ (where $\mathcal{M}_{f_0}(B) = \{\tau \in M_{H^\infty(B)} : \tau|_{A_u(B)} = \delta_{f_0}\}$).*

As a consequence of Proposition 2.5, it is obtained that the cluster value theorem of $H^\infty(B)$ over $A_u(B)$ at 0 is equivalent to the cluster value theorem of $H^\infty(B)$ over $A_u(B)$ at every $f_0 \in B$, when $X = C(K)$ [**26**].

COROLLARY 2.1. *If X is a Banach space $C(K)$, then the cluster value theorem of $H^\infty(B)$ over $A_u(B)$ at 0 is equivalent to the cluster value theorem of $H^\infty(B)$ over $A_u(B)$ at every $f_0 \in B$.*

Let us note that, from the Gelfand Representation Theorem, this corollary holds for any nonzero unital commutative C^*-algebra.

The previous result is a reduction of a cluster value problem at any point in the ball of $C(K)$ to the origin. In the next section we draw attention to a reduction of a cluster value problem for any separable Banach space to a space with a more specific structure.

2.4. The cluster value problem for separable spaces. In [**27**] it is proven that for any separable Banach space Y, a cluster value problem for $H(B_Y)$ ($H = H^\infty$ or $H = A_u$) can be reduced to a cluster value problem for $H(B_X)$ for some Banach space X that is an ℓ_1-sum of a sequence of finite-dimensional spaces. The proof relies on the construction of an isometric quotient map from a suitable X to Y that induces an isometric algebra homomorphism from $H(B_Y)$ to $H(B_X)$ with 1-complemented range, where the projection mapping is built using ultrafilter techniques. Other tools include the infinite-dimensional version of Schwarz' Lemma, as well as such familiar one complex variable results as Cauchy's inequality and Montel's theorem. This is the content of the following two lemmas [**27**]:

LEMMA 2.1. *Let Y be a separable Banach space and $Y_1 \subset Y_2 \subset Y_3 \subset \dots$ an increasing sequence of finite dimensional subspaces whose union is dense in Y. Set $X = (\sum Y_n)_1$. Then the isometric quotient map $Q : X \to Y$ defined by*

$$Q(z_n)_n := \sum_{n=1}^{\infty} z_n$$

induces an isometric algebra homomorphism $Q^\# : H(B_Y) \to H(B_X)$, where H denotes either the algebra A_u or the algebra H^∞.

The idea behind is to use the little open mapping theorem to get $Q(B_X) = B_Y$. Then a left inverse to $Q^\#$, that allows to go back to $H(B_Y)$, is found.

LEMMA 2.2. *Under the assumptions of the previous lemma, there is a norm one algebra homomorphism $T : H(B_X) \to H(B_Y)$ so that $T(X^*) \subset Y^*$ and $T \circ Q^\# = I_{H(B_Y)}$.*

THEOREM 2.3. *Let Y be a separable Banach space and $Y_1 \subset Y_2 \subset Y_3 \subset \dots$ an increasing sequence of finite dimensional subspaces whose union is dense in Y. Set $X = (\sum Y_n)_1$. Let H denote either the algebra A_u or the algebra H^∞. If $H(B_X)$ satisfies the cluster value theorem at every $x^{**} \in \overline{B_X}^{**}$ then $H(B_Y)$ satisfies the cluster value theorem at every $y^{**} \in \overline{B_Y}^{**}$.*

A clear consequence pointed out in [**27**] is that if ℓ_1 satisfies a cluster value theorem, then so does L_1. It is also shown in [**27**] that if every Banach space with an unconditional basis satisfies the cluster value theorem for an algebra H, then so does every separable Banach lattice.

3. Some open $\bar\partial$ and cluster value problems

The ideas of McDonald in [**32**] suggest that a solution to a $\bar\partial$ problem for a strongly pseudoconvex unit ball B may help solve the respective cluster value problem if the solution is weakly continuous. The $\bar\partial$ problem for the open unit ball of a Banach space is open, but has been solved positively (under certain conditions) for the unit ball of ℓ_1 by Lempert in [**31**]. Let us describe the $\bar\partial$ problem in open subsets of Banach spaces:

Let X and Y denote complex Banach spaces, and $X_\mathbb{R}$ and $Y_\mathbb{R}$ denote the respective previous spaces seen as real Banach spaces. For every $m \in \mathbb{N}$, $L(^m X_\mathbb{R}, Y_\mathbb{R})$ denotes the continuous m-linear mappings $A : X_\mathbb{R}^m \to Y_\mathbb{R}$, while $L^a(^m X_\mathbb{R}, Y_\mathbb{R})$ denotes the continuous m-linear mappings $A : X_\mathbb{R}^m \to Y_\mathbb{R}$ that are *alternating*, i.e.

$$A(x_{\sigma(1)}, \cdots, x_{\sigma(m)}) = (-1)^\sigma A(x_1, \cdots, x_m), \ \forall \sigma \in S_m \text{ and } x_1, \cdots, x_m \in X.$$

Also, given $m \in \mathbb{N}$ and $p, q \in \mathbb{N}_0$ such that $p+q=m$, $L^a({}^{p,q}X_{\mathbb{R}}, Y_{\mathbb{R}})$ is the subspace of $A \in L^a({}^m X_{\mathbb{R}}, Y_{\mathbb{R}})$ such that

$$A(\lambda x_1, \cdots, \lambda x_m) = \lambda^p \bar{\lambda}^q A(x_1, \cdots, x_m), \ \forall \lambda \in \mathbb{C} \text{ and } x_1, \cdots, x_m \in X;$$

while $L^{apq}({}^m X_{\mathbb{R}}, Y_{\mathbb{R}})$ denotes the subspace of all $A \in L({}^m X_{\mathbb{R}}, Y_{\mathbb{R}})$ which are alternating in the first p variables and are alternating in the last q variables.

The following definition can be found in [**33**, p.107].

DEFINITION 3.1. *Let U be an open subset of the complex Banach space X and let $f : U \to Y$ be an $\mathbb{R}$-differentiable mapping. Let $Df(a)$ denote the real differential of f at a. Define $D'f(a)$ and $D''f(a)$ by*

$$D'f(a)(t) = 1/2[Df(a)(t) - iDf(a)(it)],$$

$$D''f(a)(t) = 1/2[Df(a)(t) + iDf(a)(it)],$$

for every $t \in X$. Note that $D'f(a)$ is $\mathbb{C}$-linear while $D''f(a)$ is $\mathbb{C}$-antilinear.

Given $A \in L({}^m X_{\mathbb{R}}, Y_{\mathbb{R}})$, define $A^a \in L^a({}^m X_{\mathbb{R}}, Y_{\mathbb{R}})$ by

$$A^a(x_1, \cdots, x_m) = \frac{1}{m!} \sum_{\sigma \in S_m} (-1)^\sigma A(x_{\sigma(1)}, \cdots, x_{\sigma(m)}), \ \forall x_1, \cdots, x_m \in X.$$

Given U an open subset of X and $p, q \in \mathbb{N}_0$, let

$$C^\infty_{p,q}(U, Y) := C^\infty(U, L^a({}^{p,q}X_{\mathbb{R}}, Y_{\mathbb{R}})).$$

Then for each $f \in C^\infty_{p,q}(U, Y)$, $\bar{\partial} f \in C^\infty_{p,q+1}(U, Y)$ is given by

$$\bar{\partial} f(x) = (m+1)[D''f(x)]^a, \ \forall x \in U.$$

REMARK 3.1. Since $D''f(x) \in L(X_{\mathbb{R}}, L^a({}^m X_{\mathbb{R}}, Y_{\mathbb{R}})) = L^{a1m}({}^{m+1}X_{\mathbb{R}}, Y_{\mathbb{R}})$, then Proposition 18.6 in [**33**] implies that $\forall t_1, \cdots, t_{m+1} \in X$,

$$\begin{aligned} \bar{\partial} f(x)(t_1, \cdots, t_{m+1}) &= (m+1)[D''f(x)]^{a1m}(t_1, \cdots, t_{m+1}) \\ &= \frac{m+1}{m+1} \sum_{\sigma \in S_{1m}} (-1)^\sigma D''f(x)(t_{\sigma(1)}, \cdots, t_{\sigma(m+1)}) \end{aligned}$$

where S_{1m} denotes the set of all permutations $\sigma \in S_{m+1}$ such that $\sigma(1) < \cdots < \sigma(m)$, so

$$\bar{\partial} f(x)(t_1, \cdots, t_{m+1}) = \sum_{j=1}^{m+1} (-1)^{j-1} D''f(x)(t_j)(t_1, \cdots, t_{j-1}, t_{j+1}, \cdots, t_{m+1}).$$

The $\bar{\partial}$ problem for $g \in C^\infty_{p,q+1}(U, Y)$ asks whether the equation $\bar{\partial} g = 0$ implies the existence of $f \in C^\infty_{p,q}(U, Y)$ such that $\bar{\partial} f = g$.

Because of the ideas of Lempert in [**31**] and of Kerzman in [**29**, pp. 342–345], it is worthwhile to solve the cluster value problem for Banach spaces whose unit ball is strongly pseudoconvex. First off, a definition of a strongly pseudoconvex domain in an infinite-dimensional Banach space is provided in [**34**].

3.1. Strong pseudoconvexity in infinite-dimensional Banach spaces. There are certain notions of pseudoconvexity in the literature ([**33**, p.274], [**25**, Theorem 2.6.12]) that hint towards a plausible extension of the definition of a strongly pseudoconvex domain to an infinite-dimensional Banach space (see [**29**] and [**39**]). The following definitions in [**34**] build on the ideas exposed in Chapter VIII of [**33**].

DEFINITION 3.2. *Let U be an open subset of a complex Banach space X.*

A function $f : U \to [-\infty, \infty)$ is said to be locally uniformly plurisubharmonic at $a \in U$ if f is upper semicontinuous and there exists $C(a) > 0$ such that

$$C(a)\|b\|^2/4 + f(a) \leq \frac{1}{2\pi}\int_0^{2\pi} f(a + e^{i\theta}b)d\theta$$

for each $b \neq 0 \in X$ such that $a + \bar{\Delta}b \subset U$.

Similarly, a function $f : U \to [-\infty, \infty)$ is said to be uniformly plurisubharmonic if f is upper semicontinuous and there exists $C > 0$ such that

$$C\|b\|^2/4 + f(a) \leq \frac{1}{2\pi}\int_0^{2\pi} f(a + e^{i\theta}b)d\theta$$

for each $a \in U$ and $b \neq 0 \in X$ such that $a + \bar{\Delta}b \subset U$.

DEFINITION 3.3. *Let U be an open and bounded subset of a complex Banach space X. U is said to be locally uniformly pseudoconvex at $a \in U$ if $-\log d_U$ is locally uniformly plurisubharmonic at a. Similarly, U is said to be uniformly pseudoconvex if $-\log d_U$ is uniformly plurisubharmonic.*

DEFINITION 3.4. *Let U be an open and bounded subset of a complex Banach space X. A function $f : U \to [-\infty, \infty)$ is said to be strictly plurisubharmonic if f is upper semicontinuous and, for every finite-dimensional subspace M of X, $f|_{U\cap M}$ is locally uniformly plurisubharmonic at each $a \in U \cap M$ with respect to a function $C(\cdot) \in C^\infty(U \cap M)$.*

DEFINITION 3.5. *Let U be an open and bounded subset of a complex Banach space X. U is said to be strictly pseudoconvex if $-\log d_U$ is strictly plurisubharmonic.*

In [**34**], some Banach spaces whose unit ball is uniformly pseudoconvex are exhibited. To begin with, they use the following definition in [**11**].

DEFINITION 3.6. *If $0 < q < \infty$ and $2 \leq r < \infty$, a continuously quasi-normed space $(X, \|\|)$ is r-uniformly PL-convex if there exists $\lambda > 0$ such that*

$$(\frac{1}{2\pi}\int_0^{2\pi} \|a + e^{i\theta}b\|^q d\theta)^{1/q} \geq (\|a\|^r + \lambda\|b\|^r)^{1/r}$$

for all a and b in X; we shall denote the largest possible value of λ by $I_{r,q}(X)$.

Davis, Garling and Tomczak-Jaegermann proved in [**11**, Propositon 3.1] that $I_{2,1}(\mathbb{C}) = 1/2$. Moreover, a simple modification of [**11**, Theorem 4.1] gives that $L_p(\Sigma, \Omega, \mu)$ is 2-uniformly PL-convex (for $q = 1$) when $p \in [1, 2]$, and actually $I_{2,1}(L_p) = I_{2,1}(\mathbb{C}) = 1/2$.

The following theorem in [**34**] then implies that $L_p(\Sigma, \Omega, \mu)$, for $p \in [1, 2]$, has a uniformly strictly pseudoconvex unit ball.

THEOREM 3.1. *If X is a 2-uniformly PL-convex Banach space for $q = 1$ then B_X is uniformly pseudoconvex.*

In contrast, the next theorem in [**34**] shows that, for $p > 2$, the unit balls of ℓ_p and L_p are not even strictly pseudoconvex (because ℓ_p and L_p contain each ℓ_p^n).

THEOREM 3.2. *If $n \geq 2$ and $p > 2$, the unit ball of ℓ_p^n is not strictly pseudoconvex.*

Thus, as pointed out in [**34**], strictly convex spaces do not necessarily have strictly pseudoconvex unit ball; locally uniformly convex spaces do not necessarily have locally uniformly pseudoconvex unit ball; and uniformly convex spaces do not necessarily have uniformly pseudoconvex unit ball.

3.2. $\bar{\partial}$ and cluster value problems for $H^\infty(B)$ when $\dim(X) < \infty$. It is also worthwhile to study the cluster value problem for $H^\infty(B)$ when B is the ball of a finite-dimensional Banach space, and in particular for the ball of ℓ_1^n. A solution to the $\bar{\partial}$ problem for the ball of a finite-dimensional space, such as $B_{\ell_1^n}$, may hint to a solution of the respective cluster value problem for ℓ_1 and even ℓ_1-sums of finite-dimensional spaces. A relevant question to investigate is whether Lemma 3.1 in [**31**] can be strengthened to a continuous solution in $\bar{B}$, as that would guarantee a solution to the respective cluster value problem for $H^\infty(B)$.

References

[1] M. D. Acosta and M. L. Lourenço, *Shilov boundary for holomorphic functions on some classical Banach spaces*, Studia Math. **179** (2007), no. 1, 27–39, DOI 10.4064/sm179-1-3. MR2291721 (2008c:46061)

[2] L. V. Ahlfors, *Complex analysis*, 3rd ed., McGraw-Hill Book Co., New York, 1978. An introduction to the theory of analytic functions of one complex variable; International Series in Pure and Applied Mathematics. MR510197 (80c:30001)

[3] F. Albiac and N. J. Kalton, *Topics in Banach space theory*, Graduate Texts in Mathematics, vol. 233, Springer, New York, 2006. MR2192298 (2006h:46005)

[4] R. M. Aron and P. D. Berner, *A Hahn-Banach extension theorem for analytic mappings* (English, with French summary), Bull. Soc. Math. France **106** (1978), no. 1, 3–24. MR508947 (80e:46029)

[5] R. M. Aron, D. Carando, T. W. Gamelin, S. Lassalle, and M. Maestre, *Cluster values of analytic functions on a Banach space*, Math. Ann. **353** (2012), no. 2, 293–303, DOI 10.1007/s00208-011-0681-0. MR2915537

[6] R. M. Aron, B. J. Cole, and T. W. Gamelin, *Spectra of algebras of analytic functions on a Banach space*, J. Reine Angew. Math. **415** (1991), 51–93. MR1096902 (92f:46056)

[7] C. Bessaga and A. Pełczyński, *Spaces of continuous functions. IV. On isomorphical classification of spaces of continuous functions*, Studia Math. **19** (1960), 53–62. MR0113132 (22 #3971)

[8] R. Braun, W. Kaup, and H. Upmeier, *On the automorphisms of circular and Reinhardt domains in complex Banach spaces*, Manuscripta Math. **25** (1978), no. 2, 97–133, DOI 10.1007/BF01168604. MR500878 (80g:32003)

[9] L. Carleson, *Interpolations by bounded analytic functions and the corona problem*, Ann. of Math. (2) **76** (1962), 547–559. MR0141789 (25 #5186)

[10] B. J. Cole, T. W. Gamelin, and W. B. Johnson, *Analytic disks in fibers over the unit ball of a Banach space*, Michigan Math. J. **39** (1992), no. 3, 551–569, DOI 10.1307/mmj/1029004606. MR1182507 (93i:46090)

[11] W. J. Davis, D. J. H. Garling, and N. Tomczak-Jaegermann, *The complex convexity of quasinormed linear spaces*, J. Funct. Anal. **55** (1984), no. 1, 110–150, DOI 10.1016/0022-1236(84)90021-1. MR733036 (86b:46032)

[12] J. Diestel, *Sequences and series in Banach spaces*, Graduate Texts in Mathematics, vol. 92, Springer-Verlag, New York, 1984. MR737004 (85i:46020)

[13] S. Dineen, *Complex analysis on infinite-dimensional spaces*, Springer Monographs in Mathematics, Springer-Verlag London, Ltd., London, 1999. MR1705327 (2001a:46043)

[14] J. Duncan and S. A. R. Hosseiniun, *The second dual of a Banach algebra*, Proc. Roy. Soc. Edinburgh Sect. A **84** (1979), no. 3-4, 309–325, DOI 10.1017/S0308210500017170. MR559675 (81f:46057)

[15] J. D. Farmer, *Fibers over the sphere of a uniformly convex Banach space*, Michigan Math. J. **45** (1998), no. 2, 211–226, DOI 10.1307/mmj/1030132179. MR1637638 (99h:46063)

[16] J. E. Fornæss and N. Sibony, *Smooth pseudoconvex domains in* $\mathbf{C}^2$ *for which the corona theorem and* L^p *estimates for* $\overline{\partial}$ *fail*, Complex analysis and geometry, Univ. Ser. Math., Plenum, New York, 1993, pp. 209–222. MR1211882 (94a:32028)

[17] T. W. Gamelin, *Localization of the corona problem*, Pacific J. Math. **34** (1970), 73–81. MR0276742 (43 #2482)

[18] T. W. Gamelin, *Iversen's theorem and fiber algebras*, Pacific J. Math. **46** (1973), 389–414. MR0343039 (49 #7783)

[19] T. W. Gamelin, *Analytic functions on Banach spaces*, Complex potential theory (Montreal, PQ, 1993), NATO Adv. Sci. Inst. Ser. C Math. Phys. Sci., vol. 439, Kluwer Acad. Publ., Dordrecht, 1994, pp. 187–233. MR1332962 (96m:46082)

[20] T. W. Gamelin, *Uniform algebras and Jensen measures*, London Mathematical Society Lecture Note Series, vol. 32, Cambridge University Press, Cambridge-New York, 1978. MR521440 (81a:46058)

[21] T. W. Gamelin, *Uniform algebras*, Prentice-Hall, Inc., Englewood Cliffs, N. J., 1969. MR0410387 (53 #14137)

[22] J. B. Garnett, *Corona problems, interpolation problems, and inhomogeneous Cauchy-Riemann equations*, Proceedings of the International Congress of Mathematicians, Vol. 1, 2 (Berkeley, Calif., 1986), Amer. Math. Soc., Providence, RI, 1987, pp. 917–923. MR934294 (89d:32010)

[23] J. B. Garnett and P. W. Jones, *The corona theorem for Denjoy domains*, Acta Math. **155** (1985), no. 1-2, 27–40, DOI 10.1007/BF02392536. MR793236 (87e:30044)

[24] K. Hoffman, *Banach spaces of analytic functions*, Prentice-Hall Series in Modern Analysis, Prentice-Hall, Inc., Englewood Cliffs, N. J., 1962. MR0133008 (24 #A2844)

[25] L. Hörmander, *An introduction to complex analysis in several variables*, 3rd ed., North-Holland Mathematical Library, vol. 7, North-Holland Publishing Co., Amsterdam, 1990. MR1045639 (91a:32001)

[26] W. B. Johnson and S. Ortega Castillo, *The cluster value problem in spaces of continuous functions*, Proc. Amer. Math. Soc. **143** (2015), no. 4, 1559–1568, DOI 10.1090/S0002-9939-2014-12190-3. MR3314069

[27] W. B. Johnson, S. Ortega Castillo, *The cluster value problem for Banach spaces*, to appear in Illinois Journal of Mathematics.

[28] S. Kakutani, *Rings of analytic functions*, Lectures on functions of a complex variable, The University of Michigan Press, Ann Arbor, 1955, pp. 71–83. MR0070060 (16,1125b)

[29] N. Kerzman, *Hölder and* L^p *estimates for solutions of* $\bar{\partial}u = f$ *in strongly pseudoconvex domains*, Comm. Pure Appl. Math. **24** (1971), 301–379. MR0281944 (43 #7658)

[30] S. G. Krantz, *The corona problem with two pieces of data*, Proc. Amer. Math. Soc. **138** (2010), no. 10, 3651–3655, DOI 10.1090/S0002-9939-10-10462-6. MR2661563 (2011i:32006)

[31] L. Lempert, *The Dolbeault complex in infinite dimensions. II*, J. Amer. Math. Soc. **12** (1999), no. 3, 775–793, DOI 10.1090/S0894-0347-99-00296-9. MR1665984 (2000e:32053)

[32] G. McDonald, *The maximal ideal space of* $H^\infty + C$ *on the ball in* $\mathbf{C}^n$, Canad. J. Math. **31** (1979), no. 1, 79–86, DOI 10.4153/CJM-1979-009-6. MR518708 (80f:46055)

[33] J. Mujica, *Complex analysis in Banach spaces*, North-Holland Mathematics Studies, vol. 120, North-Holland Publishing Co., Amsterdam, 1986. Holomorphic functions and domains of holomorphy in finite and infinite dimensions; Notas de Matemática [Mathematical Notes], 107. MR842435 (88d:46084)

[34] S. Ortega Castillo, *Strong pseudoconvexity in Banach spaces*, working notes.

[35] A. Pełczyński and Z. Semadeni, *Spaces of continuous functions. III. Spaces* $C(\Omega)$ *for* Ω *without perfect subsets*, Studia Math. **18** (1959), 211–222. MR0107806 (21 #6528)

[36] S. Sakai, *C^*-algebras and W^*-algebras*, Springer-Verlag, New York-Heidelberg, 1971. Ergebnisse der Mathematik und ihrer Grenzgebiete, Band 60. MR0442701 (56 #1082)

[37] I. J. Schark, *Maximal ideals in an algebra of bounded analytic functions*, J. Math. Mech. **10** (1961), 735–746. "I. J. Schark" is a pseudonym for the group: Irving Kaplansky, John Wermer, Shizuo Kakutani, R. Creighton Buck, Halsey Royden, Andrew Gleason, Richard Arens and Kenneth Hoffman. MR0125442 (23 #A2744)

[38] N. Sibony, *Problème de la couronne pour des domaines pseudoconvexes à bord lisse* (French), Ann. of Math. (2) **126** (1987), no. 3, 675–682, DOI 10.2307/1971364. MR916722 (88k:32012)

[39] N. Sibony, *Some aspects of weakly pseudoconvex domains*, Several complex variables and complex geometry, Part 1 (Santa Cruz, CA, 1989), Proc. Sympos. Pure Math., vol. 52, Amer. Math. Soc., Providence, RI, 1991, pp. 199–231. MR1128526 (92g:32034)

[40] J.-P. Vigué, *Le groupe des automorphismes analytiques d'un domaine borné d'un espace de Banach complexe. Application aux domaines bornés symétriques*, Ann. Sci. École Norm. Sup. (4) **9** (1976), no. 2, 203–281. MR0430335 (55 #3340)

Department of Mathematics, Mailstop 3368, Texas A&M University, College Station, Texas 77843-3368

Current address: CIMAT A. C., Guanajuato, Guanajuato, México 36240

E-mail address: `sofia.ortega@cimat.mx`

A survey on the convergence of manifolds with boundary

Raquel Perales

ABSTRACT. This survey reviews precompactness theorems for classes of Riemannian manifolds with boundary. We begin with the works of Kodani, Anderson-Katsuda-Kurylev-Lassas-Taylor and Wong. We then present new results of Knox and the author with Sormani.

1. Introduction

Given a sequence of Riemannian manifolds, one can say that the sequence converges if the manifolds are resembling more and more some metric space. This survey reviews theorems that are able to tell when some sequences of manifolds have subsequences that converge, using a variety of notions of convergence presented in Section 2. Section 3 is devoted to those theorems that require smooth boundary conditions and Section 4 states theorems for open manifolds for which the boundary is defined to be $\partial M := \bar{M} \setminus M$.

Section 2 reviews the definitions of Lipschitz, $C^{1,\alpha}$ and Gromov-Hausdorff distance. In addition, Cheeger-Gromov, Gromov and Anderson precompactness theorems for Riemannian manifolds without boundary are stated (Theorems 2.3, 2.7, 2.9, 2.14). Although there are interesting theorems proven both for manifolds with and without boundary for the intrinsic flat distance [**17**, **18**] and for ultralimits [**8**, **19**], these are not discussed here.

Section 3 surveys results concerning Riemannian manifolds that have smooth boundaries. In 1990, Kodani [**12**] proves a precompactness theorem for Riemannian manifolds with boundary using the Lipschitz topology. He assumes that the manifolds have uniformly bounded sectional curvature, nonnegative second fundamental forms uniformly bounded from above, and a uniform lower bound for the volume of the manifolds. See Subsection 3.1. In 2004, Anderson-Katsuda-Kurylev-Lassas-Taylor [**1**] proves a precompacness theorem using a different approach under completely different hypotheses. They use $C^{1,\alpha}$ convergence ($\alpha < 1$) instead of Lipschitz convergence. They assume uniform bounds on the norm of the Ricci tensor of the manifolds and their boundaries, the mean curvatures, the diameter of the manifolds, and on three different radii: injectivity, interior and boundary.

2010 *Mathematics Subject Classification.* Primary 53-02; Secondary 53C21.

Key words and phrases. Gromov-Hausdorff convergence, intrinsic flat convergence, Lipschitz convergence, $C^{n,alpha}$ convergence.

Doctoral student at Stony Brook. Partially supported by NSF DMS 10060059.

See Subsection 3.2. In 2008, Wong [20] uses Alexandrov spaces to prove two precompactness theorems in the Gromov-Hausdorff topology with the same conditions as Kodani, except that the volume bound is replaced by a diameter bound. He assumes a lower bound on sectional curvatures of the manifolds, second fundamental forms and diameter of the manifolds bounded above. Furthermore, Wong also proves that the sectional curvature bound can be replaced by a Ricci curvature bound. See Subsection 3.3. Recently, Knox with the same technique used by Anderson-Katsuda-Kurylev-Lassas-Taylor proves a precompactness theorem in the C^α, $\alpha < 1$, and $L^{1,p}$ topologies with hypothesis analogous to theirs. He assumes bounds on the secional curvatures of the manifolds and their boundaries, the mean curvatures, the diameter of the manifolds, and a uniform lower bound in the volume of the boundaries. This theorem appears in 3.4.

Section 4 surveys Gromov-Hausdorff precompactness theorems concerning classes of open manifolds where no smoothness of the boundary is required. These results appear in work with Sormani with the author [13]. Rather than proving that a sequence of manifolds converges, we study regions within the manifolds, take the Gromov-Hausdorff limits of those inner regions and then glue the limits together to create a glued limit space. We assume conditions on the Ricci curvature, volume bounded below and above, and diameter of the inner regions uniformly bounded.

I would like to thank Monica Moreno, Fernando Galaz and Noe Barcenas, organizers of Taller de Vinculacion: Matematicos Mexicanos Jovenes en el Mundo. I would like to thank my doctoral advisor, Christina Sormani, who suggested the topic of the survey and went over it, Yuri Sobolev and Ki Song who read the survey and checked the grammar.

2. Types of convergence

2.1. Lipschitz Convergence. For details about Lipschitz convergence see [10] and [3].

DEFINITION 2.1. *Let (X, d_X) and (Y, d_Y) be two metric spaces. The dilation of a Lipschitz map $f : X \to Y$ is defined by*

$$\mathrm{dil}(f) = \sup_{x \neq x' \in X} \frac{d_Y(f(x), f(x'))}{d_X(x, x')}.$$

A function $f : X \to Y$ is called bi-Lipschitz if both $f : X \to Y$ and $f^{-1} : f(X) \to X$ are Lipschitz maps.

DEFINITION 2.2. *The Lipschitz distance between two metric spaces (X, d_X) and (Y, d_Y) is defined by*

$$d_L(X, Y) = \inf_{f : X \to Y} \log(\max\{\mathrm{dil}(f), \mathrm{dil}(f^{-1})\})$$

where the infimum is taken over all bi-Lipschitz homeomorphisms $f : X \to Y$.

THEOREM 2.3 (Cheeger-Gromov). *The class of connected closed m-dimensional Riemannian manifolds M satisfying:*

$$|\sec(M)| \leq K, \ \mathrm{Vol}(M) \geq v \ \textit{and} \ \mathrm{Diam}(M) \leq D, \tag{1}$$

is precompact in the Lipschitz topology.

This theorem follows from a Gromov's precompactness theorem (Theorem 8.25 in [**9**]), in which positive uniformly bounded injectivity radii is needed, and Cheeger's doctoral dissertation [**4**], which proves that the class of manifolds satisfying (1) have positive uniformly bounded injectivity radii.

2.2. Gromov-Hausdorff Convergence. More about Gromov-Hausdorff convergence can be find in [**10**] and [**3**]. Gromov's embedding theorem appears in [**9**]. Examples and pictures about Hausdorff converging sequences can be found in [**15**].

DEFINITION 2.4 (Hausdorff). *Let (Z, d_Z) be a metric space, the Hausdorff distance between two subsets, $A_1, A_2 \subset Z$, is defined as*

$$d_H^Z(A_1, A_2) = \inf\Big\{r : A_1 \subset T_r(A_2),\ A_2 \subset T_r(A_1)\Big\}, \tag{2}$$

where $T_r(A) = \big\{x \in Z : d_Z(x, A) < r\big\}$.

Meanwhile Gromov-Hausdorff distance avoids fixing a metric space by considering isometric embeddings of two metric spaces into a common metric space:

DEFINITION 2.5. *A function $\varphi : (X, d_X) \to (Z, d_Z)$ between metric spaces is an isometric embedding if*

$$d_Z(\varphi(x_1), \varphi(x_2)) = d_X(x_1, x_2), \tag{3}$$

for all $x_1, x_2 \in X$

DEFINITION 2.6 (Gromov). *Let (X_1, d_{X_1}) and (X_2, d_{X_2}) be two compact metric spaces. The Gromov-Hausdorff distance between them is defined as*

$$d_{GH}\Big((X_1, d_{X_1}), (X_2, d_{X_2})\Big) = \inf\Big\{d_H^Z(\varphi_1(X_1), \varphi_2(X_2)) :\ \varphi_i : X_i \to Z\Big\} \tag{4}$$

where the infimum is taken over all isometric embeddings $\varphi_i : X_i \to Z$ and all metric spaces Z.

Gromov proved that the Gromov-Hausdorff distance is a distance on the space of isometry classes of compact metric spaces. In general, elements of the classes are used and the class to which they belong is never mentioned.

The most general Gromov's precompactness theorem says:

THEOREM 2.7. *[Gromov] Let $D > 0$ and $N : (0, D] \to \mathbb{N}$ a function. Then the collection $\mathcal{M}^{D,N}$, of compact metric spaces (X, d_X) with* Diam$(X) \leq D$ *that can be covered by $N(\epsilon)$ balls of radius $\epsilon > 0$, is precompact with respect to the Gromov-Hausdorff distance.*

Given ϵ and a metric space X, the minimum number of of ϵ-balls needed to cover X is the same as the maximum number of pairwise disjoint $\epsilon/2$-balls in X. Then N can be thought as a function that bounds the maximum number of pairwise disjoint balls of compact metric spaces inside a certain class.

The converse to Theorem 2.7 also holds.

THEOREM 2.8 (Gromov). *Suppose (X_j, d_j) are compact metric spaces. If there exists $\epsilon_0 > 0$ such that X_j contains at least j disjoint balls of radius ϵ_0, then no subsequence of the X_j has a Gromov-Hausdorff limit.*

Thus proving precompactness theorems with respect to the Gromov-Hausdorff distance of a certain class of compact metric spaces is "reduced" to finding a function N and uniform upper diameter bound D. For sequences of compact Riemannian manifolds with no boundary, Gromov applied the Bishop-Gromov Volume Comparison Theorem (Theorem 2.18 [**13**]) to control the volume of the balls and obtain the following precompactness theorem.

THEOREM 2.9 (Gromov). *Given $m \in \mathbb{N}$ and $D > 0$, let $\mathcal{M}^{m,D}$ be the class of compact m dimensional Riemannian manifolds M with*

$$\text{Ricci(M)} \geq 0 \quad \textit{and} \ \text{Diam(M)} \leq \text{D}. \tag{5}$$

Then $\mathcal{M}^{m,D}$ is precompact with respect to the Gromov-Hausdorff distance.

2.3. $C^{1,\alpha}$ Convergence. For a more detailed account on $C^{k,\alpha}$ convergence, consult [**14**].

DEFINITION 2.10. *Let $\{(M_i, g_i)\}$ be a sequence of m-dimensional Riemannian manifolds. The sequence converges in the $C^{1,\alpha}$ topology to a $C^{1,\alpha}$ manifold (M, g) if M is a C^∞ manifold such that for some fixed $C^{1,\alpha}$ atlas on M compatible with its C^∞ structure, g is $C^{1,\alpha}$, and there are diffeomorphisms $\varphi_i : M_i \to M$, $i = 1, 2, 3 \ldots$, for which $\varphi_i^* g_i \to g$ with the $C^{1,\alpha}$-norm.*

REMARK 2.11. *Theorem 2.3 also holds for $C^{1,\alpha}$ convergence.*

$C^{1,\alpha}$ precompactness theorems for manifolds with or without boundary have been proved using the notion of $(r, N, C^{1,\alpha})$ harmonic coordinate atlas and harmonic radius.

DEFINITION 2.12. *Let (M, g) be an m-dimensional compact Riemannian manifold. (M, g) has an adapted harmonic coordinate atlas $(r, N, C^{1,\alpha})$ if there exist $C > 1$ and $\{B(x_k, r)) \subset M\}_{k=1}^N$ such that $\{B(x_k, r/2))\}_{k=1}^N$ cover M, $\{B(x_k, r/4))\}_{k=1}^N$ is pairwise disjoint, and for each k there is an harmonic coordinate chart $u = (u_1, \ldots, u_m) : B(x_k, 10r) \to \mathbb{R}^m$ with*

$$C^{-1}\delta_{ij} \leq g_{ij} \leq C\delta_{ij} \tag{6}$$

and

$$r^{1+\alpha}||g_{ij}(x)||_{C^{1,\alpha}} \leq C \tag{7}$$

for all $x \in B(x_k, 10r)$, where $g_{i,j} = g(\nabla u_i, \nabla u_j)$ and ∇ is the Levi-Civita connection.

DEFINITION 2.13. *Let (M, g) be a compact Riemannian manifold. For $x \in M$, the $C^{1,\alpha}$ harmonic radius at x, $r_h(x)$, is the largest radius of a geodesic ball centered at x for which there is a constant $C > 1$ and a coordinate chart $v : B(x, r) \to \mathbb{R}^m$ that satisfy*

$$C^{-1}\delta_{ij} \leq g_{ij} \leq C\delta_{ij} \tag{8}$$

and

$$r^{1+\alpha}||g_{ij}(x)||_{C^{1,\alpha}} \leq C, \tag{9}$$

where $g_{i,j} = g(\nabla v_i, \nabla v_j)$.

THEOREM 2.14 (Anderson). *The class of compact, connected Riemannian m-manifolds M satisfying*

$$|\mathrm{Ricci(M)}| \leq \mathrm{R}, \quad \mathrm{injrad}(M) \geq i \textit{ and } \mathrm{Diam}(M) \leq D,$$

is precompact in the $C^{1,\alpha}$ topology.

The theorem is proven by Anderson **[2]** by showing first that the harmonic radii for manifolds in this class is uniformly bounded below. Then, for given r, N and C, he uses the fact that the class of compact Riemannian manifolds with $(r, N, C^{1,\alpha})$ atlases is precompact in the $C^{1,\alpha'}$ topology for all $0 < \alpha' < \alpha$.

3. Precompactness Theorems for Manifolds with Smooth Boundary Conditions

3.1. Kodani's Precompactness Theorem. In 1990, Kodani **[12]** proves a theorem in the same line as Theorem 2.3, except that the Riemannian manifolds that Kodani considers have boundary and the diameter bound is replaced by bounds on the second fundamental forms of the boundaries. See examples of neccesity of this replacement in **[12]**.

THEOREM 3.1 (Kodani). *Given a positive integer m, and numbers $K, \lambda, v > 0$, the class $\mathcal{M}(m, K, \lambda, v)$ of connected m-dimensional Riemannian manifolds, M, with boundary that satisfy*

$$|\mathrm{sec(M)}| \leq \mathrm{K}, \quad 0 \leq \mathrm{II} \leq \lambda, \quad \textit{and} \quad \mathrm{Vol(M)} \geq \mathrm{v}, \tag{10}$$

where II stands for the second fundamental form of ∂M, is precompact in the Lipschitz topology.

The following two definitions are needed to explain the proof of Theorem 3.1 and state Theorem 3.5.

DEFINITION 3.2. *Let M be a Riemannian manifold with boundary and p a point in its interior. Define the interior injectivity radius of p, $i_{int}(p)$, to be the supremum over all $r > 0$ such that if $\gamma : [0, t_\gamma] \to M$ is a normal geodesic with $\gamma(0) = p$, then it is minimizing from 0 to $\min\{t_\gamma, r\}$, where t_γ is the first time γ intersects ∂M. The interior injectivity radius of M is defined as $i_{int}(M) := \inf\{i_{int}(p) | p \in M\}$.*

DEFINITION 3.3. *Let M be a Riemannian manifold with boundary and p a point in ∂M. Define the boundary injectivity radius of p, $i_\partial(p)$, to be the supremum over all $r > 0$ such that there is a minimizing geodesic $\gamma : [0, r] \to M$ with $\gamma(0) = p$ normal to ∂M. The boundary injectivity radius of M is defined as $i_\partial(M) := \inf_{p \in \partial M}(i_\partial(p))$.*

Theorem 3.1 is a corollary of the following:

THEOREM 3.4 (Kodani). *Let $\mathcal{M}(m, K, \lambda, i)$ be the class of connected m-dimensional Riemannian manifolds, M, with boundary and*

$$|\mathrm{sec(M)}| \leq \mathrm{K}, \quad |\mathrm{II}| \leq \lambda, \quad \mathrm{i_{int}(M)} \geq \mathrm{i}, \text{ and } \mathrm{i_\partial(M)} \geq \mathrm{i}, \tag{11}$$

where II stands for second fundamental form of ∂M, i_{int} is the interior injectivity radius and i_∂ the boundary injectivity radius. Then

- *for all $\varepsilon > 0$, there exists $\delta > 0$ for which if $M, N \in \mathcal{M}(m, K, \lambda, i)$ and $d_{GH}(M, N) < \delta$ then $d_L(M, N) < \varepsilon$. Thus sequences in $\mathcal{M}(m, K, \lambda, i)$ that converge in Gromov-Hausdorff sense also converge in Lipschitz sense.*

- $\mathcal{M}(m, K, \lambda, v) \subset \mathcal{M}(m, K, \lambda, i)$
- $\mathcal{M}(m, K, \lambda, v)$ *is precompact in the Gromov-Hausdorff topology.*

Proving $\mathcal{M}(m, K, \lambda, v) \subset \mathcal{M}(m, K, \lambda, i)$ involves finding lower bounds for i_∂ and i_{int}, which is done by looking at the conjugate radius of M and the length of simple closed geodesics in M, and looking at the focal radius of ∂M and the length of geodesics whose endpoints are orthogonal to ∂M, respectively.

The fact that $\mathcal{M}(m, K, \lambda, v)$ is precompact in the Gromov-Hausdorff topology comes from applying volume comparison theorems. First, he shows that for all $M \in \mathcal{M}(m, K, \lambda, v)$, $\mathrm{Vol}(M) \leq V$. Second, he shows that if $M \in \mathcal{M}(m, K, \lambda, i)$, $p \in M$ and $\varepsilon > 0$ then $\mathrm{Vol}(B(p, \varepsilon)) \leq C$ where $C > 0$ is a constant that only depends on ε and K.

Wong proves later [**20**] that II does not have to be nonnegative and the volume condition can be replaced by a diameter condition. See Subsection 3.3.

3.2. Anderson-Katsuda-Kurylev-Lassas-Taylor's Precompactness Theorem. This subsection reviews the precompactness theorem that appear in [**1**] which extend Theorem 1.1 [**2**] of Anderson to manifolds with boundary.

THEOREM 3.5 (Anderson-Katsuda-Kurylev-Lassas-Taylor). *Let $\mathcal{M}(m, R, i, H_0, D)$ be the class of compact, connected Riemannian m-manifolds with boundary M satisfying*

$$\begin{aligned} &|\mathrm{Ricci(M)}| \leq \mathrm{R}, \quad |\mathrm{Ricci}(\partial \mathrm{M})| \leq \mathrm{R} \\ &\mathrm{injrad}(M) \geq i, \quad i_{int}(M) \geq i, \quad i_\partial(M) \geq 2i \\ &\mathrm{Diam}(M) \leq D, \quad |H|_{Lip(\partial M)} \leq H_0 \end{aligned}$$

where $i_\partial(M)$ denotes the boundary injectivity radius of M, and H is the mean curvature of ∂M in M. Then $\mathcal{M}(m, R, i, H_0, D)$ is precompact in the $C^{1,\alpha}$ topology.

Theorem 3.5 is proved by showing that a larger class of manifolds is precompact in the $C^{1,\alpha'}$ topology for each $0 < \alpha' < 1$. The second step is to show that $\mathcal{M}(m, R, i, H_0, D)$ is contained in the larger class. This part relies completely on the use of harmonic coordinates and harmonic radii for manifolds with boundary.

3.3. Wong's Precompactness Theorem. This theorem, which appears in [**20**], is an improvement of Theorem 3.1. Unlike Theorem 3.5, the hypotheses do not assume any type of injectivity radius, do not require any bound on the Ricci curvature of the boundary, and the condition on the mean curvature vector is replaced by a condition on the second fundamental form.

THEOREM 3.6 (Wong). *The class $\mathcal{M}(m, r-, \lambda^{\pm}, D)$ of n-dimensional Riemannian manifolds with boundary with*

$$\mathrm{Ricci(M)} \geq \mathrm{r}-, \quad \lambda^- \leq \mathrm{II} \leq \lambda^+, \quad \text{and} \quad \mathrm{Diam(M)} \leq \mathrm{D}, \tag{12}$$

where II denotes the second fundamental form of ∂M, is precompact in the Gromov-Hausdorff topology.

The proof consists of applying Theorem 2.7 (Gromov Compactness Theorem). To show that the maximum number of disjoint ε-balls, $N(\varepsilon, M)$, for any $M \in \mathcal{M}(m, r^-, \lambda^{\pm}, D)$ and $\varepsilon > 0$ is bounded above by some constant $N(\varepsilon)$. Clearly, $N(r, M) \leq N(cr, M)$ for $c < 1$. Wong shows that there is an isometric extension $\tilde{M}$ of M that is an Alexandrov space. Then he proves that there are constants $c < 1$

and $\tilde{D} > 0$ such that for all $M \in \mathcal{M}(m, K^-, \lambda^\pm, D)$, $N(cr, M) \leq N(cr, \tilde{M})$ and $\mathrm{Diam}(\tilde{M}) \leq \tilde{D}$. Then by volume comparison in $\tilde{M}$, $N(cr, \tilde{M}) \leq \tilde{N}(cr)$.

3.4. Knox's Precompactness Theorem. The following precompactness theorem appears in [**11**]. The approach taken to prove it is similar to Theorem [**1**] of Anderson-Katsuda-Kurylev-Lassas-Taylor. Unlike Theorem 3.5, there are no conditions on any type of injectivity radius, but the Ricci curvature is replaced by sectional curvature and a lower bound on the volume is added. Note that this theorem is not an extension of an existing theorem for manifolds without boundary because it requires a lower bound on $\mathrm{Vol}(\partial M)$.

THEOREM 3.7 (Knox). *Let $\mathcal{M}(m, K, H_0, D, v_\partial)$ be the class of compact connected Riemannian m-manifolds with connected boundary satisfying*

$$|\sec(M)| \leq K,\ |\sec(\partial M)| \leq K$$
$$0 < 1/H_0 < H < H_0$$
$$\mathrm{Diam}(M) \leq D,\ \mathrm{Vol}(\partial M) \geq v_\partial,$$

where H is the mean curvature. Then $\mathcal{M}(m, K, H_0, D, v_\partial)$ is precompact in the C^α and weak $L^{1,p}$ topologies, for any $0 < \alpha < 1$ and any $p < \infty$.

Knox notes that if $0 < 1/H_0 < H < H_0$ is replaced by a bound on the Lipschitz norm of H, then C^2_* convergence can also be obtained. The proof of Theorem 3.7 follows once it is shown that $\mathcal{M}(m, K, H_0, D, v_\partial)$ satisfies the hypothesis of Theorem 3.8.

THEOREM 3.8 (Knox). *If $\{(M_i, g_i)\}$ is a sequence of Riemannian manifolds with boundary such that*

$$r_h(g_i) \geq r_0 \text{ and } \mathrm{Diam}(M_i) \leq D \tag{13}$$

where $r_h(g_i)$ is the $L^{k,p}$ harmonic radius. Then there is a subsequence of $\{(M_i, g_i)\}$ that converges in weak $L^{k,p}$ topology to a manifold with boundary whose metric is in $L^{k,p}$.

The harmonic radius, $r_h(g)$, of a Riemannian manifold with boundary, (M, g), depends on the harmonic radius of points in the interior of M and the harmonic radius of points in ∂M. Knox deals with these two cases separately. First, by looking at the volume of cylinders whose base is in ∂M, he finds that there is a $c > 0$ that only depends on $\mathcal{M}(m, K, H_0, D, v_\partial)$ such that

$$r_h(x) \geq c d_M(x, \partial M) \tag{14}$$

for all $x \in M \setminus \partial M$, where $(M, g) \in \mathcal{M}(m, K, H_0, D, v_\partial)$. Then he shows that

$$r_h(x) \geq r \tag{15}$$

for all $x \in \partial M$ where r is a constant that only depends on $\mathcal{M}(m, K, H_0, D, v_\partial)$. Thus, by definition of harmonic radius of a manifold with boundary, $r_h(g)$ has a lower bound that depends only on the class $\mathcal{M}(m, K, H_0, D, v_\partial)$.

4. Precompactness Theorems for Manifolds without Boundary Conditions

This section presents a joint work between the author with Sormani appearing in [**13**]. We make no assumptions on the boundary. In fact, we consider open Riemannian manifolds, (M^m, g), endowed with the length metric, d_M. The boundary of M is defined as $\partial M := \bar{M} \setminus M$ where $\bar{M}$ is the metric completion of M. Here the boundary is avoided by considering δ inner regions. Precompactness theorems are proven for these inner regions, then their Gromov-Hausdorff limit spaces are glued together into a single metric space (Theorem 4.4). At the end, the Hausdorff dimension of this single metric space is obtained and it is shown that its Hausdorff measure has positive lower density everywhere. Analogous theorems for constant sectional curvature are proven in [**13**].

4.1. δ-Inner Regions and their Limits. Let (M^m, g) be an open Riemannian manifold and $\delta > 0$, the δ inner region of M is defined by $M^\delta := \Big\{x \in M : d_M(x, \partial M) > \delta\Big\}$. There are two metrics in M^δ: the restricted metric d_M, and the induced length metric d_{M^δ}. If M^δ is not path connected, then the distance between points in two different path components is defined to be infinity. In general, $d_M(x,y) \le d_{M^\delta}(x,y)$ for all $x, y \in M^\delta$.

DEFINITION 4.1. *Given $m \in \mathbb{N}$, $\delta > 0$, $D > 0$, $V > 0$, and $\theta > 0$, set $\mathcal{M}_\theta^{m,\delta,D,V}$ to be the class of m-dimensional open Riemannian manifolds, M, with boundary, with*

$$\mathrm{Ricci(M)} \ge 0, \ \ \mathrm{Vol(M)} \le \mathrm{V}, \ \ \textit{and} \ \mathrm{Diam(M^\delta, d_{M^\delta})} \le \mathrm{D}, \tag{16}$$

that are noncollapsing at a point:

$$\exists q \in M^\delta \ \textit{such that} \ \mathrm{Vol}(B_q(\delta)) \ge \theta\delta^m. \tag{17}$$

THEOREM 4.2 (P–Sormani). *If $(M_j, g_j) \subset \mathcal{M}_\theta^{m,\delta,D,V}$, then there is a subsequence $\{M_{j_k}\}$ and a compact metric space (Y^δ, d) such that $(\bar{M}^\delta_{j_k}, d_{M_{j_k}}) \xrightarrow{GH} (Y^\delta, d_Y)$.*

Note that even though D bounds the diameter of the inner regions with respect to the induced length metric, the convergence is guaranteed endowing the inner regions with the restricted metric.

Replacing $\delta > 0$ in the above theorem by a decreasing sequence, $\delta_i \to 0$, and adding bounds on the diamater of δ_i-inner regions the following can be proved.

THEOREM 4.3 (P–Sormani). *Given $m \in \mathbb{N}$, a decreasing sequence, $\delta_i \to 0$, $D_i > 0$, $i = 0, 1, 2 \ldots$, $V > 0$, $\theta > 0$. Suppose that $\{(M_j, g_j)\} \subset \mathcal{M}_\theta^{m,\delta_0,D_0,V}$ and*

$$\sup\Big\{ \mathrm{Diam}\left(M_j^{\delta_i}, d_{M_j^{\delta_i}}\right) : j \in \mathbb{N}\Big\} < D_i \qquad \forall i \in \mathbb{N}. \tag{18}$$

Then there is a subsequence $\{M_{j_k}\}$, and there are compact metric spaces $(Y^{\delta_i}, d_{Y^{\delta_i}})$ such that $(\bar{M}^{\delta_i}_{j_k}, d_{M_{j_k}}) \xrightarrow{GH} (Y^{\delta_i}, d_{Y^{\delta_i}})$ for all i.

4.2. Constructing a Glued Limit Space. By constructing isometric embeddings between the limit spaces, $\varphi_{\delta_{i+1},\delta_i} : Y^{\delta_i} \to Y^{\delta_{i+1}}$, it is possible to define a metric space into which all the limit spaces isometrically embed. Set $\varphi_{\delta_{i+j},\delta_i} = \varphi_{\delta_{i+j},\delta_{i+j-1}} \circ \cdots \circ \varphi_{\delta_{i+1},\delta_i}$. Define

$$Y := Y(\{\delta_i\}, \{\varphi_{\delta_{i+1},\delta_i}\}) = Y^{\delta_0} \sqcup \left(\sqcup_{i=1}^\infty \left(Y^{\delta_{i+1}} \setminus \varphi_{\delta_{i+1},\delta_i}\left(Y^{\delta_i}\right)\right)\right) \tag{19}$$

and

$$d_Y(x,y) := \begin{cases} d_{Y^{\delta_0}}(x,y) & \text{if } x,y \in Y^{\delta_0}, \\ d_{Y^{\delta_{i+1}}}(x,y) & \text{if } x,y \in Y^{\delta_{i+1}} \setminus \varphi_{\delta_{i+1},\delta_i}\left(Y^{\delta_i}\right), \\ d_{Y^{\delta_{i+1}}}\left(x, \varphi_{\delta_{i+1},\delta_0}(y)\right) & \text{if } x \in Y^{\delta_{i+1}} \setminus \varphi_{\delta_{i+1},\delta_i}\left(Y^{\delta_i}\right) \\ & \text{for some } i \in \mathbb{N} \text{ and } y \in Y^{\delta_0}, \\ d_{Y^{\delta_{i+j+1}}}\left(x, \varphi_{\delta_{i+j+1},\delta_{i+1}}(y)\right) & \text{if } x \in Y^{\delta_{i+j+1}} \setminus \varphi_{\delta_{i+j+1},\delta_{i+j}}\left(Y^{\delta_{i+j}}\right) \\ & \text{and } y \in Y^{\delta_{i+1}} \setminus \varphi_{\delta_{i+1},\delta_i}\left(Y^{\delta_i}\right) \\ & \text{for some } i,j \in \mathbb{N} \end{cases}$$

THEOREM 4.4 (P–Sormani). *Under the hypothesis of Theorem 4.3. There exists a metric space (Y, d_Y) such that for all $\delta \in (0, \delta_0]$, there is a subsequence of $\{(\bar{M}^\delta_{j_k}, d_{M_{j_k}})\}$ that Gromov-Hausdorff converges to some compact metric space (Y^δ, d_{Y^δ}). For any such Y^δ, there exists an isometric embedding*

$$F_\delta = F_{\delta,\{\delta_i\}} : Y^\delta \to Y. \tag{20}$$

If $\delta = \delta_i$ for some i, then

$$F_{\delta_i}(Y^{\delta_i}) \subset F_{\delta_{i+1}}(Y^{\delta_{i+1}}). \tag{21}$$

If β_j is any sequence decreasing to 0, then

$$Y = \bigcup_{i=1}^{\infty} F_{\beta_j}(Y^{\beta_j}). \tag{22}$$

This glued limit space may exist even when (M_j, d_j) has no Gromov-Hausdorff limit.

Hausdorff measures and topologies of the Gromov-Hausdorff limit spaces of noncollapsing sequences of manifolds have been studied by Cheeger, Colding, Naber, Wei and Sormani (c.f. [**5**], [**6**], [**7**] and [**16**]). Applying some of these results, we are able to prove the following.

THEOREM 4.5 (P–Sormani). *Suppose that Y is a glued limit constructed as in Theorem 4.4. Then Y has Hausdorff dimension m, $\mathcal{H}^m(Y) \leq V_0$ and every point in Y has positive lower density.*

References

[1] Michael Anderson, Atsushi Katsuda, Yaroslav Kurylev, Matti Lassas, and Michael Taylor, *Boundary regularity for the Ricci equation, geometric convergence, and Gel'fand's inverse boundary problem*, Invent. Math. **158** (2004), no. 2, 261–321, DOI 10.1007/s00222-004-0371-6. MR2096795 (2005h:53051)

[2] Michael T. Anderson, *Convergence and rigidity of manifolds under Ricci curvature bounds*, Invent. Math. **102** (1990), no. 2, 429–445, DOI 10.1007/BF01233434. MR1074481 (92c:53024)

[3] Dmitri Burago, Yuri Burago, and Sergei Ivanov, *A course in metric geometry*, Graduate Studies in Mathematics, vol. 33, American Mathematical Society, Providence, RI, 2001. MR1835418 (2002e:53053)

[4] Jeff Cheeger, *COMPARISON AND FINITENESS THEOREMS FOR RIEMANNIAN MANIFOLDS*, ProQuest LLC, Ann Arbor, MI, 1967. Thesis (Ph.D.)–Princeton University. MR2616706

[5] Jeff Cheeger and Tobias H. Colding, *On the structure of spaces with Ricci curvature bounded below. I*, J. Differential Geom. **46** (1997), no. 3, 406–480. MR1484888 (98k:53044)

[6] Tobias H. Colding, *Ricci curvature and volume convergence*, Ann. of Math. (2) **145** (1997), no. 3, 477–501, DOI 10.2307/2951841. MR1454700 (98d:53050)

[7] Tobias Holck Colding and Aaron Naber, *Characterization of tangent cones of noncollapsed limits with lower Ricci bounds and applications*, Geom. Funct. Anal. **23** (2013), no. 1, 134–148, DOI 10.1007/s00039-012-0202-7. MR3037899

[8] M. Gromov, *Asymptotic invariants of infinite groups*, Geometric group theory, Vol. 2 (Sussex, 1991), London Math. Soc. Lecture Note Ser., vol. 182, Cambridge Univ. Press, Cambridge, 1993, pp. 1–295. MR1253544 (95m:20041)

[9] Mikhael Gromov, *Groups of polynomial growth and expanding maps*, Inst. Hautes Études Sci. Publ. Math. **53** (1981), 53–73. MR623534 (83b:53041)

[10] Misha Gromov. *Metric structures for Riemannian and non-Riemannian spaces*, volume 152 of *Progress in Mathematics*. Birkhäuser Boston Inc., Boston, MA, 1999. Based on the 1981 French original [MR0682063 (85e:53051)], With appendices by M. Katz, P. Pansu and S. Semmes, Translated from the French by Sean Michael Bates.

[11] Kenneth Steven Knox, *Compactness Theorems for Riemannian Manifolds with Boundary and Applications*, ProQuest LLC, Ann Arbor, MI, 2013. Thesis (Ph.D.)–State University of New York at Stony Brook. MR3167242

[12] Shigeru Kodani, *Convergence theorem for Riemannian manifolds with boundary*, Compositio Math. **75** (1990), no. 2, 171–192. MR1065204 (92b:53066)

[13] Raquel Perales and Christina Sormani, *Sequences of open Riemannian manifolds with boundary*, Pacific J. Math. **270** (2014), no. 2, 423–471, DOI 10.2140/pjm.2014.270.423. MR3253689

[14] Peter Petersen, *Riemannian geometry*, 2nd ed., Graduate Texts in Mathematics, vol. 171, Springer, New York, 2006. MR2243772 (2007a:53001)

[15] C. Sormani. How riemannian manifolds converge. *Proceedings of the International Conference on Metric and Differential Geometry in Tianjing and Beijing*, 2010.

[16] Christina Sormani and Guofang Wei, *Hausdorff convergence and universal covers*, Trans. Amer. Math. Soc. **353** (2001), no. 9, 3585–3602 (electronic), DOI 10.1090/S0002-9947-01-02802-1. MR1837249 (2002e:53057)

[17] Christina Sormani and Stefan Wenger, *Weak convergence of currents and cancellation*, Calc. Var. Partial Differential Equations **38** (2010), no. 1-2, 183–206, DOI 10.1007/s00526-009-0282-x. With an appendix by Raanan Schul and Wenger. MR2610529 (2011c:49092)

[18] Christina Sormani and Stefan Wenger, *The intrinsic flat distance between Riemannian manifolds and other integral current spaces*, J. Differential Geom. **87** (2011), no. 1, 117–199. MR2786592 (2012f:53072)

[19] L. van den Dries and A. J. Wilkie, *Gromov's theorem on groups of polynomial growth and elementary logic*, J. Algebra **89** (1984), no. 2, 349–374, DOI 10.1016/0021-8693(84)90223-0. MR751150 (85k:20101)

[20] Jeremy Wong, *An extension procedure for manifolds with boundary*, Pacific J. Math. **235** (2008), no. 1, 173–199, DOI 10.2140/pjm.2008.235.173. MR2379775 (2008m:53099)

SUNY at Stony Brook
E-mail address: praquel@math.sunysb.edu

Convexity is a local property in CAT(κ) spaces

Carlos Ramos-Cuevas

ABSTRACT. In this note we show that a connected, closed and locally convex subset (with an extra assumption on the diameter with respect to the induced length metric if $\kappa > 0$) of a CAT(κ) space is convex.

1. Introduction

The study of convex subsets of geodesic metric spaces is a very natural and interesting geometric question. In this note we are interested in metric spaces with curvature bounded from above in the sense of Alexandrov, that is, in CAT(κ) spaces. Convex subsets play a special role in the study of the geometry of such spaces.

For instance, in **[9]** Kleiner and Leeb study convex subsets of symmetric spaces of noncompact type (which are examples of simply connected Riemannian manifolds of nonpositive curvature, in particular, CAT(0) spaces) and their boundaries at infinity under the actions of certain groups of isometries. These boundaries at infinity endowed with the Tits metric are examples of spherical buildings, which are in turn CAT(1) spaces. Motivated by their analysis in **[9]** the authors ask if the circumradius of a convex subset of a spherical building must be $\leq \frac{\pi}{2}$ or it is itself a spherical building. A weaker version of this question is Tits' Center Conjecture, which is concerned with fixed points of groups of isometries acting on convex subcomplexes of spherical buildings (see **[11]**, **[10]** **[15]**, **[12]** for a complete answer to this conjecture, we refer also to **[3]** for a related result).

In the case of CAT(0) spaces there is a *folklore* question asking if the closed convex hull of a finite subset in a CAT(0) space is compact. Clearly, the interesting case is when the space is not locally compact. See e.g. mathoverflow.net/questions/6627/ for a discussion on this problem.

If we regard unbounded subsets instead of finite subsets, we can ask now how the asymptotic geometry is affected by taking the convex hull. In **[8]** they give an example of a subset S of a CAT(-1) space, such that the boundary at infinity of the convex hull $CH(S)$ is bigger than the boundary at infinity of S. In contrast, they prove that if S is the union of finitely many convex subsets, then its convex hull has the same boundary at infinity as S. This result is not true in general in the CAT(0) case (e.g. consider the Euclidean space).

2010 *Mathematics Subject Classification.* Primary 51K10; Secondary 53C21.
Key words and phrases. CAT(κ) spaces, convex sets, spherical buildings.

Consider the following related question. Which convex subsets B of the ideal boundary $\partial_T X$ (with the Tits metric) of a CAT(0) space X can be realized as ideal boundaries of convex subsets of X? In [**9**] they address the case when X is a symmetric space or a Euclidean building and $B \subset \partial_T X$ is a top-dimensional subbuilding. In [**2**] Balser studies the case of Euclidean buildings X of type A_2 and characterizes 0-dimensional convex subsets $B \subset \partial_T X$ arising as ideal boundaries of convex subsets of X.

All these problems concern themselves in one way or another with the constructions of convex subsets of CAT(κ) spaces. It is, therefore, useful to have a tool to decide whether a given subset is convex or not. One can construct convex subsets of CAT(0) spaces by considering points, geodesics, horoballs and taking tubular neighborhoods and intersections of them. In [**2**] the author also considers unions of these objects. In general unions of convex sets will not be convex, hence, in order to verify that the constructed subsets are convex, he proves and uses the following result. Let $C \subset X$ be a closed connected subset of a CAT(0) space X and suppose that there is an $\varepsilon > 0$ such that C is ε-locally convex (i.e. $B_p(\varepsilon) \cap C$ is convex for all $p \in C$), then C is convex. That is, one must verify the convexity of C only locally. One can actually drop the assumption that the ε must be chosen uniformly in C as is required in [**2**], this result is a consequence of the Cartan-Hadamard Theorem [**5**, Thm. 4.1]. Indeed, the assumption implies that C is a locally CAT(0) space (in the restricted metric) and the Cartan-Hadamard Theorem says that in this situation the universal cover $\tilde{C}$ of C is a CAT(0) space. Now we claim that $C = \tilde{C}$. If there is a non null-homotopic loop in C, then its lift to $\tilde{C}$ is a non closed curve. Since $\tilde{C}$ is CAT(0), this lift is homotopic to the unique geodesic segment joining its endpoints. Projecting this homotopy to C shows that every non null-homotopic loop in C is homotopic to a unique locally geodesic loop, but there are no locally geodesic loops in X, hence, C is simply connected and a CAT(0) space. In particular C is a geodesic space and since geodesics in X are unique we conclude that C is convex. (c.f. [**7**, Sec. 24], [**6**].)

In the case of CAT(κ) spaces for $\kappa > 0$ there is an analogous result in [**4**] to Cartan-Hadamard for locally compact spaces. It says that a locally compact, locally CAT(1) space is CAT(1) if and only if every loop of length $< 2\pi$ can be homotoped to a point by loops of length $< 2\pi$ (this property being the analogous of simply connected in the CAT(0) case). One could use again this result as above to show that a closed, connected, locally convex subset (with an extra assumption on the diameter with respect to the induced length metric, see Theorem 1.1 below) of a locally compact CAT(1) space is convex. Another argument for the locally compact case using the Hopf-Rinow Theorem can be found in [**6**]. In this note we observe, that in order to show this result for CAT(κ) spaces, we do not need the full strength of the Cartan-Hadamard Theorem and we can use instead the fact that our locally CAT(κ) space already lives in an ambient CAT(κ) (and in particular, geodesic) space. This allows us to give a unified proof for arbitrary κ without the assumption of local compactness.

Let D_κ be $\pi/\sqrt{\kappa}$ for $\kappa > 0$ and ∞ for $\kappa \leq 0$ (see Section 2). The main result of this note is:

THEOREM 1.1. *Let $C \subset X$ be a closed, connected subset of a* CAT(κ) *space* (X, d). *Suppose that for every point $p \in C$ there is an $\varepsilon = \varepsilon(p) > 0$ such that*

$B_p(\varepsilon) \cap C$ is convex, that is, C is locally convex. Denote with ℓ the induced length metric in C. Then it holds:

(1) *If for two points $x, y \in C$ holds $\ell(x, y) < D_\kappa$, then $xy \subset C$. In particular, $\ell(x, y) \le D_\kappa$ implies $\ell(x, y) = d(x, y)$.*
(2) *If the diameter* $\mathrm{diam}_\ell(C)$ *of C with respect to the length metric is $\le D_\kappa$, then C is a convex subset.*
(3) *(C, ℓ) is a* CAT(κ) *space.*

Observe that in the case $\kappa \le 0$ the assumption in (2) on the diameter is always satisfied and C is always a convex subset. If $\kappa > 0$ then it is not true in general that $d(x, y) = \ell(x, y)$, consider e.g. a segment of length $\pi + \epsilon$ on the unit circle, it is locally convex but not convex. Nevertheless, it is certainly CAT(1) with respect to the length metric (cf. with the result in Section 4).

Theorem 1.1(2) was first shown in the case of Euclidean spaces by Tietze and Nakajima in [**16**], [**13**]. For locally compact Busemann spaces it has been proven by Papadopoulos in [**14**, Sec. 8.3] (we note that our proof of Theorem 1.1(1-2) also works unchanged for general Busemann spaces). Bux and Witzel proved the cases of CAT(0) and locally compact CAT(κ) spaces in [**6**].

In Section 4 we use Theorem 1.1 to show that in the case of subsets of dimension ≥ 2 of spherical buildings, we do not need the assumption on the diameter, thus, connected, closed, locally convex subsets are always convex. The condition on the dimension is necessary as explained in the example above.

I would like to thank S. Witzel for pointing me out some missing references and drawing my attention to their preprint [**6**].

2. Preliminaries

In this section we will recall some definitions and fix the notation. We refer to [**5**] for more information on CAT(κ) spaces.

2.1. Metric spaces. Let (X, d) be a metric space. For $r > 0$ and $x \in X$ we denote with $B_x(r)$ the *open metric ball* of radius r centered at x.

A *geodesic* segment is a curve whose length realizes the distance between its endpoints. If there is a unique geodesic between two points $x, y \in X$ we denote it with xy. X is said to be a *geodesic metric space* if there exists geodesics between any two points and it is *d-geodesic* if this is true for any two points at distance $< d$.

A *midpoint* between $x, y \in X$ is a point z, such that $d(x, z) = d(z, y) = \frac{1}{2} d(x, y)$. If the midpoint is unique we denote it with $m(x, y)$.

The *length metric* associated to d is defined as follows: the distance between two points is the infimum of the lengths (measured with respect to d) of rectifiable curves connecting them.

2.2. CAT(κ) **spaces.** Let M_κ be the complete, simply connected Riemannian surface of constant sectional curvature κ and let D_κ denote its diameter. Then $D_\kappa = \pi/\sqrt{\kappa}$ for $\kappa > 0$ and $D_\kappa = \infty$ for $\kappa \le 0$.

Recall that a complete D_k-geodesic metric space (X, d) is said to be CAT(κ) if all geodesic triangles of perimeter $< 2D_\kappa$ are not thicker than the corresponding comparison triangles on the model space M_κ.

In a CAT(κ) space, points at distance $< D_\kappa$ are joined by a unique geodesic, and these geodesics vary continuously with their endpoints. A local geodesic of length $< D_\kappa$ is a geodesic.

We say that a subset C of a CAT(κ) is *convex* if for every pair of points in C at distance $< D_\kappa$ the unique geodesic segment between them is contained in C. A closed convex subset of a CAT(κ) space is itself CAT(κ).

3. Proof of the main result

In this section all geometric objects like geodesics, metric balls, etc. will be taken with respect to the metric d if not said otherwise.

First notice that the induced length metric ℓ in C is finite. Indeed, the set of points that can be connected to a given point with a rectifiable curve in C is clearly open and closed in C, because of the local convexity of C.

Let $x \in C$ and define C_x as the subset of C of points y such that $d(x,y) < D_\kappa$ and the geodesic segment xy is contained in C. In particular $\ell(x,y) = d(x,y)$ for all $y \in C_x$.

LEMMA 3.1. *C_x is open in C.*

PROOF. Let $y \in C_x$ and write $D := d(x,y) < D_\kappa$. By local convexity, we can choose an ε with $D_\kappa/2 > \varepsilon > 0$ and such that $B_{y'}(\varepsilon) \cap C$ is convex for all $y' \in xy$. We follow the proof of [**5**, Lemma 4.3(1)] (see also [**1**]) and modify it to work in our setting. The proof is some kind of induction. Consider the following statement in the real number $d \geq 0$:

There exists a positive number $\epsilon_d > 0$ depending only on $d \leq D$ such that if $x', y' \in xy \subset C$ and $d(x',y') \leq d$, then for any $\bar{x} \in B_{x'}(\epsilon_d) \cap C$ and $\bar{y} \in B_{y'}(\epsilon_d) \cap C$ holds $d(\bar{x},\bar{y}) < D_\kappa$ and the geodesic $\bar{x}\bar{y}$ lies in C.

Notice that if the statement is true for some d, then it is true for any $d' < d$. If it is true for D then we are done: the open set $B_y(\epsilon_D) \cap C$ is then contained in C_x.

Induction basis: If $d \leq \varepsilon/2$ then the statement is true: Take $\epsilon_d := \varepsilon/2$. Then any $\bar{x}, \bar{y}$ as in the statement lie in $B_{x'}(\varepsilon) \cap C$, which is convex by assumption. Further, $d(\bar{x},\bar{y}) \leq 3\varepsilon/2 < 3D_\kappa/4$.

Induction step: Now assume that the statement is true for some $d \leq D$, then we want to show, that it is true for $d' := \min\{3d/2, D\}$. Let $\delta := D_\kappa - D > 0$.

By triangle comparison and Proposition A.1, there exists a constant $K = K(D) < 1$ such that if $x, y, z \in X$ are the vertices of a geodesic triangle in X with perimeter $< 2D_\kappa$ and $d(x,y), d(x,z) \leq \frac{2}{3}(D + \delta/2) < \frac{2}{3}D_\kappa$ then $d(m(x,y), m(x,z)) \leq K d(y,z)$.

Let $\epsilon_{d'} := (1-K)\min\{\epsilon_d, \delta/3\}$. If $x', y', \bar{x}, \bar{y} \in C$ are points as in the statement, then $d(\bar{x},\bar{y}) \leq d(x',y') + 2\epsilon_{d'} \leq D + 2(1-K)\delta/3 < D + 2\delta/3 < D_\kappa$. This shows the first conclusion of the statement.

Let $a_0, b_0 \in x'y' \subset xy$ be two points such that $d(x',a_0) = d(a_0,b_0) = d(b_0,y')$, that is, a_0, b_0 cut the segment $x'y'$ in thirds. Since $d(x',b_0) = \frac{2}{3}d(x',y') \leq \frac{2}{3}d' \leq d$ and $\bar{x} \in B_{x'}(\epsilon_{d'}) \cap C \subset B_{x'}(\epsilon_d) \cap C$, by the induction hypothesis it follows that $\bar{x}b_0 \subset C$. Analogously, $a_0\bar{y} \subset C$. We define inductively $a_{n+1} := m(\bar{x}, b_n)$ and $b_{n+1} := m(\bar{y}, a_n)$ (see Figure 1). That they are well defined (i.e. $d(\bar{x}, b_n), d(\bar{y}, a_n) < D_\kappa$) is a consequence of (iii) below. We want to prove the following.

(i) $d(a_{n-1}, a_n), d(b_{n-1}, b_n) \leq K^n \epsilon_{d'}$.
(ii) $\bar{x}b_n, \bar{y}a_n \subset C$.
(iii) $d(\bar{x}, b_n), d(\bar{y}, a_n) \leq \frac{2}{3}(D + \delta/2)$.

This would finish the proof of the induction step. Indeed, by (i) the sequences $\{a_n\}, \{b_n\}$ are Cauchy and because X is complete and C is closed, there are points

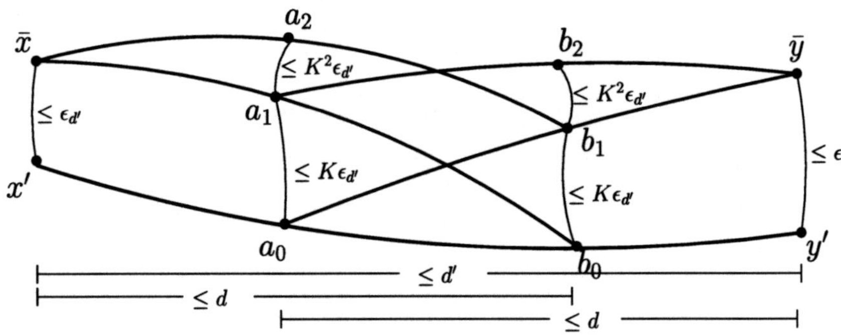

FIGURE 1. Construction of the sequences $\{a_n\}, \{b_n\}$

$a, b \in C$ with $a_n \to a$ and $b_n \to b$. The geodesics $\bar{x}b_n, \bar{y}a_n \subset C$ converge to geodesics $\bar{x}b, \bar{y}a \subset C$, because geodesics of length $< D_\kappa$ in $CAT(\kappa)$ spaces vary continuously with their endpoints. The midpoint of the geodesic $\bar{x}b$ is a and the midpoint of the geodesic $a\bar{y}$ is b, hence, the union of $\bar{x}b$ and $a\bar{y}$ is a local geodesic between $\bar{x}$ and $\bar{y}$. This local geodesic must coincide with $\bar{x}\bar{y}$ because it has length $d(\bar{x}, a) + d(a, \bar{y}) = \frac{1}{2}d(\bar{x}, b) + d(a, \bar{y}) \leq \frac{1}{3}(D + \delta/2) + \frac{2}{3}(D + \delta/2) < D_\kappa$, by (iii). It follows that $\bar{x}\bar{y} \subset C$.

(ii) follows from (i) and the induction hypothesis for d: We have that $d(x', b_0)$, $d(a_0, y') \leq \frac{2}{3}d' \leq d$ and $\bar{x} \in B_{x'}(\epsilon_{d'}) \cap C \subset B_{x'}(\epsilon_d) \cap C$, respectively, $\bar{y} \in B_{y'}(\epsilon_{d'}) \cap C \subset B_{y'}(\epsilon_d) \cap C$. By (i), it holds $d(a_0, a_n) \leq \sum_{i=1}^{n} K^i \epsilon_{d'} < \epsilon_{d'}/(1 - K) \leq \epsilon_d$. Analogously, $d(b_0, b_n) < \epsilon_d$. Thus, if $a_n, b_n \in C$ then by the induction hypothesis, $\bar{x}b_n, \bar{y}a_n \subset C$, which in turn implies that $a_{n+1}, b_{n+1} \in C$. Now (ii) follows inductively.

It remains to prove (i) and (iii). Suppose (i) is true for all $n \leq m$. Then for $n \leq m$ holds $d(\bar{y}, a_n) \leq d(\bar{y}, y') + d(y', a_0) + \sum_{i=1}^{n} d(a_{i-1}, a_i) \leq \epsilon_{d'} + \frac{2}{3}d' + \sum_{i=1}^{n} K^i \epsilon_{d'} \leq \frac{2}{3}d' + \frac{\epsilon_{d'}}{1-K} \leq \frac{2}{3}(D + \frac{\delta}{2})$. Analogously, $d(\bar{x}, b_n) \leq \frac{2}{3}(D + \frac{\delta}{2})$. Thus, (iii) is true for all $n \leq m$. Consider the triangle $(\bar{y}, a_{m-1}, a_m)$. Then, by the definition of the constant K above (given by Proposition A.1), it follows that $d(b_m, b_{m+1}) = d(m(\bar{y}, a_{m-1}), m(\bar{y}, a_m)) \leq Kd(a_{m-1}, a_m) \leq K^{m+1}\epsilon_{d'}$. Analogously, $d(a_m, a_{m+1}) \leq K^{m+1}\epsilon_{d'}$. Hence, (i) is true for all $n \leq m + 1$. Now (i) and (iii) follow again inductively. □

Let now $y \in C$ such that $\ell(x, y) < D_\kappa$. Let γ be a curve in C connecting x and y of length $L(\gamma)$ with $\ell(x, y) \leq L(\gamma) < D_\kappa$. By Lemma 3.1, the set $C_x \cap \gamma$ is open in γ. It is also closed because $L(\gamma) < D_\kappa$ and geodesics of length $< D_\kappa$ in CAT(κ) spaces vary continuously with their endpoints. Hence, $y \in C_x$ and $xy \subset C$. This proves the first assertion of (1).

If $\ell(x, y) = D_\kappa$, then we can choose points $y_n \in C$ with $\ell(x, y_n) < D_\kappa$ and $\ell(y_n, y) \to 0$. The second assertion of (1) follows by the first part and continuity. (2) is a direct consequence of (1).

To show (3), first notice that by (1), (C,ℓ) is D_κ-geodesic. Let $x_0, x_1, x_2 \in C$ be the vertices of a geodesic (with respect to ℓ) triangle Δ of perimeter $< 2D_\kappa$. In particular, each side of the triangle has length $< D_\kappa$. Then by (1) it follows that $\ell(x_i, x_j) = d(x_i, x_j)$ and $x_i x_j \subset C$. Hence Δ is also a geodesic triangle with respect to d. Let a, b be two points on the sides of Δ, they split the triangle in two curves in C and at least one of them has length $< D_\kappa$. This implies that $\ell(a,b) < D_\kappa$ and therefore $\ell(a,b) = d(a,b)$. That is, (Δ, ℓ) is isometric to (Δ, d). The CAT(κ) property for (C,ℓ) follows from the CAT(κ) property of (X,d). □

4. Locally convex subsets of spherical buildings

We apply now Theorem 1.1 in the case of spherical buildings and obtain the following result.

THEOREM 4.1. *Let C be a connected, closed, locally convex subset of a spherical building B. Suppose that* $\dim(C) \geq 2$, *then C is convex.*

PROOF. Suppose there are two points $x, y \in C$ with $\ell(x,y) > \pi$ and let γ be a curve in C connecting them. There must be a point $\hat{x} \neq y$ in γ with $\ell(x,\hat{x}) = \pi$ and therefore $d(x,\hat{x}) = \pi$ by Theorem 1.1. That is, the curve γ contains antipodes of x. The antipodes of x have the same type (i.e. the same image in the model Weyl chamber under the natural projection). Recall that the distances between points of the same type in a spherical building have finitely many different possible values. This implies that γ contains only finitely many antipodes $\hat{x}_1, \ldots, \hat{x}_k$ of x. We can also assume that γ meets each one of these antipodes only once.

The next step is to observe that since $\dim(C) \geq 2$ we can change the curve γ to avoid each one of the $\hat{x}_1, \ldots, \hat{x}_k$ and obtain a curve connecting x and y that does not meet any antipode of x. Let us make this observation more precise. Let $r > 0$ be such that $\overline{B_{\hat{x}_1}(r)}$ (the closed ball of radius r centered at $\hat{x}_1$) does not contain any antipode of x and $\overline{B_{\hat{x}_1}(r)} \cap C$ is convex. Then $(\overline{B_{\hat{x}_1}(r)} \cap C) \setminus \{\hat{x}_1\}$ is connected. Let $x_1, y_1 \in C$ be the first, respectively the last, point of γ in $\overline{B_{\hat{x}_1}(r)} \cap C$. Hence we can replace the part of γ between x_1 and y_1 with a curve in $(\overline{B_{\hat{x}_1}(r)} \cap C) \setminus \{\hat{x}_1\}$. Repeating this procedure, we obtain a curve in C connecting x and y avoiding any antipodes of x. This contradicts our first observation at the beginning of the proof. It follows that $\mathrm{diam}_\ell(C) \leq \pi$ and Theorem 1.1 implies that C is convex. □

Appendix A. A computation in spherical trigonometry

PROPOSITION A.1. *Consider a triangle in the 2-dimensional unit sphere S^2 with vertices x, y, z and sides lengths $a = d(x,y), b = d(x,z), c = d(y,z)$. Let y', z' be the midpoints of the segments xy and xz respectively and let $c' = d(y',z')$. Suppose that $a, b \leq C < \frac{2\pi}{3}$. Then there exists a constant $K = K(C) < 1$, such that $c' \leq Kc$.*

PROOF. Let $\theta := \angle_x(y,z)$ be the angle in x, then the spherical law of cosines for the sides c and c' imply

$$\begin{aligned}
(\cos\frac{a}{2}\cos\frac{b}{2})\cos c' &= (\cos\frac{a}{2}\cos\frac{b}{2})(\cos\frac{a}{2}\cos\frac{b}{2} + \sin\frac{a}{2}\sin\frac{b}{2}\cos\theta)\\
&= \cos^2\frac{a}{2}\cos^2\frac{b}{2} + \frac{\sin a}{2}\frac{\sin b}{2}\cos\theta\\
&= \cos^2\frac{a}{2}\cos^2\frac{b}{2} + \frac{1}{4}(\cos c - \cos a\cos b)\\
&= \frac{1}{4}(4\cos^2\frac{a}{2}\cos^2\frac{b}{2} + \cos c - (2\cos^2\frac{a}{2} - 1)(2\cos^2\frac{b}{2} - 1))\\
&= \frac{1}{4}(\cos c - 1 + 2(\cos^2\frac{a}{2} + \cos^2\frac{b}{2})).
\end{aligned}$$

Suppose first that $a = b$ and write $\alpha := \cos\frac{a}{2}$. Then we have $4\alpha^2(\cos c' - 1) = \cos c - 1$, or equivalently, $\sin^2\frac{c'}{2} = \frac{1}{4\alpha^2}\sin^2\frac{c}{2}$. Notice that $\frac{1}{4\alpha^2} \le \frac{1}{4\cos^2 C/2} < 1$. Since $c, c' < \pi$, this implies that there is a constant $K = K(C) < 1$ such that $c' \le Kc$ proving the proposition in this case.

Now we may assume w.l.o.g. that $b \le a$. Let $\beta := \cos\frac{b}{2} \ge \cos\frac{a}{2} = \alpha > \frac{1}{2}$. We fix a, c and consider c' as a function of b, or equivalently, as a function $f(\beta) := c'$ of β. We can derive the equation above with respect to β and obtain

$$\alpha(\cos f(\beta) - \beta f'(\beta)\sin f(\beta)) = \beta.$$

Hence $f'(\beta) \le 0$ if and only if $\alpha\cos f(\beta) - \beta \le 0$. The latter can be seen easily using again the equation above for $\cos c'$:

$$\begin{aligned}
\alpha\cos f(\beta) - \beta &= \frac{4\alpha\beta\cos f(\beta) - 4\beta^2}{4\beta} = \frac{\cos c - 1 + 2(\alpha^2 + \beta^2) - 4\beta^2}{4\beta}\\
&= \frac{(\cos c - 1) + 2(\alpha^2 - \beta^2)}{4\beta} \le 0.
\end{aligned}$$

Thus, f is monotone non-increasing and in particular, $c' = f(\beta) \le f(\alpha) \le Kc$ by the first case considered above. □

References

[1] Stephanie B. Alexander and Richard L. Bishop, *The Hadamard-Cartan theorem in locally convex metric spaces*, Enseign. Math. (2) **36** (1990), no. 3-4, 309–320. MR1096422 (92c:53044)

[2] Andreas Balser, *Convex rank 1 subsets of Euclidean buildings (of type A_2)*, Geom. Dedicata **131** (2008), 123–158, DOI 10.1007/s10711-007-9221-1. MR2369196 (2008k:22033)

[3] Andreas Balser and Alexander Lytchak, *Centers of convex subsets of buildings*, Ann. Global Anal. Geom. **28** (2005), no. 2, 201–209, DOI 10.1007/s10455-005-7277-4. MR2180749 (2006g:53049)

[4] B. H. Bowditch, *Notes on locally* CAT(1) *spaces*, Geometric group theory (Columbus, OH, 1992), Ohio State Univ. Math. Res. Inst. Publ., vol. 3, de Gruyter, Berlin, 1995, pp. 1–48. MR1355101 (97e:53070)

[5] Martin R. Bridson and André Haefliger, *Metric spaces of non-positive curvature*, Grundlehren der Mathematischen Wissenschaften [Fundamental Principles of Mathematical Sciences], vol. 319, Springer-Verlag, Berlin, 1999. MR1744486 (2000k:53038)

[6] Kai-Uwe Bux and Stefan Witzel, *Local convexity in CAT(κ)-spaces*, arXiv preprint arXiv:1211.1871v1 (2012).

[7] M. Gromov, CAT(κ)*-spaces: construction and concentration* (English, with Russian summary), Zap. Nauchn. Sem. S.-Peterburg. Otdel. Mat. Inst. Steklov. (POMI) **280** (2001), no. Geom. i Topol. 7, 100–140, 299–300, DOI 10.1023/B:JOTH.0000008756.15786.0f; English transl., J. Math. Sci. (N. Y.) **119** (2004), no. 2, 178–200. MR1879258 (2002j:53045)

[8] Christoph Hummel, Urs Lang, and Viktor Schroeder, *Convex hulls in singular spaces of negative curvature*, Ann. Global Anal. Geom. **18** (2000), no. 2, 191–204, DOI 10.1023/A:1006698910715. MR1744590 (2002d:53047)

[9] Bruce Kleiner and Bernhard Leeb, *Rigidity of invariant convex sets in symmetric spaces*, Invent. Math. **163** (2006), no. 3, 657–676, DOI 10.1007/s00222-005-0471-y. MR2207236 (2006k:53064)

[10] Bernhard Leeb and Carlos Ramos-Cuevas, *The center conjecture for spherical buildings of types F_4 and E_6*, Geom. Funct. Anal. **21** (2011), no. 3, 525–559, DOI 10.1007/s00039-011-0118-7. MR2810858 (2012i:51012)

[11] Bernhard Mühlherr and Jacques Tits, *The center conjecture for non-exceptional buildings*, J. Algebra **300** (2006), no. 2, 687–706, DOI 10.1016/j.jalgebra.2006.01.011. MR2228217 (2007e:51018)

[12] Bernhard Mühlherr and Richard M. Weiss, *Receding polar regions of a spherical building and the center conjecture* (English, with English and French summaries), Ann. Inst. Fourier (Grenoble) **63** (2013), no. 2, 479–513. MR3112519

[13] Sôji Nakajima, *Über konvexe Kurven und Flächen*, Tôhoku Math. J. **29** (1928), 227–230.

[14] Athanase Papadopoulos, *Metric spaces, convexity and nonpositive curvature*, IRMA Lectures in Mathematics and Theoretical Physics, vol. 6, European Mathematical Society (EMS), Zürich, 2005. MR2132506 (2005k:53042)

[15] Carlos Ramos-Cuevas, *The center conjecture for thick spherical buildings*, Geom. Dedicata **166** (2013), 349–407, DOI 10.1007/s10711-012-9799-9. MR3101173

[16] Heinrich Tietze, *Über Konvexheit im kleinen und im großen und über gewisse den Punkten einer Menge zugeordnete Dimensionszahlen* (German), Math. Z. **28** (1928), no. 1, 697–707, DOI 10.1007/BF01181191. MR1544985

Mathematisches Institut, Ludwig-Maximilians-Universität, Theresienstr. 39, 80333, Munich, Germany

E-mail address: cramos@mathematik.uni-muenchen.de

Contemporary Mathematics
Volume 657, 2016
http://dx.doi.org/10.1090/conm/657/13098

An introduction to the half-infinite wedge

Rodolfo Ríos-Zertuche

Abstract. After a quick review of the representation theory of the symmetric group, we give an exposition of the tools brought about by the so-called half-infinite wedge representation of the infinite symmetric group. We show how these can be applied to find the limit shapes of several distributions on partitions. We also briefly review the variational methods available to compute these limit shapes.

Contents

1. Introduction

The representation theory of the infinite symmetric group is located at the crossroads of a number of subjects that, together, give light to its nature and permit its application for the solution of a wide range of problems. These subjects include combinatorics, probability, arithmetic, algebraic geometry, and variational analysis. In this review article, we have tried to compile the main facts and techniques developed in relation with this theory, with the aim of reducing the steepness of the learning curve for these useful tools.

Although for many parts of the theory there exist very good introductions, such as **[18, 32, 34]**, there are important tools for which the standard references are still quite hard to drudge through. A good example is the half-infinite wedge fermionic Fock space, for which the standard reference **[15**, Chapter 14] is admittedly inadequate for the beginner who does not want to become a specialist in a

2010 *Mathematics Subject Classification.* Primary 20C32.
Key words and phrases. Symmetric group, partition, limit shape.

much broader theory. Other good sources, like **[21]** or **[25**, Section 2], do not go into enough depth because they pursue different aims. Here we try to explain the tool and its techniques, and we also give examples of its application.

We feel that the increased production of research papers that leverage the usefulness of these tools (see for example **[7, 8, 19, 20, 25, 27, 33]**) justifies this new effort for their exposition.

We start with a quick review of the theory, including general facts of the representations of finite groups applied to the symmetric group in Section 2.

In Section 3, we review the different available formulas for the dimension of the irreducible representations, and what is known about their asymptotics. We also include some words on the useful Okounkov-Olshanski formula for the dimension of skew Young diagrams.

Then we explain the computation of the characters of the symmetric group through the Murnaghan-Nakayama rule (Section 4) and through the theory of symmetric functions (Section 5).

All comes into play in the half-infinite wedge fermionic Fock space, which we explain in Section 6. Because they are so useful, we have included a few words on computations with vertex operators in Section 7. We also give a quick review of the theory of the simplest theta functions, which are quite prone to appear in this type of calculation.

Because it is useful for the examples we want to give, we also talk about the basic variational method to deal with the asymptotics of products and quotients of hooks. This is the content of Section 8.

Finally, we give several examples in Section 9. These concentrate on the issue of finding limit shapes of different distributions.

The reader who is in a hurry to learn how to use the half-infinite wedge fermionic Fock space can skip Sections 2.1, 3, 7.3, and 8, and is encouraged to work out the example in Section 9.1 carefully.

Part of the material of this article was originally written for the author's thesis **[31]**. It has been significantly expanded, carefully revised, and published in the hope that others will find it useful. Needless to say, this introduction is by no means exhaustive in any direction.

Acknowledgements. The author is deeply indebted to his adviser, Professor Andrei Okounkov, who taught him virtually all he knows in relation to the subject of this paper. He is also very grateful to Dr. Noé Bárcenas Torres, Dr. Fernando Galaz García, and Dr. Mónica Moreno Rocha, organizers of the workshop "Taller de vinculación: matemáticos mexicanos jóvenes en el mundo," for a wonderful conference, for their encouragement to get this piece written up, and for making its publication possible. The author is also very grateful for the comments of the anonymous referees that have helped to greatly improve this paper.

2. General representation theory of the symmetric group

The *symmetric group* $S(n)$ in n elements is the set of bijections of the set of n elements $\{1, 2, \ldots, n\}$. Its elements are known as *permutations*. The group operation is the composition of permutations.

For simplicity, we will work throughout over the field of complex numbers, which we denote $\mathbb{C}$, and we will implicitly assume that all vector spaces are defined over this field.

A *representation* (V, ρ) of the symmetric group $S(n)$ consists of a vector space V and a homomorphism $\rho : S(n) \to \mathrm{GL}(V)$ of $S(n)$ into the multiplicative group $\mathrm{GL}(V)$ of invertible linear transformations on V. A representation is *reducible* if there are nontrivial subspaces $U, W \subset V$ such that $V = U \oplus W$ and ρ factors through $\mathrm{GL}(U) \times \mathrm{GL}(W)$, that is, there exist representations $\rho_U : S(n) \to \mathrm{GL}(U)$ and $\rho_W : S(n) \to \mathrm{GL}(W)$ such that $\rho(\sigma)|_U = \rho_U(\sigma)$ and $\rho(\sigma)|_W = \rho_W(\sigma)$ for all $\sigma \in S(n)$. If two such subspaces fail to exist, we say that (V, ρ) is an *irreducible representation* of $S(n)$.

The *dimension* of the representation (V, ρ) is simply the dimension of the underlying vector space V.

Two representations (U, ρ_U) and (V, ρ_V) are *isomorphic* if there exists a linear bijection $\phi : U \to V$ such that $\phi \circ \rho_V = \rho_U$. The existence of an isomorphism of representations implies that the underlying vector spaces U and V are of the same dimension.

For an arbitrary matrix A, we denote its trace by $\operatorname{tr} A$. The *character* $\chi_V : S(n) \to \mathbb{C}$ of a representation (V, ρ) is the function $\chi_V(\sigma) = \operatorname{tr} \rho(\sigma)$. The character is invariant under conjugation: for every $\sigma, \tau \in S(n)$,

$$\chi_V(\tau^{-1}\sigma\tau) = \operatorname{tr} \rho(\tau^{-1}\sigma\tau) = \operatorname{tr} \left(\rho(\tau)^{-1}\rho(\sigma)\rho(\tau)\right) = \operatorname{tr} \rho(\sigma) = \chi_V(\sigma).$$

It is thus constant over conjugacy classes.

The classification of the finite-dimensional representations of the symmetric group can be easily achieved using the standard theory of group characters; for details we refer the reader to [**11**, **32**], for example.

In order to explain this classification, we must introduce some combinatorial objects. A *partition* of a positive integer n is a way to write n as a sum $n = a_1 + a_2 + \cdots + a_k$ of positive integers $a_1, a_2, \ldots, a_k$, and is usually denoted $(a_1, a_2, \ldots, a_k)$. Conventionally, we order the parts so that $a_1 \geq a_2 \geq \cdots \geq a_k$. For example, the partitions of the number 5 are (5), $(4,1)$, $(3,2)$, $(3,1,1)$, $(2,2,1)$, $(2,1,1,1)$, and $(1,1,1,1,1)$. It is also standard to abbreviate repetitions using exponents. For example, $(3,3,1,1,1,1,1)$ can be written $(3^2, 1^5)$. We will use the notation $|\lambda| = \sum_i \lambda_i = n$, and $\ell(\lambda)$ will denote the number of parts of λ.

The irreducible representations of the symmetric group $S(n)$ can be indexed by partitions of n, that is, for each partition λ of n there exists an irreducible representation $(V_\lambda, \rho_\lambda)$ of $S(n)$. We will denote by χ^λ the character of the representation corresponding to the partition λ.

In the case of the symmetric group $S(n)$, the conjugacy classes are also indexed by partitions: the lengths of the cycles of an element of $S(n)$ are invariant under conjugation, they completely determine the conjugacy class, and they correspond to a partition of n. For example, the permutation in $S(5)$ that takes $1 \mapsto 2$, $2 \mapsto 5$, $5 \mapsto 1$, $3 \mapsto 4$, $4 \mapsto 3$ is denoted in cycle notation by $(1,2,5)(3,4)$; it is composed of two cycles, $(1,2,5)$ and $(3,4)$, of lengths 3 and 2 respectively. Its conjugacy type corresponds to the partition $(3,2)$, and contains all other permutations with one cycle of length 3 and one of length 2, such as $(1,2,3)(4,5)$ and $(3,4,5)(1,2)$. The partition λ that determines the conjugacy class C_λ of a permutation is also known as its *cycle type*. The number of elements $|C_\lambda|$ in a conjugacy class is equal to $|S(n)|/\mathfrak{z}(\lambda)$, where

$$\mathfrak{z}(\lambda) = \prod_{i=1}^{\infty} i^{m_i(\lambda)} m_i(\lambda)!$$

is the size of the centralizer of C_λ, and $m_i(\lambda)$ is the number of parts in λ that are equal to i.

The only permutation of cycle type $(1, 1, \dots, 1)$ is the identity. A permutation σ of cycle type $(2, 2, \dots, 2, 1, 1, \dots, 1)$ is an involution, that is, $\sigma \circ \sigma$ is the identity. A permutation of cycle type $(2, 2, \dots, 2)$ is an involution without fixed points.

Since the characters are constant over conjugacy classes, we usually denote the character of a permutation of cycle type μ in the irreducible representation corresponding to the partition λ by $\chi^\lambda(\mu)$. In the following sections, we study some related objects that provide ways to compute the values of these functions more or less explicitly, and that relate them to other combinatorial constructions.

Because of their connection with the classification of irreducible representations and of conjugacy classes, partitions play a crucial role in the theory. We remark that to a partition $\lambda = (\lambda_1, \lambda_2, \dots, \lambda_k)$, where $\lambda_1 \geq \lambda_2 \geq \cdots \geq \lambda_k \geq 0$, we can assign a *Young diagram* consisting of a drawing of rows of boxes stacked on each other, with row i consisting of of λ_i square boxes for $i = 1, 2, \dots, k$. For example, if our partition is $(5, 4, 4, 2)$, the diagram looks like this:

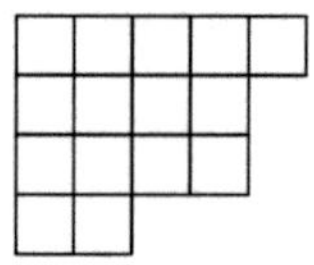

In this diagram, each square is called a *cell*.

We recall two more general facts about group representations. The first one is the decomposition of the group algebra $\mathbb{C}G$ of any finite group G, which is itself a representation of G. The group algebra $\mathbb{C}G$ is the algebra obtained by forming the vector space over $\mathbb{C}$ whose generators are the elements of G and whose product operation is inherited from the operation of G. It turns out to contain each of the irreducible representations of G exactly as many times as their dimensions, that is,

$$\mathbb{C}G = \bigoplus_i V_i^{\oplus \dim V_i},$$

where the subspaces V_i range over the different irreducible representations of G. See, for example, [**32**, Section 1.10].

The second one is *Schur's lemma*: Let V and W be two irreducible representations of G. If $\phi : V \to W$ is a G-homomorphism (i.e., ϕ commutes with the action of G), then either ϕ is a G-isomorphism, or ϕ is the zero map. See, for example, [**32**, Section 1.6].

2.1. The branching rule and Frobenius reciprocity. The restriction of a representation $\rho : S(n) \to \mathrm{GL}(V)$ of $S(n)$ to a smaller symmetric group $S(m)$, $m < n$, induces a restriction of the corresponding representation, which we denote $\mathrm{Res}_{S(m)} V$. If ρ is irreducible and corresponds, say, to the partition λ of n, $\mathrm{Res}_{S(m)} V$ is, in general, reducible. The *branching rule* gives its decomposition into irreducible representations. It states that, if $n = m + 1$ and V^λ is the irreducible representation corresponding to the partition λ, then

$$\mathrm{Res}_{S(m)} V^\lambda = \bigoplus_{\lambda = \mu \cup \square} V^\mu,$$

where the direct sum is taken over all partitions μ of m from which λ can be obtained by adjoining one cell $\square$. For example,

$$\operatorname{Res}_{S(11)} V^{(4,3,3,2)} = V^{(4,3,3,1)} \oplus V^{(4,3,2,2)} \oplus V^{(3,3,3,2)}.$$

Note that, iterating, this gives a complete description of the decomposition corresponding to any pair $m < n$. For a proof, see for example [**32**, Section 2.8] or [**10**, Section 7.3]. If χ is the character of ρ, then we will denote by $\operatorname{Res}_{S(m)} \chi$ the character of $\operatorname{Res}_{S(m)} V$.

A representation $\rho : S(m) :\to \mathrm{GL}(V)$ induces a representation $\operatorname{Ind}_{S(n)} V$ of $S(n)$ (see [**32**] for a full discussion). There is a version of the branching rule in this case:

$$\operatorname{Ind}_{S(n)} V^{\lambda} = \bigoplus_{\lambda \cup \square = \mu} V^{\mu},$$

where the sum is taken over all partitions μ that can be formed by adjoining a cell $\square$ to λ. Let χ be the character of ρ and denote by $\operatorname{Ind}_{S(n)} \chi$ the character of the induced representation. There is a way to compute $\operatorname{Ind}_{S(n)} \chi$ using Frobenius reciprocity, described next.

The vector space of real-valued functions on $S(n)$ contains a subspace generated by the characters on $S(n)$. This subspace has an inner product

$$\langle \chi_1, \chi_2 \rangle_{S(n)} = \frac{1}{n!} \sum_{\sigma \in S(n)} \chi_1(\sigma)\chi_2(\sigma).$$

The characters of irreducible representations turn out to be orthonormal with respect to this inner product. The relation

$$\langle \operatorname{Ind}_{S(n)} \chi_1, \chi_2 \rangle_{S(n)} = \langle \chi_1, \operatorname{Res}_{S(m)} \chi_2 \rangle_{S(m)},$$

which is true for all characters χ_1 and χ_2 of representations of $S(m)$ and $S(n)$, $m < n$, respectively, is known as *Frobenius reciprocity*. For a proof see [**11**, Section 3.3], [**32**, Theorem 1.12.6].

3. Formulas for the dimension

3.1. The hook formula. The formula obtained in 1954 by J. Frame, G. Robinson, and R. Thrall [**9**] to calculate the dimension of the irreducible representation of $S(n)$ corresponding to the partition λ, or equivalently, to calculate the character $\chi^{\lambda}(1, 1, \ldots, 1)$, has become known as the *hook formula* due to the combinatorial objects it involves.

To each cell in a Young diagram we can assign several quantities. For our purposes, the most important one is the *hook length*, which we now define. For a given cell, the *hook* consists of the cell itself, together with all the cells to the right of it and all the cells below it. So, for example, in this diagram we have marked with bullets the hook of cell $(2, 2)$:

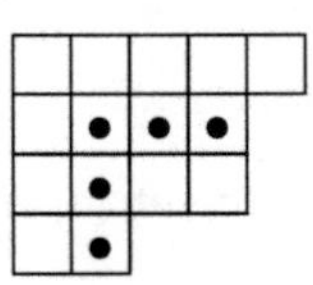

The *hook length* is the number of cells in the hook. In the example of the last diagram, the hook length is 5. In the following diagram we have filled in each cell with the corresponding hook length:

8	7	5	4	1
6	5	3	2	
5	4	2	1	
2	1			

Now let λ be a partition of n, and let $h(\lambda)$ be the product of the hook lengths of λ. For the example above, $\lambda = (5,4,4,2)$, $n = 15$, and $h(\lambda) = 16\,128\,000$. Then the hook formula, discovered by Frame, Robinson, and Thrall, says that

$$\boxed{\dim \lambda := \chi^{\lambda}(1,1,\ldots,1) = \frac{n!}{h(\lambda)}.} \tag{1}$$

Here, $(1,1,\ldots,1)$ stands for the partition with n parts equal to 1 and corresponds to the conjugacy class of the identity.

In the example above, formula (1) gives

$$\dim(5,4,4,2) = \chi^{(5,4,4,2)}(1,1,\ldots,1) = \frac{15!}{h((5,4,4,2))} = 81\,081.$$

A proof of the hook formula can be found in the original paper [**9**], and also in [**18**, Examples I.1.1 and I.7.6], [**32**, Section 3.10], and [**34**, Corollary 7.21.6].

3.2. Other formulas. There is an alternative way to express this as a function of the coordinates of the partition λ [**18**, Example I.1.1]:

$$\boxed{\dim \lambda = \frac{|\lambda|!\prod_{i<j}(\lambda_i - \lambda_j - i + j)}{\prod_i(\lambda_i + \ell(\lambda) - i)!}.} \tag{2}$$

Here $\ell(\lambda)$ denotes the number of parts of λ.

Yet another formula can be obtained if one uses instead the so-called Frobenius coordinates for a partition. For each partition λ, there is a *transposed partition* λ' that is obtained by exchanging rows by columns in the Young diagram; for example, if $\lambda = (5,4,4,2)$, then $\lambda' = (4,4,3,3,1)$:

$\lambda =$ 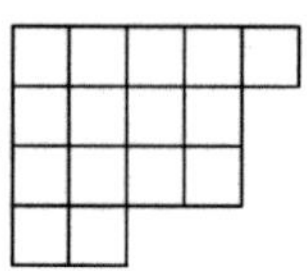 $\lambda' =$ 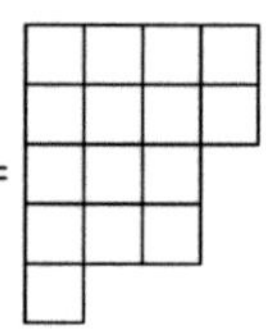

The *Frobenius coordinates* are defined as follows: for a partition $\lambda = (\lambda_1, \lambda_2, \ldots, \lambda_k)$, let d be the largest integer $\leq k$ such that $\lambda_d \geq d$. In other words, d is the size of the diagonal of λ. Then let $p_i = \lambda_i - i$ and $q_i = \lambda'_i - i$ for $i = 1,2,\ldots,d$, where the numbers λ'_i are the parts of the transposed partition λ'. Of course, $p_i, q_i \geq 0$. For $\lambda = (5,4,4,2)$ as above, the Frobenius coordinates are $p = (4,2,1)$, $q = (3,2,0)$. We have [**28**, eq. (2.7)]

$$\boxed{\dim \lambda = \frac{\prod_{i<j}(p_i - p_j)(q_i - q_j)}{\prod_{i,j}(p_i + q_j + 1)\prod_i p_i!\, q_i!}.}$$

This can also be written as a determinant [**4**, eq. (2.2)]:

$$\frac{\dim \lambda}{|\lambda|!} = \det\left[\frac{1}{(p_i + q_j + 1)p_i!\, q_i!}\right]_{1\leq i,j\leq d}.$$

3.3. Combinatorial interpretation. Combinatorially, $\dim\lambda$ is the number of ways in which one can construct the Young diagram of λ by adding one cell at a time, starting from the upper left corner and making sure that, at each step, the diagram obtained corresponds to some partition. So for example, for the partition $(3,2)$, one could add the cells in the following orders:

$$\begin{array}{|c|c|c|}\hline 1&2&3\\\hline 4&5\\\cline{1-2}\end{array}\quad\begin{array}{|c|c|c|}\hline 1&2&4\\\hline 3&5\\\cline{1-2}\end{array}\quad\begin{array}{|c|c|c|}\hline 1&2&5\\\hline 3&4\\\cline{1-2}\end{array}\quad\begin{array}{|c|c|c|}\hline 1&3&4\\\hline 2&5\\\cline{1-2}\end{array}\quad\begin{array}{|c|c|c|}\hline 1&3&5\\\hline 2&4\\\cline{1-2}\end{array}$$

In each row and column the numbers must be increasing. We conclude that $\dim(3,2) = 5$. This interpretation follows immediately from the branching rule (see Section 2.1) or from the Murnaghan-Nakayama rule (see Section 4).

3.4. Asymptotic formula. Vershik and Kerov [**37**] obtained noteworthy asymptotic estimates for the maximal dimension of the irreducible representation corresponding to the partition λ as $|\lambda| \to \infty$.

They found that there exist positive constants c and c' such that

$$\sqrt{n!}\exp\left(-\frac{c}{2}\sqrt{n}\right) \le \max_{\lambda\in S(n)} \dim\lambda \le \sqrt{n!}\exp\left(-\frac{c'}{2}\sqrt{n}\right).$$

Since the actual values of c and c' are not known explicitly, the true value of this formula lies on the description it gives of the asymptotic situation as $n \to \infty$.

It is also possible to obtain asymptotics for the typical dimensions for some probability distributions; see Section 9 for examples.

3.5. The Okounkov-Olshanski formula. In this section we will state a formula due to A. Okounkov and G. Olshanski [**24**] for the dimension $\dim(\lambda/\mu)$ of a skew diagram.

Let λ and μ be two partitions, and assume that $\mu_i \le \lambda_i$, $i = 1, 2, \dots$ A *skew Young diagram* λ/μ, consists of the cells of the Young diagram of λ that would not belong to the Young diagram of μ if we were to overlap them. For example, let $\mu = (3,1)$ and $\lambda = (5,4,4,2)$. Then

$\lambda/\mu =$ 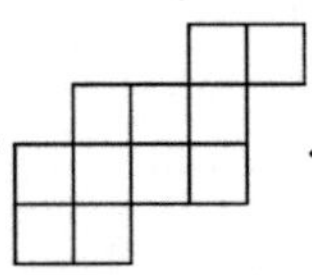 .

The *dimension* of λ/μ is defined as a straight-forward generalization of the combinatorial principle explained in Section 3.3: it is the number of ways in which one can assign the numbers $1, 2, \dots, |\lambda| - |\mu|$ to the cells λ/μ in such a way that in each row the numbers increase as one moves downwards, and in each column the numbers increase as one moves to the right.

To simplify notations, let

$$(x \downharpoonleft k) = x(x-1)(x-2)\cdots(x-k+1).$$

This is known as the *Pochhammer symbol* and another common notation for it is $(x)_k$.

We define the *shifted Schur polynomials* in n variables, indexed by the partition μ, as the following ratio of two $n \times n$ determinants:

$$\mathbf{s}_\mu(x_1,\dots,x_n) = \frac{\det[(x_i+n-i \downharpoonleft \mu_j+n-j)]}{\det[(x_i+n-i \downharpoonleft n-j)]}, \quad 1 \le i,j \le n.$$

These polynomials satisfy [**24**] a stability condition

$$\mathbf{s}_\mu(x_1, \dots, x_n, 0) = \mathbf{s}_\mu(x_1, \dots, x_n),$$

which allows us to take inverse limits, just as in the definition of symmetric functions (see Section 5). The resulting objects are known as *shifted Schur functions* and we will denote them by $\mathbf{s}_\mu(x_1, x_2, \dots)$. These are also sometimes called *Frobenius Schur functions* (see for example [**29**]). The algebra Λ^* they generate coincides with the algebra generated by the shifted-symmetric power functions $\mathbf{p}_k$.

We then have [**24**]:

(3) $$\boxed{\dim(\lambda/\mu) = \frac{\mathbf{s}_\mu(\lambda)\,\dim\lambda}{(|\lambda| \downarrow |\mu|)}.}$$

4. The Murnaghan-Nakayama rule

The Murnaghan-Nakayama rule gives a combinatorial algorithm to compute the value of the character $\chi^\lambda(\mu)$ of an irreducible representation of the symmetric group $S(n)$ corresponding to the partition λ and evaluated at an element of cycle type μ.

In order to state the rule, we need to define some combinatorial objects. Recall the definition of a skew Young diagram from Section 3.5. A special kind of Young diagram that we will be interested in is the *strip*, which is defined to be a connected skew Young diagram that does not contain a 2×2 block:

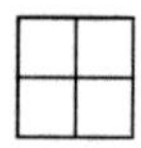

A skew Young diagram is connected if we can get from any cell to any other cell jumping on *adjacent* cells. For example, the following are strips:

(4)

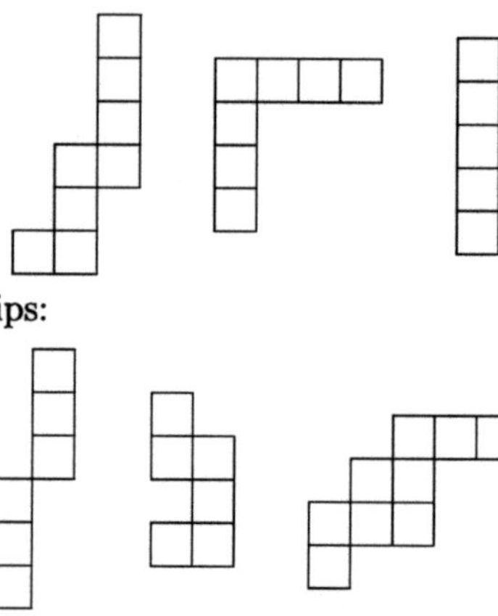

The following are *not* strips:

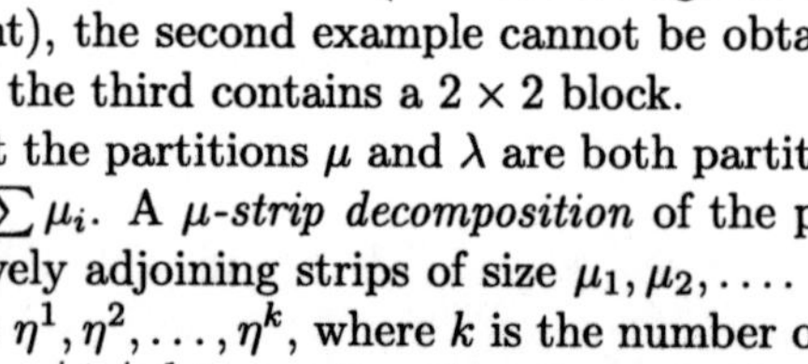

Here, the first example is disconnected (it is not enough to have a common corner for cells to be adjacent), the second example cannot be obtained as a difference of Young diagrams, and the third contains a 2×2 block.

Now suppose that the partitions μ and λ are both partitions of the same number, that is, $\sum \lambda_i = \sum \mu_i$. A *$\mu$-strip decomposition* of the partition λ is a way to recover λ by successively adjoining strips of size $\mu_1, \mu_2, \dots$. More precisely, it is a sequence of partitions $\eta^1, \eta^2, \dots, \eta^k$, where k is the number of parts in μ, such that η^1 is a strip of size μ_1, η^i/η^{i-1} is a strip of length μ_i, $i = 2, \dots, k$, and $\eta^k = \lambda$. For example, if $\lambda = (5, 4, 4, 2)$ and $\mu = (5, 3, 2, 2, 2, 1)$, a μ-strip decomposition of λ can be represented as follows: we will use number 1 to represent the strip of size 5, then number 2 to represent the strip of size 3, and so on, always assigning the number i to the strip of length μ_i:

$$
\begin{array}{|c|c|c|c|c|}\hline 1&1&1&3&6\\\hline 1&2&2&3\\\cline{1-4} 1&2&5&5\\\cline{1-4} 4&4\\\cline{1-2}\end{array}
\qquad
\begin{array}{|c|c|c|c|c|}\hline 1&1&2&3&3\\\hline 1&2&2&5\\\cline{1-4} 1&4&4&5\\\cline{1-4} 1&6\\\cline{1-2}\end{array}
\qquad
\begin{array}{|c|c|c|c|c|}\hline 1&1&2&4&4\\\hline 1&2&2&5\\\cline{1-4} 1&3&3&5\\\cline{1-4} 1&6\\\cline{1-2}\end{array}
\tag{5}
$$

Note that the second and third examples are different only in the order of the third and fourth strips.

The *height* of a strip λ/μ is equal to the *vertical* distance between the center of a cell in the lowest row of the strip to the center of a cell in the highest row of the strip, assuming that each cell has side length 1. For example, the strips in (4) have heights 5, 3, and 4, respectively. The *height of a μ-strip decomposition* equals the sum of the heights of the involved strips. For example, the height of the first decomposition in (5) is $2+1+1+0+0+0=4$, and the heights of the other two are both equal to $3+1+0+0+1+0=5$.

The *Murnaghan Nakayama rule* is the simple assertion that

$$
\boxed{\chi^\lambda(\mu) = \sum_{\{\eta_i\}} (-1)^{\mathrm{ht}\{\eta^i\}},}
\tag{6}
$$

where the sum is taken over all the μ-strip decompositions $\{\eta^i\}$ of λ. For example, if $\lambda = (4,3,2)$ and $\mu = (6,2,1)$, then there are exactly two μ-strip decompositions of λ, namely (in the same notations as above),

$$
\begin{array}{|c|c|c|c|}\hline 1&1&1&1\\\hline 1&2&3\\\cline{1-3} 1&2\\\cline{1-2}\end{array}
\quad\text{and}\quad
\begin{array}{|c|c|c|c|}\hline 1&1&1&1\\\hline 1&2&2\\\cline{1-3} 1&3\\\cline{1-2}\end{array}
$$

The heights of these decompositions are $2+1+0=3$ and $2+0+0=2$, respectively, so the Murnaghan-Nakayama rule implies that

$$
\chi^{(4,3,2)}((6,2,1)) = (-1)^3 + (-1)^2 = 0.
$$

Proofs of the Murnaghan-Nakayama rule can be found in [**18**, Exercises I.3.11 and I.7.5], [**32**, Section 4.10], or [**34**, Theorem 7.17.3].

5. Symmetric and shifted-symmetric functions

5.1. Schur functions and the characters of the symmetric group. For variables $x_1, x_2, \dots, x_n$, a *symmetric polynomial* is a polynomial $p(x_1, x_2, \dots x_n)$ with coefficients in the rational numbers $\mathbb{Q}$ such that p is invariant by any reordering of the variables:

$$
p(x_1, x_2, \dots, x_n) = p(x_{\sigma(1)}, x_{\sigma(2)}, \dots, x_{\sigma(n)})
$$

for all $\sigma \in S(n)$. Denote by P_n the set of all symmetric polynomials in n variables.

Let $\psi : P_{n+1} \to P_n$ be the specialization to n variables, given by

$$
\psi(p)(x_1, x_2, \dots, x_{n+1}) = p(x_1, x_2, \dots, x_n, 0).
$$

Using the identification provided by ψ, we can form the inverse limit

$$
\Lambda = \varprojlim P_n
$$

whose elements are known as *symmetric functions*, and consist of formal sums of monomials of the form

$$a x_{i_1} x_{i_2} \cdots x_{i_k}$$

in countably many variables $x_1, x_2, \ldots$, for $a \in \mathbb{Q}$, $i_1, \ldots, i_k \in \mathbb{N}$, where the numbers i_j may be repeated. By construction, symmetric functions are invariant by any finite permutation. They form a ring, denoted Λ, that is obviously also a vector space over $\mathbb{Q}$.

A basis of the symmetric functions Λ is given by the *symmetric power functions* p_μ. To define them, we let

$$p_k(x_1, x_2, \ldots) = x_1^k + x_2^k + \cdots, \quad k = 1, 2, \ldots,$$

and then extend this definition to partitions $\mu = (\mu_1, \mu_2, \ldots, \mu_k)$ by letting

$$p_\mu = p_{\mu_1} p_{\mu_2} \cdots p_{\mu_k}.$$

In other words, Λ is algebraically generated by the functions p_k [**18**].

While the symmetric power functions are by far the simplest linear basis of the symmetric functions Λ, there are a few other bases of Λ that are of interest. Among these is the one composed of the so-called *Schur functions* $s_1, s_2, \ldots$, which can be defined in many equivalent ways. One way is as the sum [**18**]

$$\boxed{s_\lambda = \sum_\rho \frac{\chi^\lambda(\rho)}{\mathfrak{z}(\rho)} p_\rho,} \tag{7}$$

where the sum is over all partitions $\rho = (\rho_1, \rho_2, \ldots, \rho_k)$, $\mathfrak{z}(\rho) = \prod_i i^{m_i(\rho)} m_i(\rho)!$, $m_i(\rho)$ is the number of parts ρ_j of ρ that are equal to i, and $\chi^\lambda(\rho)$ stands for the character of the representation of the symmetric group corresponding to the partition λ evaluated at an element of cycle type ρ.

A second way to define the Schur functions is by their characterization as being the symmetric functions that specialize (by setting the variables $x_{n+1}, x_{n+2}, \ldots$ to zero) to the *Schur polynomials*, defined by quotients of determinants:

$$s_\lambda(x_1, x_2, \ldots, x_n, 0, 0, \ldots) = \frac{\det(x_i^{\lambda_j + n - j})_{i,j=1,\ldots,n}}{\det(x_i^{n-j})_{i,j=1,\ldots,n}}. \tag{8}$$

This form of the definition, together with equation (7) gives a way to compute the characters $\chi^\lambda(\rho)$. Other ways to define s_λ are described in the following sections.

We want to mention an extremely important generalization that will come up in Section 7.1. Endow the space of symmetric functions with the inner product induced by

$$(s_\lambda, s_\mu) = \delta_{\lambda\mu}.$$

Define the *skew Schur functions* by

$$(s_{\lambda/\mu}, s_\nu) = (s_\lambda, s_\mu s_\nu).$$

In other words, if we have the expansion of the product in the basis given by the Schur functions s_λ,

$$s_\mu s_\nu = \sum_\lambda c_{\mu\nu}^\lambda s_\lambda,$$

then we define

$$s_{\lambda/\mu} = \sum_\nu c_{\mu\nu}^\lambda s_\nu.$$

The functions $s_{\lambda/\mu}$ vanish unless the Young diagram of μ is completely contained inside the Young diagram of λ (see [**18**, (I.5.7)]). The numbers $c^{\lambda}_{\mu\nu}$ have a nice combinatorial interpretation known as the *Littlewood-Richardson rule* that is out of the scope of this exposition; see [**32**, Section 4.9], [**18**, Section I.9].

Apart from the symmetric power functions and the Schur functions, there are other bases of the space of symmetric functions that appear frequently in the subject. The first one is the one one given by the *elementary symmetric functions*,

$$e_\lambda = \prod_i e_{\lambda_i}, \quad e_r = \sum_{i_1<i_2<\cdots<i_r} x_{i_1}x_{i_2}\cdots x_{i_r}, \quad r \in \mathbb{Z}_+.$$

There is also the basis given by the *monomial symmetric functions*,

$$m_\lambda = \sum_\alpha x_1^{\alpha_1}x_2^{\alpha_2}\cdots,$$

where the sum is over all $\alpha \in \mathbb{Z}^\infty$ such that $\alpha_i = 0$ for all but finitely many i, and the α_i that do not vanish are a permutation of the parts λ_i of the partition λ. These directly generate the space of symmetric functions.

One last basis we need to mention is the one given by the *complete symmetric functions* h_r, given by

$$h_r = \sum_{|\lambda|=r} m_\lambda.$$

Many interesting facts are known about the interaction of these bases and of their generating functions; the reader is encouraged to look at [**18**, Chapter I]. We mention only two relations that will come up in Section 7.1:

$$s_{\lambda/\mu} = \det\left(h_{\lambda_i-\mu_j+j-i}\right), \quad \sum_{r=1}^\infty h_r z^r = \exp\sum_{k=1}^\infty \frac{p_k z^k}{k}. \tag{9}$$

The reference for these is [**18**, pages 25 and 70].

5.2. Shifted-symmetric functions. Parallel to the symmetric functions, we have the *shifted-symmetric functions*, which are symmetric on the shifted variables $x_i - i + \frac{1}{2}$. Section 6 will clarify why these coordinates are very natural.

The shifted-symmetric Schur functions were already defined in Section 3.5. The *shifted symmetric power functions* are given by

$$\mathbf{p}_\lambda = \prod_i \mathbf{p}_{\lambda_i}, \quad \mathbf{p}_k(x) = \sum_{j=1}^\infty \left[\left(x_j - j + \tfrac{1}{2}\right)^k - \left(-j+\tfrac{1}{2}\right)^k\right] + \left(1 - \frac{1}{2^k}\right)\zeta(-k),$$

where ζ is the Riemann zeta function. The last summand appears due to the *zeta-function regularization* of the divergent sum $\sum_i (x_i - i + \frac{1}{2})^k$; see [**8**, Remark 2.14]. It will be useful to record their exponential generating function [**3**, eq. (0.18)], [**8**, eq. (2.8)]

$$\sum_k \frac{z^k}{k!}\mathbf{p}_k(x) = \sum_i e^{\left(x_i - i + \frac{1}{2}\right)z}. \tag{10}$$

Moments of the limit shape. Let $\Omega : \mathbb{R}_+ \to \mathbb{R}_+$ be a non-increasing continuous function such that

$$\int_{\mathbb{R}_+} \Omega(x)\, dx = 1.$$

Let λ^n, $n = 1, 2, \ldots$, be a sequence of partitions such that $|\lambda^n| \to \infty$ as $n \to \infty$. We say that λ^n *converges* to Ω if

$$\limsup_{n\to\infty} \max_j \left| \frac{1}{\sqrt{n}} \lambda_j^n - \Omega\left(\frac{j}{\sqrt{n}}\right) \right| = 0.$$

Let $L_\infty : \mathbb{R} \to \mathbb{R}$ be the function whose graph corresponds to the graph of Ω rotated 45° in the counter-clockwise direction (or 135° if we think of the graph of Ω drawn upside down, resembling the lower-right contour of the Young diagram of a very large partition, rescaled to have area 1). Also, after similarly rotating the Young diagram of λ^n, let L_n denote the function whose graph is the upper border of the rotated Young diagram of λ^n rescaled by $1/\sqrt{n}$. See the diagrams on page 210.

Then if λ^n converge to Ω, we also have that L_n converges to L_∞ in the topology of the supremum norm.

We want to give an interpretation of the shifted power functions $\mathbf{p}_k(\lambda^n)$: they converge to the moments of the contour of the limit shape L_∞. From the definition of $\mathbf{p}_k$ it is clear that as $n \to \infty$,

$$\left| n^{-\frac{k+1}{2}} \mathbf{p}_k(\lambda^n) - \int_{\mathbb{R}_+} \left((\Omega(x) - x)^k - (-x)^k \right) dx \right| \to 0.$$

Now,

$$\begin{aligned} \int_{\mathbb{R}_+} \left((\Omega(x) - x)^k - (-x)^k \right) dx &= \int_0^\infty \int_0^{\Omega(x)} \frac{d}{dy} (y - x)^k \, dy\, dx \\ &= \int_0^\infty \int_0^{\Omega(x)} k(y - x)^{k-1} \, dy\, dx = 2 \int_{-\infty}^\infty \int_{|s|}^{L_\infty(s)} k s^{k-1} \, dt\, ds \\ &= 2k \int_{-\infty}^\infty s^{k-1} \left(L_\infty(s) - |s| \right) ds \qquad (11) \end{aligned}$$

where $s = y - x$ and $t = y + x$. The change of variables is done as per the following diagram, in which the curve Γ depicts both the graph of $\Omega(x)$ and the graph of $L_\infty(s)$, and the integration domain is shaded in gray.

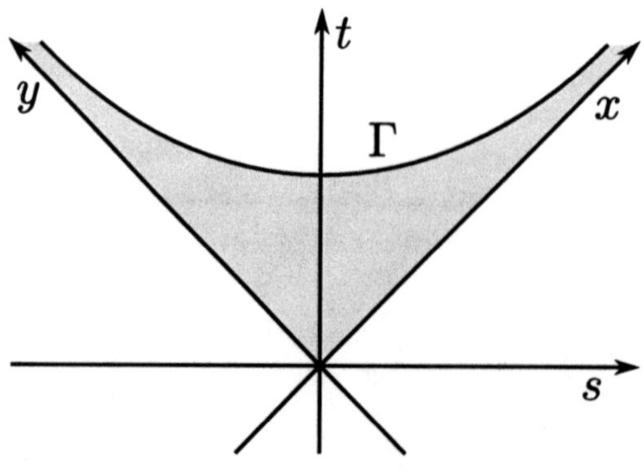

In other words, if we consider $L_\infty(s) - |s|$ as a density with moments μ_n,

$$\mu_n = \int_{\mathbb{R}} s^n (L_\infty(s) - |s|)\, ds,$$

then

$$\boxed{|\lambda^n|^{-\frac{k+1}{2}} \mathbf{p}_k(\lambda^n) \to 2k\mu_{k-1} \quad \text{as} \quad n \to \infty.}$$

Another derivation of the result of this section can be found in [**13**].

6. The half-infinite wedge fermionic Fock space

In this section we describe an algebraic framework to work with partitions and their connections to the symmetric group. Other good sources for this material are [**15**, Chapter 14], [**21**], and [**25**, Section 2].

Let $\lambda = (\lambda_1, \lambda_2, \dots, \lambda_k)$ be a partition. For convenience, we identify this vector representation with an infinite sequence by extending it with zeros:

$$\lambda = (\lambda_1, \lambda_2, \dots, \lambda_k, 0, 0, \dots),$$

so that $\lambda_i = 0$ for $i > k$. Now let $\xi_i = \lambda_i - i + \frac{1}{2}$ be the *shifted coordinates* of λ. Obviously, for $i > k$, $\xi_i = -i + \frac{1}{2}$.

We start with a countably-infinite-dimensional vector space V. We take a (countable) basis of V, and we denote its elements by $\dots, \underline{\frac{5}{2}}, \underline{\frac{3}{2}}, \underline{\frac{1}{2}}, \underline{-\frac{1}{2}}, \underline{-\frac{3}{2}}, \underline{-\frac{5}{2}}, \dots$; thus, this is a bidirectionally-infinite sequence of vectors that form a basis of V. The strange notation (using $\underline{\frac{2k+1}{2}}$ instead of something more normal like u_k) is well motivated by the geometric picture that we will now explain.

The *0-charge sector of fermionic Fock space* $\Lambda_0^{\frac{\infty}{2}} V$ (also known as the *half-infinite wedge* $\Lambda_0^{\frac{\infty}{2}}$) is a vector space generated by the basis of vectors v_λ indexed by partitions $\lambda = (\lambda_1, \lambda_2, \dots)$. These are defined by:

$$v_\lambda = \underline{\xi_1} \wedge \underline{\xi_2} \wedge \underline{\xi_3} \wedge \cdots = \underline{\lambda_1 - \tfrac{1}{2}} \wedge \underline{\lambda_2 - \tfrac{3}{2}} \wedge \underline{\lambda_3 - \tfrac{5}{2}} \wedge \cdots \wedge \underline{\lambda_i - i + \tfrac{1}{2}} \wedge \cdots.$$

Here, $\lambda_1 \geq \lambda_2 \geq \lambda_3 \geq \cdots \geq 0$ and the λ_i eventually become zero. The symbol $\wedge$ indicates that adjacent factors are anti-commutative. For example, for $\lambda = (5, 4, 4, 2)$,

$$\begin{aligned} v_\lambda &= \underline{\tfrac{9}{2}} \wedge \underline{\tfrac{5}{2}} \wedge \underline{\tfrac{3}{2}} \wedge \underline{-\tfrac{3}{2}} \wedge \underline{-\tfrac{9}{2}} \wedge \underline{-\tfrac{11}{2}} \wedge \underline{-\tfrac{13}{2}} \wedge \cdots. \\ &= -\underline{\tfrac{9}{2}} \wedge \underline{\tfrac{3}{2}} \wedge \underline{\tfrac{5}{2}} \wedge \underline{-\tfrac{3}{2}} \wedge \underline{-\tfrac{9}{2}} \wedge \underline{-\tfrac{11}{2}} \wedge \underline{-\tfrac{13}{2}} \wedge \cdots, \end{aligned}$$

since the factors $\underline{\frac{3}{2}}$ and $\underline{\frac{5}{2}}$ have been exchanged.

The relation between v_λ and λ can be best visualized as follows. We rotate the picture of the Young diagram of λ by 135° counter-clockwise, and we rescale it by $\sqrt{2}$. We then place black pebbles at each point ξ_i on the reversed x-axis, and white ones in the remaining half-integers. For example, if $\lambda = (5, 4, 4, 2)$ we start with the usual Young diagram,

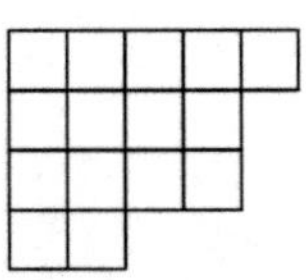

and then we rotate to get the following diagram.

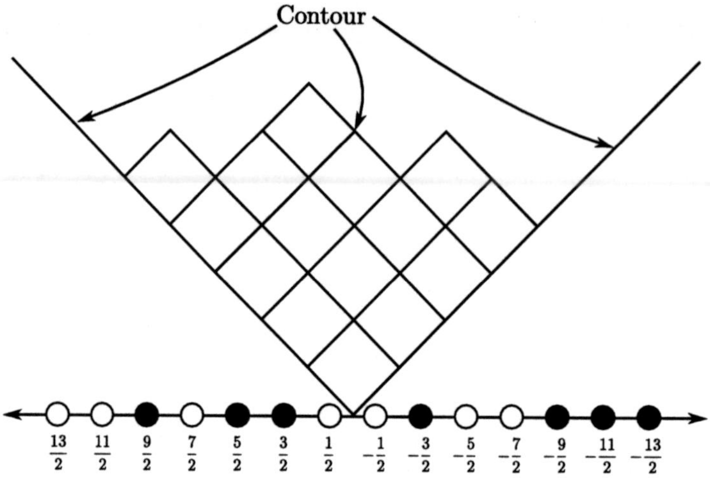

In the diagram we have also drawn the rotated axes, which extend to infinity. The *contour* L_λ of λ is the graph of the continuous function that describes the piecewise-linear curve starting (in the picture) with the diagonal axis on the left, running along the top border (or rim) of the diagram of the partition and ending with the diagonal axis on the right. It is clear that the contour of λ has slope -1 wherever there is a white pebble, and has slope $+1$ in the intervals where there is a black pebble, that is, on the segments corresponding to the shifted coordinates ξ_i of λ. The association of sequences of pebbles to a Young diagram is known as a *Maya diagram.* Observe that in such a diagram there are always as many black pebbles to the left of zero as there are white pebbles to the right of zero.

If we assign the number 0 to the white pebbles and the number 1 to the black ones, we get a sequence $\{c_i\}_{i\in\mathbb{Z}}$, such that $c_i \in \{0,1\}$, $c_i = 0$ for all $i \ll 0$, and $c_i = 1$ for all $i \gg 0$. In the case of the example above the sequence equals:

$$\ldots, 0, 0, 0, 1, 0, 1, 1, 0, 0, 1, 0, 0, 1, 1, 1, \ldots$$

The sequence completely determines the partition λ. It coincides with the sequence one gets if one assigns the number 1 to each vertical segment and the number 0 to each horizontal segment in the contour, as follows:

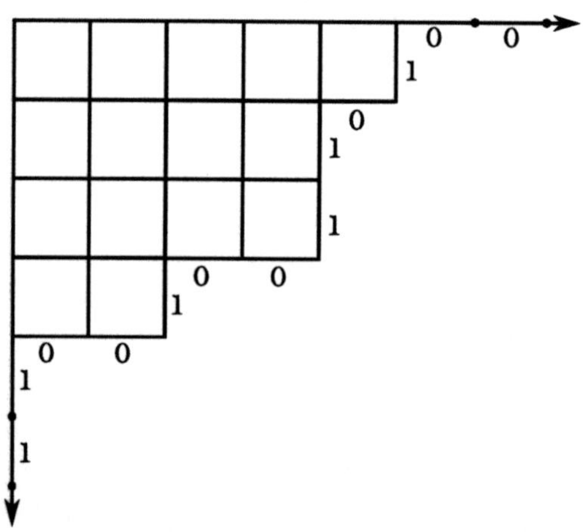

Note that two sequences $\{a_i\}_{i\in\mathbb{Z}}$ and $\{b_i\}_{i\in\mathbb{Z}}$ of zeros and ones with the above properties determine the same partition if they are translates $a_i = b_{i-n}$ of each other, for some $n \in \mathbb{Z}$. Both points of view —Maya diagrams and binary sequences— are equivalent. In what follows, we will prefer the language of pebbles placed under the rotated diagram.

The space $\Lambda_0^{\frac{\infty}{2}} V$ is contained inside *fermionic Fock space* $\Lambda^{\frac{\infty}{2}} V$, the space generated by *all* wedge products of the form

$$v_{\{s_i\}} = \underline{s_1} \wedge \underline{s_2} \wedge \underline{s_3} \wedge \cdots, \tag{12}$$

where s_i is a half-integer contained in $\mathbb{Z} + \frac{1}{2}$, $s_1 \geq s_2 \geq s_3 \geq \cdots$, and $s_i - 1 = s_{i+1}$ for i large enough. The 0-charge sector is simply the subspace whose generators can be identified with partitions. The basis of $\Lambda_0^{\frac{\infty}{2}}$ given by vectors of the form (12) is characterized by containing only those generators for which there are as many positive numbers s_i present as there are negative numbers s_i missing from the wedge.

The fermionic Fock space is a formalization of the quantum-mechanical model known as *Dirac's sea of electrons.* In this model, the ground state corresponds to the vacuum, represented by the vector

$$v_{\varnothing} = \underline{-\tfrac{1}{2}} \wedge \underline{-\tfrac{3}{2}} \wedge \underline{-\tfrac{5}{2}} \wedge \cdots.$$

We say that the state represented by this vector corresponds to the vacuum because it has no electrons (factors $\underline{i}$ with $i > 0$) and no positrons (missing factors $\underline{i}$ with $i < 0$). Electrons can be created by exchanging a white pebble with a black pebble; a black pebble on a positive number represents an electron excited to the energy given by that number, and a white pebble on a negative number represents a positron.[1] A vector in $\Lambda_0^{\frac{\infty}{2}}$ corresponds to a state with equal number of electrons and positrons, and has zero net charge. The number of cells in the corresponding partition can be defined to be equal to the total energy of the system; see below.

6.1. Basic operators. Let k be a number in $\mathbb{Z} + \frac{1}{2}$ and let ψ_k be the *creation operator* in $\Lambda^{\frac{\infty}{2}} V$ defined by

$$\psi_k(v) = \underline{k} \wedge v.$$

Note that $\underline{k}$ is added at the very beginning of the sequence, so the anti-commutativity may produce a sign (when comparing with the elements of the basis). For example,

$$\begin{aligned}\psi_{\frac{3}{2}}(v_{(3,1)}) &= \underline{\tfrac{3}{2}} \wedge \left(\underline{\tfrac{5}{2}} \wedge \underline{-\tfrac{1}{2}} \wedge \underline{-\tfrac{5}{2}} \wedge \underline{-\tfrac{7}{2}} \wedge \cdots\right) \\ &= -\left(\underline{\tfrac{5}{2}} \wedge \underline{\tfrac{3}{2}} \wedge \underline{-\tfrac{1}{2}} \wedge \underline{-\tfrac{5}{2}} \wedge \underline{-\tfrac{7}{2}} \wedge \cdots\right)\end{aligned}$$

The anti-commutativity also implies that if the factor $\underline{k}$ is there already, the result is zero. For instance, $\psi_{\frac{5}{2}}(v_{(3,1)}) = 0$. We define an inner product in $\Lambda^{\frac{\infty}{2}} V$,

$$\langle v_{\{s_i\}}, v_{\{t_i\}} \rangle = \delta_{\{s_i\},\{t_i\}},$$

[1] I am very grateful to the anonymous referee who has clarified some history to me: "Dirac originally identified the holes as protons. Dirac himself said many times that this error was due to his lack of imagination. This does not work at all. Later the holes were identified as prositrons and the idea of anti-particle was introduced for the first time in physics. [...] That works better, but is not considered nowadays to be the best way to think of positrons." Apparently the identification was not due to Dirac.

where $\delta_{\{s_i\},\{t_i\}}$ is 1 if both sequences s_i and t_i are equal, and 0 otherwise. Then ψ_k has an adjoint operator ψ_k^*, known as the *annihilation operator*. The idea is that, if the spot k is empty (i.e., occupied by a white pebble), ψ_k adds an electron (represented by a black pebble) there, and ψ_k^* removes it. These operators satisfy the anti-commutation relations

$$\psi_i\psi_j^* + \psi_j^*\psi_i = \delta_{ij},$$

$$\psi_i\psi_j + \psi_j\psi_i = 0 = \psi_i^*\psi_j^* + \psi_j^*\psi_i^*.$$

Define the *normally ordered product*, a device invented to make it easy to keep track of the convergence of infinite sums of products of the operators ψ_i and ψ_i^*, by

$$:\psi_i\psi_j^*: = \begin{cases} \psi_i\psi_j^*, & j > 0, \\ -\psi_j^*\psi_i, & j < 0. \end{cases}$$

If $i > j$, the net effect of applying $:\psi_i\psi_j^*:$ to a vector v_λ, where λ is some partition, is to add a strip to λ if possible, and to add a sign according to the parity of the height of the strip (this should be reminiscent of equation (6)). For example,

$$:\psi_{\frac{11}{2}}\psi^*_{-\frac{3}{2}}: v_{(5,4,4,2)} = -v_{(6,6,5,5)}.$$

This is illustrated in the following picture, which should be compared to the diagram on page 210:

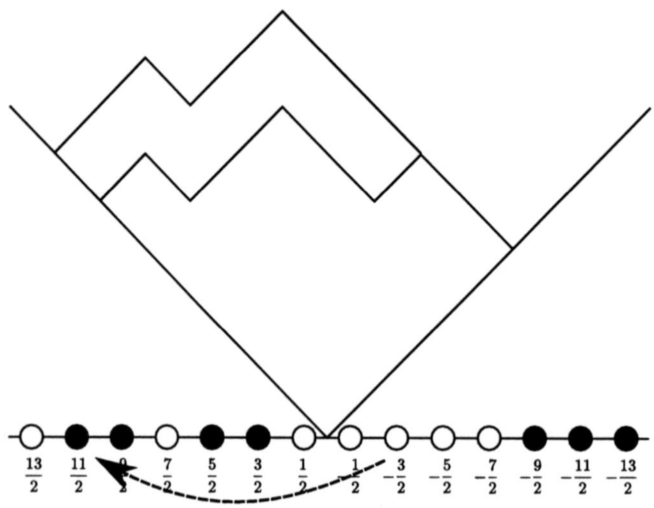

The height of the strip $(6,6,5,5)/(5,4,4,2)$ added in this procedure is 3, so the sign is indeed $(-1)^3 = -1$:

$$:\psi_{\frac{11}{2}}\psi^*_{-\frac{3}{2}}: v_{(5,4,4,2)} = -v_{(6,6,5,5)}.$$

If the addition of the strip is not possible, the result is zero. If $i < j$, the effect is precisely the opposite: a strip is removed and again a sign is added according to the parity of the height of the strip; if the strip cannot be removed, the result is zero.

Let us give some examples to show how useful a device the normally ordered product turns out to be. Right away, one can define the *energy operator* H by

$$H = \sum_{k\in\mathbb{Z}+\frac{1}{2}} k :\psi_k\psi_k^*:$$

and it is a quick exercise to convince oneself that $Hv_\lambda = |\lambda| v_\lambda$. For any given partition λ, the sum involved in Hv_λ is actually finite (and hence there is no doubt that it converges) because of the way the normally ordered product works. The operator H is often called the *Hamiltonian*, whence the notation H.

Another interesting operator is the *charge operator* C given by

$$C = \sum_{k \in Z + \frac{1}{2}} :\psi_k \psi_k^*:$$

whose action should be considered in the larger space $\Lambda^{\frac{\infty}{2}}$:

$$C(\underline{s_1} \wedge \underline{s_2} \wedge \cdots) = [(\#\text{ of present positive } s_i) \\ - (\#\text{ of missing negative } s_i)]\underline{s_1} \wedge \underline{s_2} \wedge \cdots .$$

In particular, the vectors in the 0-charge sector $\Lambda_0^{\frac{\infty}{2}}$ are characterized by being annihilated by C, as their name suggests.

Let $n > 0$ be an integer. We want to define operators α_{-n} and α_n that will, loosely speaking, respectively increase and decrease the energy of an electron by n in all possible ways or, equivalently, they will respectively add and remove all possible strips of length n. We let, for n in $\mathbb{Z}$,

$$\alpha_{-n} = \sum_{k \in \mathbb{Z} + \frac{1}{2}} :\psi_{k+n} \psi_k^*:$$

For example,

$$\alpha_{-3}\, v_{(3,1)} = v_{(6,1)} - v_{(3,2,2)} + v_{(3,1,1,1,1)},$$

which graphically corresponds to:

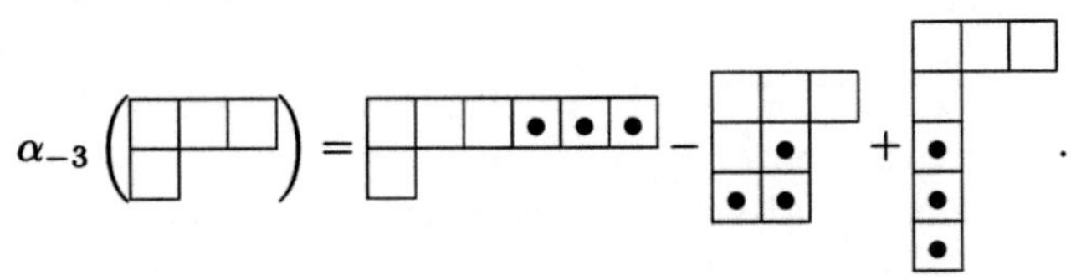

The bullets • mark the cells corresponding to the added strips. Note that the signs correspond to the parities of the heights of these strips. Similarly,

$$\alpha_3\, v_{(5,4,3)} = v_{(5,4)} - v_{(3,3,3)} - v_{(5,2,2)},$$

or

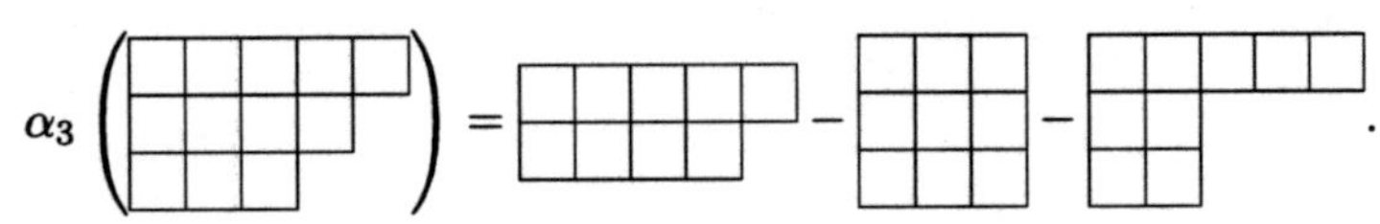

Strips of the forms 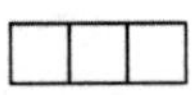 and 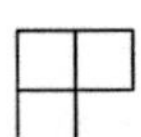 have been removed, and again the signs correspond to the parity of their heights.

Let $\varnothing$ stand for the empty partition. The vector $v_\varnothing = \underline{-\frac{1}{2}} \wedge \underline{-\frac{3}{2}} \wedge \underline{-\frac{5}{2}} \wedge \cdots$ is known as the *ground state*.

Let μ be a partition. The following important identity is an immediate consequence of the Murnaghan-Nakayama rule, equation (6):

(13)
$$\boxed{\alpha_{-\mu_1} \alpha_{-\mu_2} \cdots \alpha_{-\mu_k} v_\varnothing = \sum_\lambda \chi^\lambda(\mu)\, v_\lambda .}$$

The sum is over all partitions λ, but of course the character $\chi^\lambda(\mu)$ equals zero for all but finitely many of them. Since the characters χ^λ are linearly independent from each other, the vectors $\prod_i \alpha_{-\mu_i} v_\varnothing$ form a basis of the entire space $\Lambda_0^{\frac{\infty}{2}} V$.

The space $\Lambda_0^{\frac{\infty}{2}} V$ is almost[2] a Hilbert space when endowed with the inner product $\langle \cdot, \cdot \rangle$ induced by

$$\langle v_\lambda, v_\mu \rangle = \delta_{\lambda\mu},$$

where $\delta_{\lambda\mu}$ is 1 when the partitions λ and μ are equal, and vanishes otherwise. With this inner product, the adjoint α_n^* of α_n satisfies

$$\alpha_n^* = \alpha_{-n},$$

and we have the commutation relations

$$[\alpha_k, \alpha_r] := \alpha_k \alpha_r - \alpha_r \alpha_k = k\,\delta_{-k,r}. \tag{14}$$

6.2. p-cores and p-quotients. The Young diagram of the partition $(3, 2, 2, 2)$ can be constructed out of 3-strips, but the Young diagram of the partitions $(4, 2, 2, 1)$ cannot. The theory of p-quotients and p-cores gives a framework to understand how this works, and the infinite wedge clarifies it further. We first illustrate with the $p = 2$ case, and we comment on the general case at the end of the section. We mainly follow [**34**, Excercise 7.59], [**18**, Example I.1.8].

It is easy to identify, given the Maya diagram of a partition, the sites from which it is possible to remove a p-strip: one only needs to look for a site with a black pebble such that in the site p units to the right there is a white pebble. In the example above involving partition $(6, 6, 5, 5)$, since we have a black pebble at the site $\frac{11}{2}$, we can remove strips of sizes 2, 5, 6, 7, 8, and 9, corresponding respectively to the white pebbles at the sites $\frac{7}{2}$, $\frac{1}{2}$, $-\frac{1}{2}$, $-\frac{3}{2}$, $-\frac{5}{2}$, and $-\frac{7}{2}$.

Note that there is an immediate correspondence between the strips we can remove and the hooks of the partition. (The hooks and their lengths are defined in Section 3.1.) Indeed, if instead of joining the left-most and right-most cells of the strip along the rim, we look at the hook whose two ends correspond to these two cells, it is easy to see that the lengths of the hook and the strip are exactly the same. In the following diagram, we see the same strip as before, and the corresponding hook shaded in grey.

[2]The topological completion of $\Lambda_0^{\frac{\infty}{2}}$ with respect to the norm induced by this inner product is indeed a Hilbert space, but $\Lambda_0^{\frac{\infty}{2}}$ is not complete.

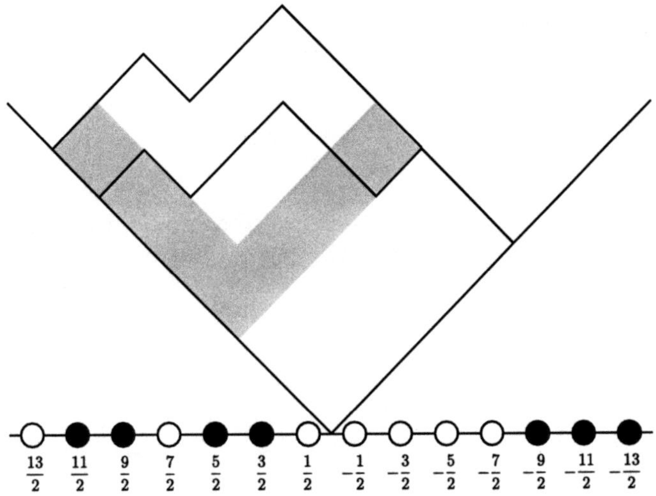

Both the strip and the hook have length $7 = \frac{11}{2} - \left(-\frac{3}{2}\right)$.

2-quotients. Given the sequence of black and white pebbles corresponding to any partition λ, we can split it into two sequences by picking the sites alternatingly. Namely, we assign to one of these sequences all the pebbles on the sites $2n + \frac{1}{2}$, $n \in \mathbb{Z}$, and to the other, the pebbles on the sites $2n - \frac{1}{2}$, $n \in \mathbb{Z}$. The resulting sequences determine two partitions α and β which are together known as the *2-quotient* (α, β) of λ, and they carry information on how λ can be built by adjoining a *2-core*, to be defined below, and a bunch of *2-dominoes*, which are simply 2-strips, so they are of the form 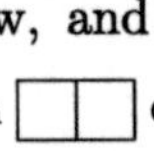 or 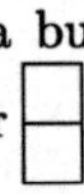. For example, in the case of the partition $(5, 4, 4, 2)$, we split the pebbles as follows:

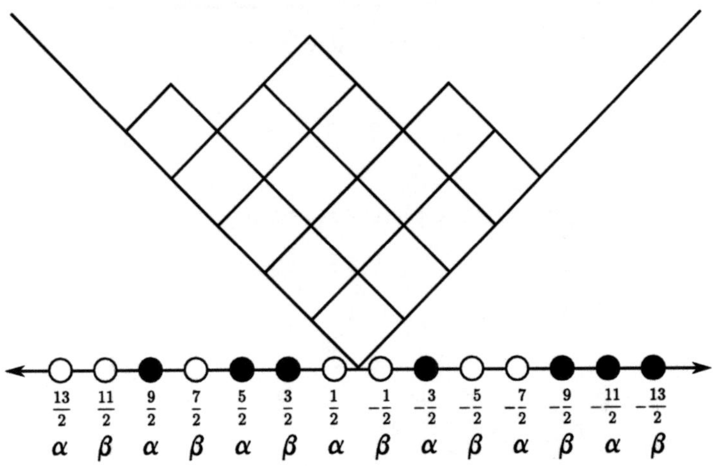

and we obtain the following two sequences of black and white pebbles:

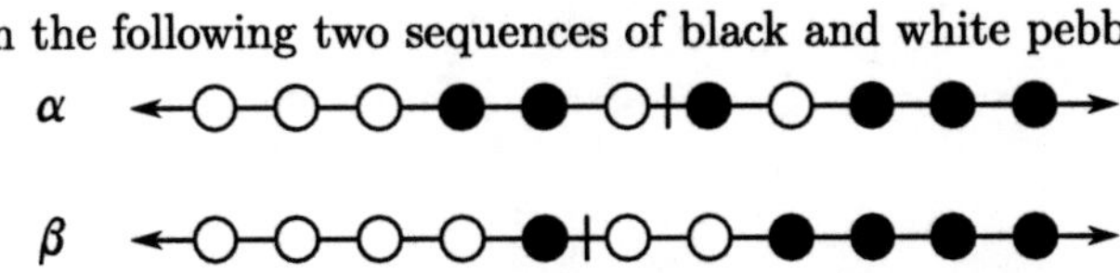

(Here, the position of zero has been recorded with a vertical line. It is off-center in the sense that the number of black pebbles to the left of it does not equal the number of white pebbles to the right of it. We will discuss this below.) These in turn correspond to the following partitions:

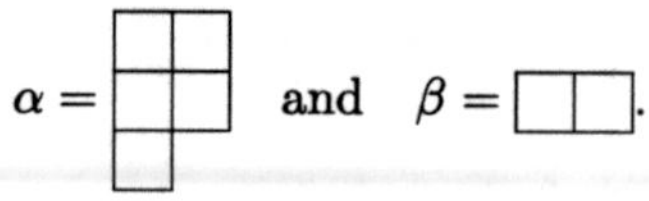

Every time we remove a 2-domino (which is the same as a strip of length 2) from λ, we are removing a cell from one of the components α and β of the 2-quotient. This is because we are exchanging two pebbles of opposite color which are exactly 2 units apart, and hence belong in the same component of the 2-quotient.

2-cores. In the process of removing 2-dominoes, we may get stuck before removing all the cells in the partition λ. The resulting partition is known as the *2-core* and it is always shaped as a staircase, that is, it is of the form $(n, n-1, n-2, \ldots, 2, 1)$, $n = 1, 2, \ldots$. No 2-dominoes can be removed from a 2-core. The following are the first few 2-cores:

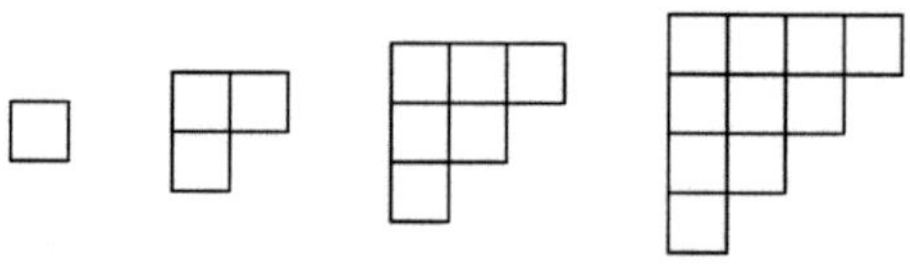

It is easy to see, for example, that the 2-core of the partition $(5, 4, 4, 2)$ is (1). For instance, we may remove its 2-dominoes in the order suggested by the following diagram:

	6	6	3	3
7	5	2	2	
7	5	1	1	
4	4			

These moves in turn correspond to the following removals in the 2-quotients:

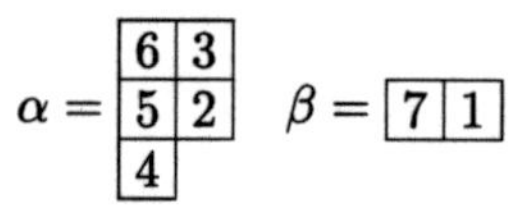

In terms of pebble exchanges, the picture is as follows:

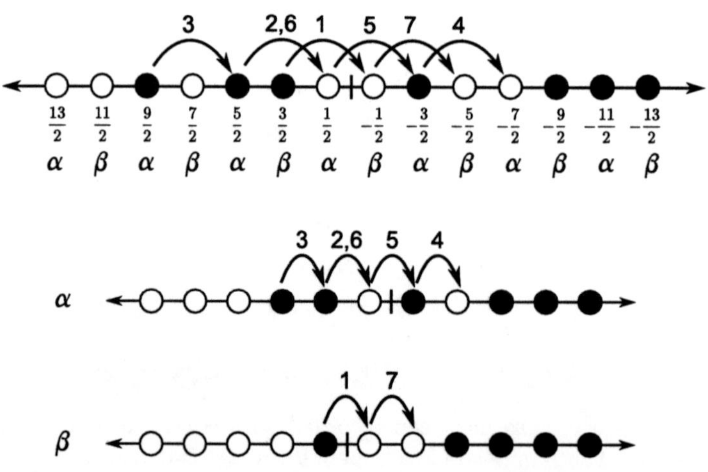

As we noted before, the vertical bar denoting the original position of zero with respect to the sequences of pebbles for α and β is off-centered; this is just a consequence of the presence of the non-trivial 2-core $\square$. A larger 2-core produces a larger translation of the components of the 2-quotient.

The general case. In the general case in which p is any positive integer ≥ 2, one uses the same principle: produce p different partitions by taking the chain of pebbles and distributing it into p sets that alternate in the same periodic order throughout $\mathbb{Z} + \frac{1}{2}$. The p-quotient is composed by these partitions.

The part of the partition that remains after removing all possible p-strips is the p-core. It is reflected in the Maya diagram in the form of translations of the chains of pebbles corresponding to the p-quotient away from the center. In other words, the charges of the Maya diagrams of the partitions of the p-quotient are what produces the p-core. We remark that p-cores do not in general look like staircases, as in the $p = 2$ case.

An application of this is the following formula for the characters of the representation corresponding to the partition λ evaluated at elements with cycle type $(p, p, \ldots, p)$: if $(\lambda^1, \lambda^2, \ldots, \lambda^p)$ is the p-quotient of λ and μ is its p-core, then

$$\left|\chi^\lambda(p, p, \ldots, p)\right| = \begin{cases} 0, & \mu \neq \varnothing, \\ \binom{|\lambda|/p}{|\lambda^1|, |\lambda^2|, \ldots, |\lambda^p|} \prod_i \dim \lambda^i, & \mu = \varnothing. \end{cases}$$

Here,

$$\binom{|\lambda|/p}{|\lambda^1|, |\lambda^2|, \ldots, |\lambda^p|} = \frac{(|\lambda|/p)!}{|\lambda^1|!\,|\lambda^2|! \cdots |\lambda^p|!}.$$

The formula follows from the Muraghan-Nakayama rule and from the picture outlined above: adding a p-strip is equivalent to adding a single cell to one of the components of the p-quotient, so the factors $\dim \lambda^i$ appear. The multinomial factor appears because one has to choose which component of the p-quotient to affect with each added p-strip; in other words, that factor is there because the order matters.

7. Vertex operators

In this section we explore some of the ways known to work with the so-called vertex operators in the half-infinite wedge fermionic Fock space. These operators have found important applications in the field of enumerative combinatorics, and they most likely will continue to do so.

We have also included Section 7.3 on theta functions and quasimodular forms because these objects tend to appear naturally in connection with the traces of vertex operators and we will need some of their properties in Section 9.

To keep things simple, we will consider only vertex operators of the form

$$\Gamma_- \Gamma_+ \tag{15}$$

where

$$\Gamma_\pm = \exp\left(\sum_{\pm n > 0} c_n \alpha_n\right),$$

$c_n \in \mathbb{C}$, acting on the space $\Lambda^{\frac{\infty}{2}} V$ defined in Section 6. Of course, this is just notation, and what we really mean is the expansion of the series of the exponential functions. Note that the order in equation (15) is not arbitrary: the operators that decrease charge must act first in order to avoid getting diverging sums.

These operators are not necessarily bounded, but they are well-defined for all vectors in $\Lambda^{\frac{\infty}{2}}V$ and, for each such vector, they give a vector in a completion of $\Lambda^{\frac{\infty}{2}}V$ where we allow sums of infinitely many generators. In our applications the topology of this completion will not be important, so for simplicity let us be a little sloppy and ignore it.

In a sense, what a vertex operator does is to simultaneously consider all the possible products of characters of the symmetric group (compare equation (13)) with special weights given by the numbers c_n. These vertex operators turn out to be quite common: they are appear naturally when one is dealing with an operator with some prescribed commutation relations, as the following lemma shows.

LEMMA 7.1 ([**15**, Lemma 14.5]). *Let D be a linear operator in $\Lambda^{\frac{\infty}{2}}$. Then:*

a. if $[\alpha_{-n}, D] = c_{-n}D$ for all $n = 1, 2, \ldots$, then $D = c\Gamma_+$ for some $c \in \mathbb{C}$,
b. if $[\alpha_n, D] = c_n D$ for all $n = 1, 2, \ldots$, then $D = c\Gamma_-$ for some $c \in \mathbb{C}$, and
c. if $[\alpha_n, D] = c_n D$ for all $n \in \mathbb{Z} \setminus \{0\}$, then $D = c\Gamma_-\Gamma_+$ for some $c \in \mathbb{C}$.

An important special case of a vertex operator is given by the following useful consequence of the so-called *boson-fermion correspondence.* The motivation for the fancy name is rather lengthy and has already been explained very nicely for example in [**21**, Chapter 5]; the idea is essentially that the left-hand-side of equation (16) below is the generating function for a sequence of operators that have the commutation properties of bosons, while the operators α_n have the commutation properties of fermions. Examples of applications of formula (16) can be found in Sections 9.1 and 9.3. Let

$$\psi(x) = \sum_{k\in\mathbb{Z}+\frac{1}{2}} x^k \psi_k \quad \text{and} \quad \psi^*(x) = \sum_{k\in\mathbb{Z}+\frac{1}{2}} x^{-k}\psi_k^*,$$

where ψ_k, ψ_k^* are the creation and annihilation operators defined in Section 6. Then:

$$\psi(xy)\psi^*(y) = \frac{1}{x^{1/2} - x^{-1/2}} \times \exp\left(\sum_{n>0} \frac{(xy)^n - y^n}{n} \alpha_{-n}\right) \exp\left(\sum_{n>0} \frac{y^{-n} - (xy)^{-n}}{n} \alpha_n\right). \tag{16}$$

This identity can be checked easily using Lemma 7.1, for it is not a hard exercise to check the appropritate commutation relations of $D = \psi(xy)\psi^*(y)$ with the operators α_n, and it can be checked that the coefficient c turns out to be

$$c = \frac{1}{x^{1/2} - x^{-1/2}} = \sum_{k=0}^{\infty} x^{-k-1/2}$$

by computing the inner product

$$\begin{aligned}
c = \langle v_\varnothing, Dv_\varnothing\rangle &= \langle v_\varnothing, \psi(xy)\textstyle\sum_{k\in\mathbb{Z}+\frac{1}{2}} y^{-k}\psi_k^* v_\varnothing\rangle \\
&= \sum_{k=-\frac{1}{2},-\frac{3}{2},\dots} \langle v_\varnothing, \psi(xy)y^{-k}\psi_k^* v_\varnothing\rangle \\
&= \sum_{\ell\in\mathbb{Z}+\frac{1}{2}} \sum_{k=-\frac{1}{2},-\frac{3}{2},\dots} \langle v_\varnothing, (xy)^\ell y^{-k}\psi_\ell\psi_k^* v_\varnothing\rangle \\
&= \sum_{k=-\frac{1}{2},-\frac{3}{2},\dots} \langle v_\varnothing, (xy)^k y^{-k}\psi_k\psi_k^* v_\varnothing\rangle = \sum_{k=-\frac{1}{2},-\frac{3}{2},\dots} x^k.
\end{aligned}$$

For the full proof of equation (16), the reader can also see [**15**, Theorem 14.10].

7.1. The value of a vertex operator. The matrix entries of $\Gamma_\pm$ can be found as follows [**7**, Section 3.1], [**27**, Section 2.2.4].

Note that, from the definition of α_n, we see that it can also be applied directly to the vectors $\underline{k}$ in the basis of V, $k \in \mathbb{Z}+\frac{1}{2}$. Their action is

$$\alpha_n \underline{k} = \underline{k-n}, \quad n \in \mathbb{Z}$$

and it follows that Γ_+ and Γ_- also have an action on $\underline{k}$.

Fix a partition λ. Choose N large enough to have $\lambda_{2N+1} = 0$. Then, when acting on the vectors $\underline{\lambda_i - i + \frac{1}{2}}$, the operators Γ_+ and Γ_- are lower and upper triangular, respectively, and their diagonal elements are all equal to 1. For this reason, the vectors $\underline{\lambda_i - i + \frac{1}{2}}$ for $i > 2N$ end up being irrelevant for the evaluation of $\langle \Gamma_-\Gamma_+ v_\lambda, v_\mu\rangle$. On account of this, it is enough to study what $\Gamma_-\Gamma_+$ does in the exterior power of the finite-dimensional vector space generated by

$$e_k = \underline{-2N + k + \tfrac{1}{2}}, \quad k = 0, 1, \dots, \lambda_1 + 2N.$$

For convenience, we will use below the same notation e_k for any $k \in \mathbb{Z}$, with a similar definition.

Consider first the action of the operator Γ_+, which is encoded in the following generating function

$$\sum_{k,\ell\geq 0} x^k y^\ell \,(\Gamma_+ e_k, e_\ell) = \sum_{k,\ell} x^k y^\ell \left(\left[\prod_{n=1}^{\infty} \exp(c_n\alpha_n)\right] e_k, e_\ell\right) \tag{17}$$

Since the action of each of these exponentials goes like

$$\exp(c_n\alpha_n)e_k = \sum_{j=0}^{\infty} \frac{c_n^j}{j!}\alpha_n^j e_k = \sum_{j=1}^{\infty} \frac{c_n^j}{j!} e_{k-nj}$$

we see that, letting $m_i(\nu)$ be the number of parts of length i in the partition ν,

$$\left[\prod_n \exp(c_n\alpha_n)\right] e_k = \sum_\nu \left(\prod_{i=1}^{\infty} \frac{c_i^{m_i(\nu)}}{m_i(\nu)!}\right) e_{k-|\nu|},$$

where the sum is taken over all partitions ν. But notice also that

$$\sum_{|\nu|=n} \prod_{i=1}^{\infty} \frac{c_i^{m_i(\nu)}}{m_i(\nu)!}$$

is the coefficient of z^n in the series expansion of

$$\phi^+(z) := \exp \sum_{k=1}^{\infty} c_k z^k.$$

So we can write the generating function (17) in the form

$$\sum_{k,\ell} x^k y^\ell \left[z^{k-\ell}\right] \phi^+(z) = \sum_\ell x^\ell y^\ell \sum_k x^k \left[z^k\right] \phi^+(z) = \sum_\ell x^\ell y^\ell \phi^+(x) = \frac{\phi^+(x)}{1-xy}.$$

Here, the notation $[z^m]$ stands for the operation of taking the coefficient of z^m in the Taylor series expansion of the function to the right of this symbol. Now let ϕ_n^+ be the coefficients of the series expansion of $\phi^+(z)$,

$$\phi^+(z) = \sum_{n=0}^{\infty} \phi_n^+ z^n.$$

In particular, $\phi_0^+ = 1$. Then

$$(\Gamma_+ e_k, e_\ell) = [x^k y^\ell] \frac{\phi^+(x)}{1-xy} = \phi_{k-\ell}^+$$

We plug this into the full half-infinite wedge to conclude that, letting

$$s_{\lambda/\mu}^+ = \det\left(\phi^+_{\lambda_i - \mu_j + j - i}\right),$$

we have

$$\langle \Gamma_+ v_\mu, v_\lambda \rangle = s_{\lambda/\mu}^+.$$

Here, $s_{\lambda/\mu}^+$ is the skew Schur function $s_{\lambda/\mu}$ (defined in Section 5.1) specialized to the case in which the homogeneous symmetric functions h_n equal ϕ_n^+ for $n \geq 0$ or, equivalently, specialized to the case in which the symmetric power functions p_n equal nc_n (compare with the relations (9)). In particular, $s_{\lambda/\mu} = 0$ if the Young diagram of μ is not completely contained in the Young diagram of λ.

A similar calculation shows that

$$\langle \Gamma_- v_\mu, v_\lambda \rangle = s_{\mu/\lambda}^-$$

where

$$s_{\mu/\lambda}^- = \det\left(\phi^-_{\mu_i - \lambda_j + j - i}\right),$$

and ϕ_n^- are the coefficients in the series expansion of

$$\phi^-(z) = \exp \sum_{n=1}^{\infty} c_{-n} \alpha_{-n}.$$

With this information in hand, it is immediate that

$$\boxed{\langle \Gamma_- \Gamma_+ v_\mu, v_\lambda \rangle = \sum_\nu s_{\lambda/\nu}^- s_{\mu/\nu}^+.}$$

As a side note, whenever one can write $\phi(z) = \phi^+(z)\phi^-(z)$, where ϕ^+ and ϕ^- are analytic and non-vanishing in some neighborhood of, respectively, the interior and the exterior of the unit disc, this expression is called the *Wiener-Hopf factorization* of the function ϕ. Clearly, this type of factorization may be important when we want to understand vertex operators. (However, the analyticity of ϕ^+ and

ϕ^- need not be true in general.) See [**27**] for an application to the enumeration of random surfaces.

7.2. Formulas for the computation of the trace of a vertex operator. Sometimes one needs to compute the trace of a vertex operator T. In such cases, there are two formulas that are quite useful.

The first one is the obvious decomposition of the space

$$\Lambda_0^{\frac{\infty}{2}} V \cong \bigotimes_{n=1}^{\infty} \bigoplus_{k=0}^{\infty} \alpha_{-n}^k v_\varnothing,$$

whose immediate consequence is that

$$\boxed{\operatorname{tr} T = \prod_{n=1}^{\infty} \operatorname{tr}\left(T|_{\bigoplus_{k=0}^{\infty} \alpha_{-n}^k v_\varnothing}\right).} \tag{18}$$

This, together with the commutation relation (14), allows for the following formula [**7**] to be used: for complex numbers A and B,

$$\boxed{\operatorname{tr} q^H e^{A\alpha_{-n}} e^{B\alpha_n}|_{\bigoplus_{k=0}^{\infty} \alpha_{-n}^k v_\varnothing} = \frac{1}{1-q^n} \exp \frac{nABq^n}{1-q^n}.} \tag{19}$$

Here, $H(v_\lambda) = |\lambda| v_\lambda$ as above, q is a parameter, and the operator q^H is defined by

$$q^H v_\lambda = |\lambda|\, v_\lambda,$$

and extended by linearity. The introduction of q^H is often a good idea because it sometimes simplifies formulas and it produces generating functions that are indexed on the size of the partitions.

To prove (19), one needs to expand both sides. Let us explain this in detail. We have

$$\begin{aligned}
\operatorname{tr} q^H e^{A\alpha_{-n}}\, e^{B\alpha_n}|_{\bigoplus_{k=0}^{\infty} \alpha_{-n}^k v_\varnothing} &= \sum_{k=0}^{\infty} \langle q^H e^{A\alpha_{-n}} e^{B\alpha_n} \alpha_{-n}^k v_\varnothing, \alpha_{-n}^k v_\varnothing \rangle \\
&= \sum_k \left\langle q^H \sum_\ell \frac{(A\alpha_{-n})^\ell}{\ell!} \sum_m \frac{(B\alpha_n)^m}{m!} \alpha_{-n}^k v_\varnothing, \alpha_{-n}^k v_\varnothing \right\rangle \\
&= \sum_{k,\ell,m} q^{n(\ell-m+k)} \frac{A^\ell B^m}{\ell!\, m!} \langle \alpha_{-n}^\ell \alpha_n^m \alpha_{-n}^k v_\varnothing, v\alpha_{-n}^k v_\varnothing \rangle
\end{aligned}$$

Note that this inner product vanishes when the partitions associated to the vectors involved do not have the same number of cells. This means that only the terms in which $m = \ell$ survive and we get

$$\sum_{k,\ell} q^{kn} \frac{(AB)^\ell}{\ell!^2} \langle \alpha_{-n}^\ell \alpha_n^\ell \alpha_{-n}^k v_\varnothing, \alpha_{-n}^k v_\varnothing \rangle .$$

One then plays a game of moving the fermionic annihilators α_n to the right, using the commutation relation (14) and the fact that $\alpha_n v_\varnothing = 0$. For example,

$$\begin{aligned}
\alpha_n \alpha_{-n} \alpha_{-n} v_\varnothing &= \alpha_{-n} \alpha_n \alpha_{-n} v_\varnothing + n\alpha_{-n} v_\varnothing = \\
&\qquad \alpha_{-n} \alpha_{-n} \alpha_n v_\varnothing + 2n\alpha_{-n} v_\varnothing = 2n\alpha_{-n} v_\varnothing .
\end{aligned}$$

In general, one gets

$$\alpha_n^a \alpha_{-n}^b v_\varnothing = \begin{cases} \frac{b!}{(b-a)!} n^a \alpha_{-n}^{b-a} v_\varnothing & \text{if } a \le b, \\ 0 & \text{otherwise.} \end{cases}$$

Therefore, the above sum becomes

$$\sum_{k=0}^{\infty} \sum_{\ell=0}^{k} q^{kn} \frac{(nAB)^\ell}{\ell!} \binom{k}{\ell}.$$

This equals the right hand side of (19); just invert the order of summation and notice that

$$\sum_{k=\ell}^{\infty} q^{kn} \binom{k}{\ell} = \frac{q^{n\ell}}{(1-q^n)^{\ell+1}}.$$

7.3. Theta functions and quasimodular forms. It is an experimental fact that, due to the form of formulas like (16) or (19), traces of vertex operators have a tendency to be expressible in terms of the Jacobi theta function. For examples, please see Section 9 or, in a more general setting, [**15**].

7.3.1. *The Jacobi theta function.* The *Jacobi theta function*

$$\vartheta(x,q) = (q^{1/2} - q^{-1/2}) \prod_{i=1}^{\infty} \frac{(1-q^i x)(1-q^i/x)}{(1-q^i)^2}, \tag{20}$$

is defined for arbitrary x in $\mathbb{C}$ and q in the unit disc $\{|q|<1\} \subset \mathbb{C}$. In this section we recall some of its properties.

The first property is its modularity. Let τ and z be defined (modulo $\mathbb{Z} \subset \mathbb{C}$) by $q = e^{2\pi i \tau}$ and $x = e^{2\pi i z}$. It is traditional to abuse notation and write $\vartheta(z,\tau)$ for $\vartheta(x,q)$, and we will embrace this tradition here. As a consequence of the Poisson summation formula, ϑ satisfies

$$\vartheta\left(\frac{z}{\tau}, -\frac{1}{\tau}\right) = -\frac{e^{\pi i z^2/\tau}}{i\tau} \vartheta(z,\tau).$$

Another property is its additive expression, which can be derived using the *Jacobi triple product*, which is the identity

$$\prod_{m=1}^{\infty} \left(1 - x^{2m}\right)\left(1 + x^{2m-1}y^2\right)\left(1 + x^{2m-1}y^{-2}\right) = \sum_{n\in\mathbb{Z}} x^{n^2} y^{2n},$$

valid for all complex numbers x and y with $|x|<1$ and $y \neq 0$. It follows that

$$\vartheta(x,q) = \eta^{-3}(q) \sum_{n\in\mathbb{Z}} (-1)^n q^{\frac{\left(n+\frac{1}{2}\right)^2}{2}} x^{n+\frac{1}{2}},$$

where η stands for the *Dedekind eta function*,

$$\eta(q) = q^{1/24} \prod_{n=1}^{\infty} (1-q^n).$$

A second property we want to record for later reference is the expression of our ϑ in terms of the more classical Jacobi thetas, namely, [**22**, page 17]

$$\begin{aligned}
\vartheta_{00}(z,\tau) &= \sum_{n\in\mathbb{Z}} \exp(\pi i n^2 \tau + 2\pi i n z) \\
\vartheta_{01}(z,\tau) &= \sum_{n\in\mathbb{Z}} \exp\left(\pi i n^2 \tau + 2\pi i n \left(z + \tfrac{1}{2}\right)\right) \\
&= \vartheta_{00}\left(z + \tfrac{1}{2}, \tau\right) \\
\vartheta_{10}(z,\tau) &= \sum_{n\in\mathbb{Z}} \exp\left(\pi i \left(n + \tfrac{1}{2}\right)^2 \tau + 2\pi i \left(n + \tfrac{1}{2}\right) z\right) \\
&= \exp\left(\frac{\pi i \tau}{4} + \pi i z\right) \vartheta_{00}\left(z + \tfrac{1}{2}\tau, \tau\right) \\
\vartheta_{11}(z,\tau) &= \sum_{n\in\mathbb{Z}} \exp\left(\pi i \left(n + \tfrac{1}{2}\right)^2 \tau + 2\pi i \left(n + \tfrac{1}{2}\right)\left(z + \tfrac{1}{2}\right)\right) \\
&= \exp\left(\frac{\pi i \tau}{4} + \pi i \left(z + \tfrac{1}{2}\right)\right) \vartheta_{00}\left(z + \tfrac{1}{2}(1+\tau), \tau\right)
\end{aligned}$$

It follows that

$$\vartheta(z,\tau) = -i\eta^{-3}(\tau)\,\vartheta_{11}(z,\tau). \tag{21}$$

We further record the modular behavior of the traditional Jacobi thetas [**22**, page 36]:

$$\begin{aligned}
\vartheta_{00}\left(\frac{z}{\tau}, -\frac{1}{\tau}\right) &= (-i\tau)^{\frac{1}{2}} \exp\left(\frac{\pi i z^2}{\tau}\right) \vartheta_{00}(z,\tau), \\
\vartheta_{01}\left(\frac{z}{\tau}, -\frac{1}{\tau}\right) &= (-i\tau)^{\frac{1}{2}} \exp\left(\frac{\pi i z^2}{\tau}\right) \vartheta_{10}(z,\tau), \\
\vartheta_{10}\left(\frac{z}{\tau}, -\frac{1}{\tau}\right) &= (-i\tau)^{\frac{1}{2}} \exp\left(\frac{\pi i z^2}{\tau}\right) \vartheta_{01}(z,\tau), \\
\vartheta_{11}\left(\frac{z}{\tau}, -\frac{1}{\tau}\right) &= -(-i\tau)^{\frac{1}{2}} \exp\left(\frac{\pi i z^2}{\tau}\right) \vartheta_{11}(z,\tau).
\end{aligned}$$

Note the additional minus sign in the rule for ϑ_{11}. These identities are useful because it is likely to find quotients of the form $\vartheta(x)/\vartheta(-x)$ involved in partition functions.

The statements of the following two lemmas are more readable if we use h instead of τ, where

$$q = e^{-h} = e^{2\pi i \tau}.$$

LEMMA 7.2 ([**8**, Proposition 4.1]). *We have*

$$\frac{\vartheta(e^u, e^{-h})}{\vartheta'(0, e^{-h})} \approx h \frac{\sin(\pi u/h)}{\pi} \exp\left(\frac{u^2}{2h}\right)$$

as $h \to +0$, uniformly in u. This asymptotic relation can be differentiated any number of times.

PROOF. We have

$$\vartheta(e^{hu}, e^{-h}) = i\sqrt{\frac{2\pi}{h}} \exp\left(\frac{hu^2}{2}\right) \vartheta\left(e^{-2\pi i u}, e^{-\frac{4\pi^2}{h}}\right),$$

so, expanding the series of ϑ,

$$\vartheta\left(e^{-2\pi i u}, e^{-\frac{4\pi^2}{h}}\right) = \sum_{n\in\mathbb{Z}} (-1)^n \exp\left(-\frac{2\pi^2\left(n+\frac{1}{2}\right)^2}{h}\right) e^{-2\pi\left(n+\frac{1}{2}\right)u}.$$

From here it is clear that, as $h \to +0$, the terms $n = 0$ and $n = -1$ dominate all others. □

LEMMA 7.3. *We have the following approximations for ϑ at $x = 1$ and $x = -1$, respectively:*

$$\vartheta(e^u, e^{-h}) \approx i\eta^{-3}(q)\sqrt{\tfrac{2\pi}{h}} \exp\left(\pi i\left(\left[\tfrac{u}{2\pi i} - \tfrac{1}{2}\right] + \tfrac{1}{2}\right) - \tfrac{2\pi^2}{h}\left\{\tfrac{u}{2\pi i} - \tfrac{1}{2}\right\}^2\right),$$
$$\vartheta(-e^u, e^{-h}) \approx i\eta^{-3}(q)\sqrt{\tfrac{2\pi}{h}} \exp\left(-\tfrac{2\pi^2}{h}\left\{\tfrac{u}{2\pi i} - \tfrac{1}{2}\right\}^2\right),$$

as $h \to 0$, where $q = e^{-h}$, $[x]$ stands for the integer closest to x, and $\{x\} = x - [x]$.

PROOF OF LEMMA 7.3. This is again a straightforward application of the modular transformation $h \mapsto -1/h$ in the expression (21). We also use the identity

$$\vartheta_{11}(-e^u, q) = -\vartheta_{10}(e^u, q).$$ □

7.3.2. *Quasimodular forms.* We follow [**7**, Section 3.3].

A *quasimodular form* for a subgroup $\Gamma \subset \mathrm{SL}_2(\mathbb{Z})$ is the holomorphic part of an almost holomorphic modular form for Γ. A function of $|q| < 1$ is called *almost holomorphic* if it is a polynomial in $(\log|q|)^{-1}$ with coefficients that are holomorphic functions of q.

Quasimodular forms constitute a graded algebra $\mathcal{QM}(\Gamma)$. It is a theorem of M. Kaneko and D. Zagier [**16**] that the space of quasimodular forms is simply

$$\mathcal{QM}(\Gamma) = \mathbb{Q}[E_2] \otimes \mathcal{M}(\Gamma),$$

where $\mathcal{M}(\Gamma)$ denotes the space of modular forms with respect to the subgroup Γ and E_2 denotes the first of the Eisenstein series, which are in general defined by

$$E_{2k}(q) = \frac{\zeta(1-2k)}{2} + \sum_{n=1}^{\infty}\left(\sum_{d|n} d^{2k-1}\right) q^n, \quad k = 1, 2, \ldots$$

In particular,

$$E_2(q), E_2(q^2), E_4(q^2) \in \mathcal{QM}(\Gamma_0(4)) \tag{22}$$

where

$$\Gamma_0(4) = \left\{\left(\begin{smallmatrix} a & b \\ c & d\end{smallmatrix}\right) : c \equiv 0 \mod 4\right\}.$$

It is well known that the three series in (22) algebraically generate all other Eisenstein series and their products, so these too are quasimodular forms. Hence, in order to show that a certain function is a quasimodular form, it is sufficient to show that it is a polynomial in the E_{2k}.

Note the close connection between the Eisenstein series and the theta function, best evidenced by the formal identities (easily checked by expanding both sides)

$$\log \frac{z}{\vartheta(e^z)} = 2\sum_{k\geq 1} \frac{z^{2k}}{(2k)!} E_{2k}(q),$$

$$\log \frac{\vartheta(-e^z)}{\vartheta(-1)} = 2\sum_{k\geq 1} \frac{z^{2k}}{(2k)!} \left[E_{2k}(q) - 2^{2k}E_{2k}(q^2)\right],$$

$$\vartheta(-1) = 2i\left(\prod_{n\geq 1}\frac{1+q^n}{1-q^n}\right)^2 = \frac{\eta(q^2)^2}{\eta(q)^4},$$

where η is the Dedekind eta function.

8. Variational methods

8.1. The entropy. When thinking about very large partitions, as we will see later, one often finds that most of the partitions in a particular distribution end up looking very much alike. In this case, they are said to have a *limit shape.*

The method to deduce what this limit shape is (if it exists) must vary according to the probability distribution being studied. Probably the easiest case comes from the case of the uniform distribution. For each number N, the uniform distribution gives equal probability weight to each of the partitions of N. For large N, most partitions look like a certain curve, which we shall now deduce.

After rotating the Young diagrams by 135° as in Section 6, each partition contour becomes a zig-zag curve, which can be interpreted as the graph of a discrete random walk that moves up or down, with slope $+1$ or -1, from left to right. For any two points in

$$\{(x,y) \in \mathbb{Z}\times\mathbb{Z} : y \geq |x|\},$$

not both on the same vertical line, there is at least one partition contour that joins them. If the two points are $m > 0$ units apart horizontally and $n \geq 0$ units apart vertically, there are, in fact

$$\binom{m}{(m+n)/2}$$

possible random walks joining them, each of them determined by the choice of $(m+n)/2$ steps upwards, for the remaining $(m-n)/2$ steps will be taken downwards. (The plot of the random walk is part of a partition's contour.) As we rescale maintaining the overall slope $p = (m+n)/2m$ constant, we find that an application of Stirling's formula gives

$$\frac{1}{m}\log\binom{m}{pm} \to S(p)$$

where

$$S(p) = -p\log p - (1-p)\log(1-p)$$

is the *Shannon entropy.*

Let us assume that there exists a limit shape, given by the function f. The parameter p, which can be interpreted as the probability of the random walk of going up, then corresponds to $(1+f'(x))/2$. For instance, if $f'(x) = 1$, then at the microscopic level the random walk must be going always upwards, so $p = 1$. And if $f'(x) = -1$, the random walk is never going up, so it is always going down and $p = 0$.

From our calculation above it follows that a heuristic count of partitions (or random walks) around the graph of f is given by

$$\frac{1}{\delta}\int_0^T S(p)\,dx = \frac{1}{\delta}\int_0^T S\left(\frac{1+f'(x)}{2}\right)dx, \tag{23}$$

where δ is the size of the side of the rescaled cells of the Young diagrams. The limit shape would be the curve f maximizing this count, because it would need to have as many of these zigzagging random walks near to it as possible (it must be the look-alike of most large partitions). If one variationally looks for a function f that will maximize this integral, while respecting the restriction that $\int f(x) - |x|\,dx = 1$, one ends up finding the correct limit shape for the uniform distribution; see Section 9.1.

This is really a textbook case, and it can be studied to greater depth. In particular, it is not difficult to find a full large deviation principle. See for example [**23**].

8.2. Approximating products of hooks. In order to deduce the asymptotic behavior of probability distributions in the space of Young diagrams given by products or quotients of hooks, people sometimes use variational arguments resembling the idea that S. Kerov and A. Vershik [**37**] and B. Logan and L. Shepp [**17**] used in the analysis of the Plancherel distribution. Here, we review this technique.

The idea is that we can approximate the logarithm of products of hook lengths of very large partitions by an integral.

For a non-increasing function $F : \mathbb{R}_+ \to \mathbb{R}_+$, define

$$F^{-1}(y) = \inf\{x : F(x) \leq y\},$$

and

$$h_F(x,y) = F(x) + F^{-1}(y) - x - y,$$

which gives an approximation of the hook length at (x, y). Let λ be a partition, and $n = |\lambda|$. We contract its diagram until it has area 1, rescaling by $1/\sqrt{n}$, and we associate to it a function $F(x)$ that describes its rim,

$$F(x) = \frac{1}{\sqrt{n}}\#\{\text{parts of } \lambda \text{ of size} \leq \lceil \sqrt{n}\,x \rceil\}.$$

In the following diagram, we have the example of $\lambda = (4, 3, 3, 2)$.

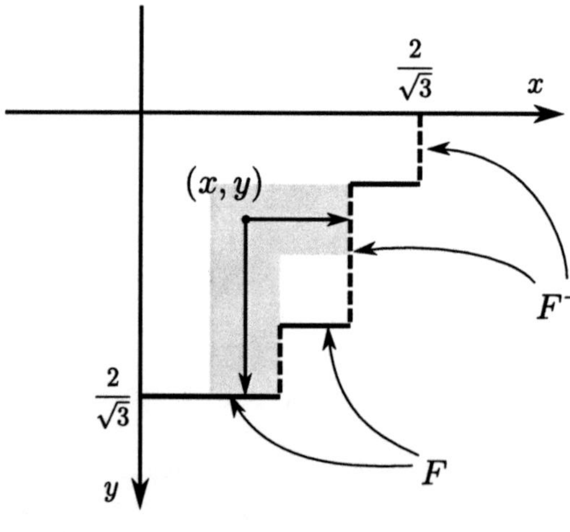

In the picture, we have inverted the orientation of the y axis to reflect the usual orientation of our Young diagrams, and we have drawn the parts of the rim of λ that are described by F with a solid line, while the parts that correspond do F^{-1} are drawn with dashed lines. The numbers on the axes reflect the rescaling by $1/\sqrt{n} = 1/\sqrt{12}$. We have also highlighted the hook corresponding to the cell $(2,2)$. We will approximate the logarithm of its length by

$$n \int_R \log\left(\sqrt{n}\, h_F(x,y)\right)\, dx\, dy, \tag{24}$$

where R is the rectangle between the points $\frac{1}{\sqrt{12}}(1,1)$ and $\frac{1}{\sqrt{12}}(2,2)$. To motivate this approximation, we have drawn, at the center of the cell $(2,2)$, the point (x,y) and an inverted L ending extending to the rescaled rim of λ; the length of this L is precisely $h_F(x,y)$: its vertical part measures $F(x) - y$, while its horizontal part measures $F^{-1}(y) - x$. Then $\sqrt{n}\, h_F(x,y) = 4$ coincides with the length of the hook of cell $(2,2)$, and the integral (24) is very close to this number as well. In general, we have

LEMMA 8.1. *Let λ be a partition and let $\square \in \lambda$ be a cell in its Young diagram, and denote by F the non-increasing function that describes the rim of the Young diagram of λ, as defined above. Let $R_\square \subseteq \{(x,y) : 0 \le y \le F(x)\}$ be the rectangular domain corresponding to $\square$. Let $h_\square$ denote the hook length of $\square$. Then*

$$\log h_\square = n \int_{R_\square} \log\left(\sqrt{n} h_F(x,y)\right) dx\, dy + c(h_\square),$$

where

$$c(x) = \frac{1}{2}\sum_{k=1}^{\infty} \frac{1}{k(k+1)(2k+1)x^{2k}}.$$

REMARK 8.2. The function c is strictly decreasing in the interval $[1, +\infty)$. It remains between $c(1) = \frac{1}{2}(3 - 4\log 2) \approx 0.113706$ and

$$\lim_{x \to +\infty} c(x) = 0.$$

This implies that the approximation of the logarithm of a product of any m hook lengths using the integral of $\log\left(\sqrt{n} h_F(x,y)\right)$ will have an error of order $\le c(1) \cdot m$. This estimate can be improved in the large-scale case; see Lemma 8.3.

PROOF. Let (x_0, y_0) be the point at the center of $R_\square$. Recall that the area of $R_\square$ is $1/n$. Then

$$\begin{aligned} n \int_{R_\square} \log\left(\sqrt{n} h_F(x,y)\right) dx\, dy &= n \int_{R_\square} \log\left(h_\square - \sqrt{n}(x - x_0) - \sqrt{n}(y - y_0)\right) dx\, dy \\ &= \log h_\square + n \int_{R_\square} \log\left(1 - \frac{(x - x_0) + (y - y_0)}{h_\square/\sqrt{n}}\right) dx\, dy \\ &= \log h_\square + \int_{\sqrt{n}y_0 - \frac{1}{2}}^{\sqrt{n}y_0 + \frac{1}{2}} \int_{\sqrt{n}x_0 - \frac{1}{2}}^{\sqrt{n}x_0 + \frac{1}{2}} \log\left(1 - \frac{(u - \sqrt{n}x_0) + (v - \sqrt{n}y_0)}{h_\square}\right) du\, dv, \end{aligned}$$

where $u = \sqrt{n}x$ and $v = \sqrt{n}y$. Now expand the Taylor series of the logarithm and integrate term by term. $\square$

It is now convenient to change variables to s and t such that $x = \frac{1}{\sqrt{2}}(L(s) - s)$ and $y = \frac{1}{\sqrt{2}}(L(t) + t)$. To explain how this change of variables works we have the following diagram.

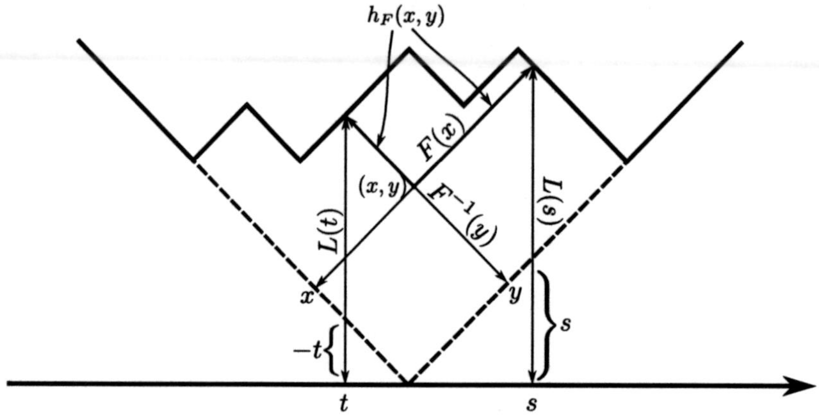

In the diagram, the point (x, y) is at the center, and through it pass two diagonals of lengths $F(x)$ and $F^{-1}(y)$ intersecting the diagonal axes at coordinates x and y, respectively. From their intersections with the rim of the rotated Young diagram, we have vertical segments of lengths $L(s)$ and $L(t)$. The transform $(s,t) \mapsto (x,y)$ is not bijective, but it is surjective, and it is injective in the region in which its Jacobian is nonzero. The determinant of its Jacobian is

$$-\frac{1}{2}\left(1 - L'(s)\right)\left(1 + L'(t)\right),$$

which is supported in a compact set because $L(x) = |x|$ for large $|x|$. Our integral becomes, after a couple of integrations by parts:

$$\log \prod_{\square\in\lambda} h_\square = \tag{25}$$

$$\boxed{\frac{n}{2}\int\!\!\int_{s>t}\left(\log\sqrt{2}(s-t)\right)(1-L'(s))(1+L'(t))\,ds\,dt + \sum_{\square\in\lambda} c(h_\square)}$$

This integral is, in fact, the square of the norm of a Sobolev space, whence it is convex and has good properties for optimization. Finally, we have:

LEMMA 8.3.

$$\sum_{\square\in\lambda} c(h_\square) = O\left(\sqrt{n}\right) \quad \textit{as } n = |\lambda| \to \infty.$$

PROOF. Since c is a decreasing function and $c(h) \to 0$ very quickly as $h \to \infty$, the worst case scenario is the case in which we maximize the number of *small* hook lengths $h_\square$. This happens in the case of the *staircase partition* $(\ell, \ell-1, \ldots, 2, 1)$, and in that case the estimate is obvious. $\square$

9. Probability distributions and limit shapes

9.1. The uniform distribution. Let $\mathcal{Y}_n$ be the set of Young diagrams corresponding to all partitions of n, rescaled by $1/\sqrt{n}$. This means that we construct the diagrams with squares of side $1/\sqrt{n}$.

It is a very interesting phenomenon that, for large n, most of the diagrams in $\mathcal{Y}_n$ are very similar. In fact, once rescaled, most of them are very close to the curve

$$e^{-\frac{\pi}{\sqrt{6}}x} + e^{-\frac{\pi}{\sqrt{6}}y} = 1. \tag{26}$$

Here is a simulation:

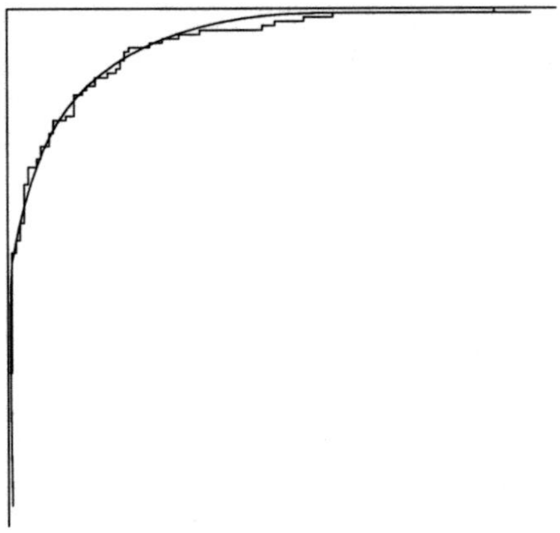

In the picture we see the graph of the curve corresponding to the limit shape, together with the contour of a random partition of very large size n rescaled by $1/\sqrt{n}$. The x-axis is horizontal and increases to the right; the y-axis is vertical and increases downwards.

This phenomenon was apparently first remarked by A. Vershik [**36**], who mentioned it informally to the Szalay and Turan after they had published a paper [**35**] from where it can be deduced immediately.

Now, let us explain how one can arrive at this conclusion. We will do this in two ways. First, let us apply the variational method explained in Section 8.1. One does standard variational calculus (see for example [**12**]) to find that the (Euler-Lagrange) equation that must be satisfied by any function f maximizing the entropy integral (23) is

$$f'' = 1 - (f')^2.$$

It is easy to see that, after a rescaling to ensure that the limit shape has area 1, the only possible candidate is then (26). To complete the argument, one must prove that the entropy integral (23) is upper-semicontinuous and from there obtain a large deviation principle. We will not go into the details, but they can be found for example in [**23**].

Our second way to deduce the form of the limit shape leverages the tools we developed in Sections 5.2, 6 and 7. This will seem like a huge roundabout, but it is perhaps the simplest illustration of how one can deal with more complicated problems. The idea is similar to the one used to prove Wigner's semicircle law: we will find the moments of the limit shape, and since they turn out to coincide with those of our candidate (26), we can apply a standard result [**2**, Chapter 30] in probability that says that two densities must coincide if they have the same moments μ_k and these satisfy the growth condition

$$\limsup_{k\to\infty} \frac{1}{2k}\mu_{2k}^{1/2k} < \infty.$$

Instead of looking at each of the layers $\mathcal{Y}_n$ of partitions Young diagrams of partitions of n separately, we introduce the *q-coupled probability distribution* m_q on the full set of Young diagrams $\bigcup_n \mathcal{Y}_n$. It will be given by

$$m_q(\lambda) = \frac{q^{|\lambda|}}{Z}, \quad Z = \sum_{\nu} q^{|\nu|} = \prod_{n=1}^{\infty} (1-q^n)^{-1}$$

for[3] $|q| < 1$, $q \in \mathbb{C}$. In the denominator we have normalized using the generating function Z of partitions, whose product form is easy to verify. In statistical mechanics the process of taking into account all possible states of all possible energies simultaneously is called 'passing to the grand canonical ensemble.' In this regime, the limit $q \nearrow 1$ corresponds to the limit $|\lambda| \to \infty$, as we will see below.

We introduce the *1-point function*

$$F(e^r) = \frac{1}{r} + \sum_{j \geq 0} \frac{r^j}{j!} \langle \mathbf{p}_j \rangle,$$

where the expectations of the shifted-symmetric power functions $\mathbf{p}_j$ are given by

$$\langle \mathbf{p}_j \rangle = \sum_{\lambda} \mathbf{p}_j(\lambda) \frac{q^{|\lambda|}}{Z}.$$

Because of the convergence of the expectations of the shifted-symmetric power functions $\mathbf{p}_k(\lambda)$ to the moments of the limit shape, as explained in Section 5.2, the asymptotics of F encode the moments of the limit shape (which we aim to compute).

There is another useful interpretation for F. Denote by $\mathfrak{s}(\lambda)$ the set of shifted coordinates $\lambda_i - i + \frac{1}{2}$ of λ. As we rescale the partition by $1/\delta$ and let $|\lambda| \to \infty$, we expect to have $f'(x)$ close to 1 in the regions where there is a high probability of finding one of these rescaled shifted coordinates $\delta(\lambda_i - i + \frac{1}{2})$ (because they correspond to the presence of a black pebble), and we expect $f'(x)$ to be close to -1 in the regions where this probability is very low. In other words,

$$f'(x) \approx \lim_{\delta \to 0} \mathbf{P}(x/\delta \in \mathfrak{s}(\lambda)).$$

If we take the generating function of these probabilities,

$$\sum_{x \in \mathbb{Z} + \frac{1}{2}} u^x \sum_{\mathfrak{s}(\lambda) \ni x} \frac{q^{|\lambda|}}{Z},$$

where the second sum is taken over all partitions λ such that x is in the set of shifted coordinates $\mathfrak{s}(\lambda)$. By equation (10), this is the same as $F(u)$, where $u = e^r$.

Using this interpretation and the language of the half-infinite wedge, we rewrite $F(u)$ as a trace,

$$F(u) = \frac{1}{Z} \sum_{x \in \mathbb{Z} + \frac{1}{2}} u^x \operatorname{tr} q^H \psi_x \psi_x^* = \frac{1}{Z} [v^0] \operatorname{tr} q^H \psi(uv) \psi^*(v).$$

Our notations are as in equation (16), and $[v^0]$ stands for the operation of "taking the constant term in the series expansion with respect to v." The idea is that the

[3]We can only interpret these as probabilities if $q \in \mathbb{R}$, but it is useful to keep in mind the complex-analytic properties of the functions involved.

composition of operators $\psi_x\psi_x^*$ will select the vectors v_λ such that x is in $\mathfrak{s}(\lambda)$. Apply formulas (16), (18) and (19) to find that

$$\begin{aligned}\frac{1}{Z}\operatorname{tr} q^H\psi(uv)\psi^*(v) &= \frac{1}{x^{1/2}-x^{-1/2}}\exp\sum_{n=1}^{\infty}\frac{((uv)^n-v^n)\left(v^{-n}-(uv)^{-n}\right)q^n}{n(1-q^n)}\\ &= \frac{1}{x^{1/2}-x^{-1/2}}\exp\sum_{n=1}^{\infty}\left(u^n+u^{-n}-2\right)\frac{\sum_{k=1}^{\infty}q^{nk}}{n}\\ &= \frac{1}{x^{1/2}-x^{-1/2}}\prod_{k=1}^{\infty}\frac{(1-q^k)^2}{(1-q^k u)(1-q^k/u)} = \frac{1}{\vartheta(u)},\end{aligned}$$

where ϑ is the Jacobi theta function defined in equation (20) and we used that

$$\log(1-a) = -\sum_{n=1}^{\infty}\frac{a^n}{n}.$$

So we have

$$F(u) = \frac{1}{\vartheta(u)}.$$

Using the same techniques as above, it is an easy exercise to see that, in the case of the uniform distribution, the generating functions of n-point correlations are given by

$$\sum_{i_1,i_2,\ldots,i_n\in\mathbb{Z}+\frac{1}{2}} u_1^{i_1}\cdots u_n^{i_n}\mathbf{P}_{\mu_q}\left(i_1,\ldots,i_n\in\mathfrak{s}(\lambda)\right) = \prod_{k=1}^{n}\frac{1}{\vartheta(u_k)}$$

(where $\mathbf{P}_{\mu_q}$ denotes the probability with respect to the distribution μ_q that assigns weights $\mu_q(\lambda)=q^{|\lambda|}$), and this is what is called the *n-point function.* This is great news, because the fact that it factors as a product of thetas implies all sorts of good quasimodularity properties; in particular, it means that the μ_q-expectations of the functions $\mathbf{p}_k$ (whose highest degree terms encode the moments of the limit shape as $q\to 1$) are quasimodular forms. (Compare with Section 7.3.2.)

We turn to the issue of understanding the asymptotic behavior of this distribution. As explained above, we want to be able to extract information about the moments of the limit shape from the n-point function.

In order to get an idea of how the moments will scale as $q\to 1$, it is insightful to compute the leading term of the asymptotics of the expectation of the size of the partition,

$$\langle|\lambda|\rangle = \frac{\sum_\lambda |\lambda| q^{|\lambda|}}{Z} = \frac{qZ'}{Z}.$$

Recall that if $q=e^{2\pi i\tau}$, then the Dedekind eta function η satisfies

$$\eta(\tau) = \sqrt{\frac{i}{\tau}}\,\eta\left(-\frac{1}{\tau}\right) = q^{1/24}\sum_{n\in\mathbb{Z}}(-q)^n q^{\frac{3n^2-n}{2}} = q^{1/24}\prod_{n=1}^{\infty}(1-q^n).$$

Hence, as $\tau \to 0 + 0i$,

$$Z = \prod_{n=1}^{\infty}(1-q^n)^{-1} = \frac{q^{1/24}}{\eta(q)} = \frac{q^{1/24}}{\sqrt{\frac{i}{\tau}}\eta\left(-\frac{1}{\tau}\right)} =$$

$$e^{\frac{\pi i\tau}{12}+\frac{\pi i}{12\tau}}\left(\sum_n(-1)^n e^{-\frac{\pi i}{\tau}(3n^2-n)}\right)^{-1}\sqrt{\frac{\tau}{i}} \approx e^{\frac{\pi i}{12\tau}}\sqrt{\frac{\tau}{i}}.$$

Plugging this approximation into the expectation, we see its highest order asymptotics are given by

$$\frac{qZ'}{Z} \approx \frac{4\pi}{\tau^2}.$$

This is the expected size of the Young diagram of a partition when q is close to 1. If we want to rescale by $1/\delta$ so that $\delta^2|\lambda| = 1$, we see that we must keep $\delta \approx \tau/i$ (up to a real constant).

To respect this scaling, we will apply Lemma (7.2) to the 1-point function rescaling also the variables, as follows:

$$F\left(e^{\tau r/i}\right) = \frac{1}{\vartheta(e^{\tau r/i}, e^{2\pi i\tau})} \approx \frac{1}{2i\tau\sin(r/2)}\exp\left(\frac{\tau r^2}{4\pi}\right)$$

The exponential term clearly tends to 1 as $\tau \to 0$. The series of the cosecant encodes exactly the moments of (26) (up to a constant due to rescaling). Let us verify this. To compute the moments, change variables $x = s+t$ and $y = s-t$ and take the same integral as in equation (11):

$$\begin{aligned}
\mu_{2k} &= \int_{\mathbb{R}} t^{2k}\left(L_\infty(t) - |t|\right)dt \\
&= \int_{\mathbb{R}} t^{2k}\left(\tfrac{\sqrt{6}}{\pi}\log\left[2\cosh\left(\tfrac{\pi}{\sqrt{6}}t\right)\right] - |t|\right)dt \\
&= \frac{6^{k+1}}{2^{2k}\pi^{2k+2}}\int_{\mathbb{R}_+} t^{2k}\log\left(1+e^{-t}\right)dt \\
&= \frac{3^{k+1}}{2^{k-1}\pi^{2n+2}}\left(1-2^{-2n-1}\right)\Gamma(2n+1)\zeta(2n+2)
\end{aligned}$$

by the Mellin transform (compare [**8**, page 99]), and $\mu_n = 0$ for odd n. One can check that the even terms in the series expansion of $1/2\sin(r/2)$ are

$$\frac{1}{(2k-1)!}\frac{1}{(2\pi)^{2k+2}}\left(1-2^{-2k-1}\right)\Gamma(2k+1)\zeta(2k+2),$$

and the difference between these two things can be traced down to the scaling, but we will not do it here.

The existence of the limit shape also follows as a special case of the Large Deviation Principle proved by A. Dembo, A. Vershik and O. Zeitouni [**6**].

Typical dimension. In order to investigate what the dimension is for a typical partition in the uniform distribution as $|\lambda| \to \infty$, we will use the hook formula

(1), Stirling's approximation for $n!$, and the approximation given by the hook integral (25):

$$\begin{aligned}\log \dim \lambda &= \log \frac{n!}{\prod_{\square\in\lambda} h_\square} \\ &\approx n\log n - n - \frac{n}{2}\iint_{s>t} \log\sqrt{2}(s-t)(1-L'_\infty(s))(1+L'_\infty(t))\, ds\, dt \\ &\approx n\log n - n\, 0.628699\ldots\end{aligned}$$

9.2. The Plancherel distribution. The *Plancherel distribution* seems to have first appeared in the search for the asymptotics of the maximal dimension of an irreducible representation of $S(n)$. It assigns the weight

$$p(\lambda) = \frac{(\dim\lambda)^2}{|\lambda|!} = \frac{|\lambda|!}{\left(\prod_{\square\in\lambda} h_\square\right)^2}$$

to the partition λ. (The equivalence of these two definitions follows from equation (1).) By Burnside's formula,

$$n! = \sum_{|\lambda|=n} (\dim\lambda)^2,$$

the weights $p(\lambda)$ define a probability on the set $\mathcal{Y}_n$ of Young diagrams of partitions of n. It was named "Plancherel" because the Fourier transform $\mathcal{F}$, understood as a map

$$\mathcal{F} : L^2\left(S(n)^{S(n)}, \mu_{\text{Haar}}\right) \to L^2\left(\widehat{S(n)}, \mu_{\text{Plancherel}}\right)$$

(where $S(n)^{S(n)}$ is the set of conjugacy classes of $S(n)$, and $\widehat{S(n)} \cong \mathcal{Y}_n$ denotes the space of complex finite-dimensional irreducible representations) is an isometry, like in the classical Plancherel theorem.

The variational methods explained in Section 8.2 were originally developed for the study of this distribution [**17**, **37**]. The maximizer of the resulting integral (25) is the curve

$$L_\infty(x) = \begin{cases} \frac{2}{\pi}\left(x\arcsin x + \sqrt{1-x^2}\right) & \text{for } |x|\le 1, \\ |x| & \text{for } |x|\ge 1.\end{cases}$$

For any other curve Ω, let $h(x) = \Omega(x) - L_\infty(x)$ be the variation. We will assume that it is compactly supported. In this case one has that the hook integral (25) has the form

$$\begin{aligned} -\frac{n}{2}\iint_{\mathbb{R}^2} &\log\sqrt{2}|s-t| h'(s)h'(t)\, ds\, dt \\ &+ \frac{n}{2}\iint_{|s|,|t|>1} \log\sqrt{2}|s-t|\,(h'(s)-h'(t))\, ds\, dt \\ &\qquad = -\frac{n}{4}\iint_{\mathbb{R}^2}\left(\frac{h(s)-h(t)}{s-t}\right)^2 ds\, dt.\end{aligned}$$

The second term in the left hand side disappears because of symmetry. To obtain the right hand side form the first term, we basically do two integrations by parts. We thus obtain a constant times the square of what actually is the norm in a Sobolev space, and it follows that the maximum is attained when $h = 0$. The full central limit theorem is attributed to S. Kerov, but its proof was first written up by V. Ivanov and G. Olshanski [**13**].

While the author is not aware of any methods to efficiently deal with this distribution using the half-infinite wedge fermionic Fock space, many interesting facts are known about this distribution. There are for example some papers that analyze very carefully the local correlations (e.g., [**4**, **5**]), and the papers dealing with the resolution of the Baik-Deift-Johansson conjecture [**1**] are very interesting (e.g., [**4**, **14**, **26**]).

Typical dimension. In the case of the Plancherel distribution, the typical dimension turns out to be the same as the maximal dimension [**37**].

9.3. The pillowcase distribution. The *pillowcase distribution* initially appeared in [**7**] in connection with the computation of the volumes of the moduli spaces of quadratic differentials, and was further studied in [**30**, **31**]. Given a parameter q, $0 < |q| < 1$, it assigns to each partition λ the weight $q^{|\lambda|}\mathsf{w}(\lambda)/Z$, where[4]

$$\mathsf{w}(\lambda) = \left(\frac{\prod \text{odd hook lengths of } \lambda}{\prod \text{even hook lengths of } \lambda}\right)^2$$

to partitions λ with empty 2-core, and no weight to any other partitions. The normalization constant Z is known to be

$$Z = \prod_{n=1}^{\infty} (1 - q^{2n})^{-1/2}.$$

In the case of the pillowcase distribution, one can apply the variational methods of Section 8.2 as follows: Since the relevant partitions λ have empty 2-core, their 2-quotient (α, β) determines them completely. According to the picture drawn in Section 6.2, the odd hook lengths of λ correspond to twice the hook lengths of α and β, while the odd hooks of λ correspond to the interactions of the Maya diagrams of α and β: there will be an odd hook whenever there is a pebble on the Maya diagram of α (resp. β) with a corresponding hole on the Maya diagram of β (resp. α), to the right of it. Since the Maya diagrams of α and β are interlaced to form the Maya diagram of λ, their contours interlace too. Letting L_α and L_β be the contours of the partitions α and β, we have

$$L'_\lambda(s) = \begin{cases} L'_\alpha\left(s - \frac{k}{2\sqrt{2n}}\right), & \text{for } s \in \frac{1}{\sqrt{2n}}(k-1, k), k \in 2\mathbb{Z}, \\ L'_\beta\left(s - \frac{k}{2\sqrt{2n}}\right), & \text{for } s \in \frac{1}{\sqrt{2n}}(k, k+1), k \in 2\mathbb{Z}, \end{cases}$$

Going through the derivation of equation (11) again, one sees that the integral for the part corresponding to the even hooks is

$$-n\int_{t<s}\left(\log\sqrt{2}(s-t)\right)\left((1 - L'_\alpha(s))(1 + L'_\alpha(t)) + (1 - L'_\beta(s))\left(1 + L'_\beta(t)\right)\right)ds\,dt,$$

and the one for the odd hooks is

$$n\int_{t<s}\left(\log\sqrt{2}(s-t)\right)\left((1 - L'_\alpha(s))(1 + L'_\beta(t)) + (1 - L'_\beta(s))(1 + L'_\alpha(t))\right)ds\,dt.$$

[4]The original definition was given in [**7**] in terms of characters of near-involutions. The equivalence is given in [**30**, **31**].

Adding and simplifying, one finds

$$n \int_{\mathbb{R}^2} \left(\log \sqrt{2}|s-t|\right) \Delta'(s)\Delta'(t)\, ds\, d$$

where $\Delta(s) = L_\alpha(s) - L_\beta(s)$. After two intergrations by parts,

$$\log \mathsf{w}(\lambda) \approx t = -\frac{n}{2} \int_{\mathbb{R}^2} \left(\frac{\Delta(s)-\Delta(t)}{s-t}\right)^2 ds\, dt + O\left(\sqrt{n}\right).$$

This integral is again a multiple of the square of the norm $\|\Delta\|^2$ of Δ in a Sobolev space. Whence one gets a hint that, in the limit, the pillowcase weights induce a distribution concentrated mostly near the partitions for which $\Delta = 0$, that is, for which the 2-quotients are equal, and that in that region the distribution is more or less uniform. So one expects to get the same limit shape as for the uniform distribution, and one can indeed prove some statements in this direction [**30, 31**]. However, the term $O\left(\sqrt{n}\right)$ makes this conclusion hard to obtain using this method.

Fock space method. It was found in [**7**] that the *pillowcase operator*, given by

$$\mathfrak{W} = \exp\left(\sum_{n>0} \frac{\alpha_{-2n-1}}{2n+1}\right) \exp\left(-\sum_{n>0} \frac{\alpha_{2n+1}}{2n+1}\right),$$

has the property that its diagonal elements exactly coincide with the pillowcase weights:

$$\langle \mathfrak{W} v_\lambda, v_\lambda \rangle = \mathsf{w}(\lambda).$$

One can thus proceed in the same way as we did in Section 9.1: the 1-point function becomes

$$F(u) = \frac{1}{Z}[v^0] \operatorname{tr} q^H \mathfrak{W} \psi(uv) \psi^*(v).$$

Turning the crank as in Section 9.1, one sees that

$$F(u) = \frac{1}{\vartheta(u)}[v^0] \sqrt{\frac{\vartheta(-v)\vartheta(uv)}{\vartheta(v)\vartheta(-uv)}}.$$

Similarly, the n-point function also has a closed expression in terms of modular forms, and again we have good quasimodularity properties. Applying Lemmas 7.2 and 7.3, one obtains the desired result: essentially, in the $q \to 1$ limit, the quotient inside the square root tends to 1 and the limit shape indeed coincides with the one for the uniform distribution. The typical dimension thus also coincides with the one for the uniform distribution.

References

[1] Jinho Baik, Percy Deift, and Kurt Johansson, *On the distribution of the length of the longest increasing subsequence of random permutations*, J. Amer. Math. Soc. **12** (1999), no. 4, 1119–1178, DOI 10.1090/S0894-0347-99-00307-0. MR1682248 (2000e:05006)

[2] Patrick Billingsley, *Probability and measure*, Wiley Series in Probability and Statistics, John Wiley & Sons, Inc., Hoboken, NJ, 2012. Anniversary edition [of MR1324786]; With a foreword by Steve Lalley and a brief biography of Billingsley by Steve Koppes. MR2893652

[3] Spencer Bloch and Andrei Okounkov, *The character of the infinite wedge representation*, Adv. Math. **149** (2000), no. 1, 1–60, DOI 10.1006/aima.1999.1845. MR1742353 (2001g:11059)

[4] Alexei Borodin, Andrei Okounkov, and Grigori Olshanski, *Asymptotics of Plancherel measures for symmetric groups*, J. Amer. Math. Soc. **13** (2000), no. 3, 481–515 (electronic), DOI 10.1090/S0894-0347-00-00337-4. MR1758751 (2001g:05103)

[5] Alexander I. Bufetov, *On the Vershik-Kerov conjecture concerning the Shannon-McMillan-Breiman theorem for the Plancherel family of measures on the space of Young diagrams*, Geom. Funct. Anal. **22** (2012), no. 4, 938–975, DOI 10.1007/s00039-012-0169-4. MR2984121

[6] A. Dembo, A. Vershik, and O. Zeitouni, *Large deviations for integer partitions*, Markov Process. Related Fields **6** (2000), no. 2, 147–179. MR1778750 (2001h:11105)

[7] Alex Eskin and Andrei Okounkov, *Pillowcases and quasimodular forms*, Algebraic geometry and number theory, Progr. Math., vol. 253, Birkhäuser Boston, Boston, MA, 2006, pp. 1–25, DOI 10.1007/978-0-8176-4532-8_1. MR2263191 (2007g:14065)

[8] Alex Eskin and Andrei Okounkov, *Asymptotics of numbers of branched coverings of a torus and volumes of moduli spaces of holomorphic differentials*, Invent. Math. **145** (2001), no. 1, 59–103, DOI 10.1007/s002220100142. MR1839286 (2002g:32018)

[9] J. S. Frame, G. de B. Robinson, and R. M. Thrall, *The hook graphs of the symmetric groups*, Canadian J. Math. **6** (1954), 316–324. MR0062127 (15,931g)

[10] William Fulton, *Young tableaux*, London Mathematical Society Student Texts, vol. 35, Cambridge University Press, Cambridge, 1997. With applications to representation theory and geometry. MR1464693 (99f:05119)

[11] William Fulton and Joe Harris, *Representation theory*, Graduate Texts in Mathematics, vol. 129, Springer-Verlag, New York, 1991. A first course; Readings in Mathematics. MR1153249 (93a:20069)

[12] I. M. Gelfand and S. V. Fomin, *Calculus of variations*, Revised English edition translated and edited by Richard A. Silverman, Prentice-Hall, Inc., Englewood Cliffs, N.J., 1963. MR0160139 (28 #3353)

[13] Vladimir Ivanov and Grigori Olshanski, *Kerov's central limit theorem for the Plancherel measure on Young diagrams*, Symmetric functions 2001: surveys of developments and perspectives, NATO Sci. Ser. II Math. Phys. Chem., vol. 74, Kluwer Acad. Publ., Dordrecht, 2002, pp. 93–151, DOI 10.1007/978-94-010-0524-1_3. MR2059361 (2005d:05148)

[14] Kurt Johansson, *Discrete orthogonal polynomial ensembles and the Plancherel measure*, Ann. of Math. (2) **153** (2001), no. 1, 259–296, DOI 10.2307/2661375. MR1826414 (2002g:05188)

[15] Victor G. Kac, *Infinite-dimensional Lie algebras*, 3rd ed., Cambridge University Press, Cambridge, 1990. MR1104219 (92k:17038)

[16] Masanobu Kaneko and Don Zagier, *A generalized Jacobi theta function and quasimodular forms*, The moduli space of curves (Texel Island, 1994), Progr. Math., vol. 129, Birkhäuser Boston, Boston, MA, 1995, pp. 165–172. MR1363056 (96m:11030)

[17] B. F. Logan and L. A. Shepp, *A variational problem for random Young tableaux*, Advances in Math. **26** (1977), no. 2, 206–222, DOI 10.1016/0001-8708(77)90030-5. MR1417317 (98e:05108)

[18] I. G. Macdonald, *Symmetric functions and Hall polynomials*, 2nd ed., Oxford Mathematical Monographs, The Clarendon Press, Oxford University Press, New York, 1995. With contributions by A. Zelevinsky; Oxford Science Publications. MR1354144 (96h:05207)

[19] Sho Matsumoto, *Correlation functions of the shifted Schur measure*, J. Math. Soc. Japan **57** (2005), no. 3, 619–637. MR2139724 (2006a:60013)

[20] Davesh Maulik and Andrei Okounkov. Quantum groups and quantum cohomology. Preprint. arXiv:1211.1287.

[21] T. Miwa, M. Jimbo, and E. Date, *Solitons*, Cambridge Tracts in Mathematics, vol. 135, Cambridge University Press, Cambridge, 2000. Differential equations, symmetries and infinite-dimensional algebras; Translated from the 1993 Japanese original by Miles Reid. MR1736222 (2001a:37109)

[22] David Mumford, *Tata lectures on theta. I*, Progress in Mathematics, vol. 28, Birkhäuser Boston, Inc., Boston, MA, 1983. With the assistance of C. Musili, M. Nori, E. Previato and M. Stillman. MR688651 (85h:14026)

[23] A. Okounkov. Limit shapes, real and imaginary. Preprint. Available from http://www.math.columbia.edu/~okounkov/AMScolloq.pdf.

[24] A. Okun′kov and G. Ol′shanskiĭ, *Shifted Schur functions* (Russian, with Russian summary), Algebra i Analiz **9** (1997), no. 2, 73–146; English transl., St. Petersburg Math. J. **9** (1998), no. 2, 239–300. MR1468548 (99f:05118)

[25] A. Okounkov and R. Pandharipande, *Gromov-Witten theory, Hurwitz theory, and completed cycles*, Ann. of Math. (2) **163** (2006), no. 2, 517–560, DOI 10.4007/annals.2006.163.517. MR2199225 (2007b:14123)

[26] Andrei Okounkov, *Random matrices and random permutations*, Internat. Math. Res. Notices **20** (2000), 1043–1095, DOI 10.1155/S1073792800000532. MR1802530 (2002c:15045)
[27] Andrei Okounkov and Nikolai Reshetikhin, *Correlation function of Schur process with application to local geometry of a random 3-dimensional Young diagram*, J. Amer. Math. Soc. **16** (2003), no. 3, 581–603 (electronic), DOI 10.1090/S0894-0347-03-00425-9. MR1969205 (2004b:60033)
[28] Grigori Olshanski, *Point processes related to the infinite symmetric group*, The orbit method in geometry and physics (Marseille, 2000), Progr. Math., vol. 213, Birkhäuser Boston, Boston, MA, 2003, pp. 349–393. MR1995385 (2005f:60114)
[29] Grigori Olshanski, Amitai Regev, and Anatoly Vershik, *Frobenius-Schur functions*, Studies in memory of Issai Schur (Chevaleret/Rehovot, 2000), Progr. Math., vol. 210, Birkhäuser Boston, Boston, MA, 2003, pp. 251–299. With an appendix by Vladimir Ivanov. MR1985729 (2004f:05188)
[30] Rodolfo Ríos-Zertuche, *The pillowcase distribution and near-involutions*, Electron. J. Probab. **19** (2014), no. 116, 22, DOI 10.1214/EJP.v19-3626. MR3296532
[31] Rodolfo Antonio Rios Zertuche Rios Zertuche, *Near-Involutions, the Pillowcase Distribution, and Quadratic Differentials*, ProQuest LLC, Ann Arbor, MI, 2012. Thesis (Ph.D.)–Princeton University. MR3122034
[32] Bruce E. Sagan, *The symmetric group*, 2nd ed., Graduate Texts in Mathematics, vol. 203, Springer-Verlag, New York, 2001. Representations, combinatorial algorithms, and symmetric functions. MR1824028 (2001m:05261)
[33] Andrey Smirnov. On the instanton R-matrix. Preprint. arXiv:1302.0799.
[34] Richard P. Stanley. *Enumerative combinatorics. Vol. 2*, volume 62 of *Cambridge Studies in Advanced Mathematics.* Cambridge University Press, Cambridge, 1999. With a foreword by Gian-Carlo Rota and appendix 1 by Sergey Fomin.
[35] M. Szalay and P. Turán, *On some problems of the statistical theory of partitions with application to characters of the symmetric group. I*, Acta Math. Acad. Sci. Hungar. **29** (1977), no. 3-4, 361–379. MR0506108 (58 #21991a)
[36] A. M. Vershik, *Statistical mechanics of combinatorial partitions, and their limit configurations* (Russian, with Russian summary), Funktsional. Anal. i Prilozhen. **30** (1996), no. 2, 19–39, 96, DOI 10.1007/BF02509449; English transl., Funct. Anal. Appl. **30** (1996), no. 2, 90–105. MR1402079 (99d:82008)
[37] A. M. Vershik and S. V. Kerov, *Asymptotic behavior of the maximum and generic dimensions of irreducible representations of the symmetric group* (Russian), Funktsional. Anal. i Prilozhen. **19** (1985), no. 1, 25–36, 96. MR783703 (86k:11051)

MAX-PLANCK-INSTITUT FÜR MATHEMATIK, VIVATSGASSE 7, 53111 BONN, GERMANY
E-mail address: **rrios@mpim-bonn.mpg.de**